Photodynamic Therapy 2021

Photodynamic Therapy 2021

Editor

Serge Mordon

MDPI • Basel • Beijing • Wuhan • Barcelona • Belgrade • Manchester • Tokyo • Cluj • Tianjin

Editor
Serge Mordon
IOncoThAI
Laser Assisted Therapies and
Immunotherapies for
Oncology
IINSERM Unité U1189
University of Lille
France

Editorial Office
MDPI
St. Alban-Anlage 66
4052 Basel, Switzerland

This is a reprint of articles from the Special Issue published online in the open access journal *Pharmaceuticals* (ISSN 1424-8247) (available at: www.mdpi.com/journal/pharmaceuticals/special_issues/photodynamic_therapy_2021).

For citation purposes, cite each article independently as indicated on the article page online and as indicated below:

LastName, A.A.; LastName, B.B.; LastName, C.C. Article Title. *Journal Name* **Year**, *Volume Number*, Page Range.

ISBN 978-3-0365-5352-8 (Hbk)
ISBN 978-3-0365-5351-1 (PDF)

Cover image courtesy of Serge Mordon

Contents

About the Editor

Serge Mordon

Pr. Serge R. Mordon, PhD has been working for the French National Institute of Health and Medical Research (INSERM) from 1981 to 2022. He has been involved in the medical applications of lasers, particularly in dermatology and plastic surgery. He is an internationally recognized expert in laser-tissues interaction and laser applications in medicine. For the last fifteen years, he has focused his research on Photodynamic Therapy. He has authored over 600 articles and book chapters. Pr. Mordon is the author of seventeen issued patents. He is the co-founder of several companies. Since 2012, he holds a Master Degree in Strategy and Organization Management (University of Paris X). He was the director of INSERM U 1189 (Laser Therapies Assisted by Image and Simulation for Oncology) and the director of the Photomedicine Center (Lille University Hospital) for 12 years. He is Board Member of several professional societies and the Past-President of The French Medical Laser Society. He is an associate editor of the editorial boards for the journals *Lasers in Surgery and Medicine* and for *Phamaceuticals*. In 2015, he was nominated Finland Distinguished Professor.

Preface to "Photodynamic Therapy 2021"

Photodynamic therapy (PDT) is a light-based photochemistry process. The illumination of a photoactivatable molecule (also called photosensitizer) with visible or near infrared light produces reactive oxygen toxic species to destroy tumor cells. This treatment modality leads to highly targeted actions, because reactive oxygen species are produced only where light is applied. Light is not harmful, nor is the photoactivable molecule. Only the combination of three elements (photosensitizer, oxygen, and light) is required to induce photo-oxidation reactions. PDT has proven to be a promising modality in many medical applications including cutaneous condition, infectious diseases, and various cancers at different stages. This reprint is the compilation of original articles shedding light on the challenges and opportunities of the development of innovative solutions for photodynamic therapy.

Serge Mordon
Editor

pharmaceuticals MDPI

Article

Study of Cytotoxic and Photodynamic Activities of Dyads Composed of a Zinc Phthalocyanine Appended to an Organotin

Isabelle Toubia [1,2,†], Christophe Nguyen [3,†], Stéphane Diring [1], Marine Pays [3], Elodie Mattana [3], Philippe Arnoux [2], Céline Frochot [2,*], Magali Gary-Bobo [3,*], Marwan Kobeissi [4,*] and Fabrice Odobel [1,*]

1 CEISAM, Chimie Et Interdisciplinarité, Synthèse, Analyse, Modélisation, CNRS, UMR CNRS 6230, Université de Nantes, 2, rue de la Houssinière—BP 92208, CEDEX 3, 44322 Nantes, France; isabelle.toubia@univ-lorraine.fr (I.T.); Stephane.Diring@univ-nantes.fr (S.D.)
2 LRGP, Laboratoire Réactions et Génie des Procédés, UMR 7274 CNRS-Université de Lorraine, 1 rue Grandville, 54000 Nancy, France; philippe.arnoux@univ-lorraine.fr
3 IBMM, Université de Montpellier, CNRS, ENSCM, 34093 Montpellier, France; christophe.nguyen@umontpellier.fr (C.N.); paysmarinepro@gmail.com (M.P.); mattana.e@gmail.com (E.M.)
4 Laboratoire Rammal Rammal, Equipe de Synthèse Organique Appliquée SOA, Faculté des Sciences 5, Université Libanaise, Nabatieh, Lebanon
* Correspondence: celine.frochot@univ-lorraine.fr (C.F.); magali.gary-bobo@inserm.fr (M.G.-B.); kobeissi_marwan@hotmail.com (M.K.); Fabrice.Odobel@univ-nantes.fr (F.O.)
† These authors contributed equally to this study.

Abstract: The combination of photodynamic therapy and chemotherapy is a promising strategy to enhance cancer therapeutic efficacy and reduce drug resistance. In this study two zinc(II) phthalocyanine-tin(IV) conjugates linked by a triethylene glycol chain were synthesized and characterized. In these complexes, the zinc(II) phthalocyanine was used as a potential photosensitizer for PDT and the tin complex was selected as cytostatic moiety. The two dyads composed of zinc(II) phthalocyanine and tin complexes exhibited high cytotoxicity, in absence of light stimulation, against MCF-7 human breast cancer cells with low LC_{50} values in the range of 0.016–0.453 μM. In addition, these complexes showed superior cytotoxicity than their mixture of equimolar component, accompanied with a higher activity towards cancer cells compared to human healthy fibroblasts. However, under irradiation of the zinc phthalocyanine unit (at 650 nm) no photodynamic activity could be detected, due to the most likely quenching of zinc(II) phthalocyanine singlet excited state by the nearby tin complex according to a photoinduced electron transfer process. This study demonstrates the potential of heterometallic anticancer chemotherapeutics composed of a zinc phthalocyanine and tin complex, and it highlights that the development of such conjugates requires that the sensitizer preserves its photophysical properties and in particular its singlet oxygen sensitization ability in the conjugate in order to combine the PDT activity with the cytotoxicity of the anticancer drug.

Keywords: photodynamic therapy; chemotherapy; zinc (II) phthalocyanine; tin complex

Citation: Toubia, I.; Nguyen, C.; Diring, S.; Pays, M.; Mattana, E.; Arnoux, P.; Frochot, C.; Gary-Bobo, M.; Kobeissi, M.; Odobel, F. Study of Cytotoxic and Photodynamic Activities of Dyads Composed of a Zinc Phthalocyanine Appended to an Organotin. *Pharmaceuticals* **2021**, *14*, 413. https://doi.org/10.3390/ph14050413

Academic Editors: Serge Mordon and Jean Jacques Vanden Eynde

Received: 6 April 2021
Accepted: 19 April 2021
Published: 28 April 2021

1. Introduction

Recently, there has been considerable interest in combining photodynamic therapy (PDT) with a chemotherapeutic anticancer agent [1–6]. The combination of different therapeutic approaches that acts on distinct disease pathways has shown several advantages, such as enhanced therapeutic efficacy, reduced side effects, and drug resistance problems. There are generally three approaches to combine PDT and chemotherapy, including consecutive administration of a photosensitizer and an anticancer drug, the use of their covalent and noncovalent conjugates, and co-encapsulation of these agents in a polymeric nanocarrier [7,8]. The combination of a photosensitizer with an appended anticancer drug, cisplatin [9,10], carboplatin [11,12], doxorubicin (DOX) [13–15], and organometallic ruthenium(II) complexes [16,17] has been attempted, in order to evaluate the possibility of synergistic anticancer properties between the two entities. Interesting results were obtained showing an enhanced anticancer efficacy at lower drug doses [18] due to the dark chemostatic effect of the drug amplified by the PDT effect generated upon light excitation of the photosensitizer.

Tin organometallic derivatives are important classes of organometallic drugs, as they exhibit a wide spectrum of biological effects and have been widely studied as bactericides, fungicides, acaricides and wood preservatives. In addition, some compounds possess antitumor activity. The cytotoxic activity of these compounds results from the inhibition of macromolecular synthesis, the alteration of energy-producing metabolism in the mitochondria, and the reduction of DNA synthesis. In addition, the cytotoxicity induced also arises from their interaction with the cell membrane which increases the concentration of cytosolic Ca^{2+} and induces apoptosis. Phthalocyanines are promising second-generation photosensitizers for PDT, owing to their suitable photo-physical and photo-chemical properties, such as the strong absorption in the tissue-penetrating red visible region and high efficiency in generating singlet oxygen [19–23].

In this work, we have investigated the combination of tin complexes with a zinc phthalocyanine photosensitizer in order to take advantage of the intrinsic cytotoxicity of tin complex in the dark added to the potentially high phototoxicity of zinc phthalocyanine upon light excitation. The two entities were linked with a triethylene glycol chain, which can enhance the biocompatibility of the system (Figure 1). Although the length of this spacer might have an impact on the biological activity, this parameter was not explored in this study. Two dyads are composed of tetrasubstituted zinc phthalocyanine connected to a tin phenyl complex liganded *via* a malonate anchor. One tin organometallic compound bears two phenyl groups in its coordination sphere ($SnPh_2$), while the other is bound to three phenyls ($SnPh_3$). Phenyl was selected as organometallic moiety around tin(IV), because it was shown in previous studies that this type of complex gives the most potent anticancer activity [24]. The cytotoxicity of the three dyads was assessed on human healthy (fibroblasts) and cancer cell lines (MCF-7) to quantify their therapeutic potential. It was shown that the two dyads **ZnPc-SnPh$_2$** and **ZnPc-Sn$_2$Ph$_6$** exhibit superior cytotoxicity than their equimolar mixtures of individual components and they display higher selectivity towards MCF-7 carcinoma cell line than healthy cells. On the other hand, these two dyads showed no photodynamic activity under light irradiation.

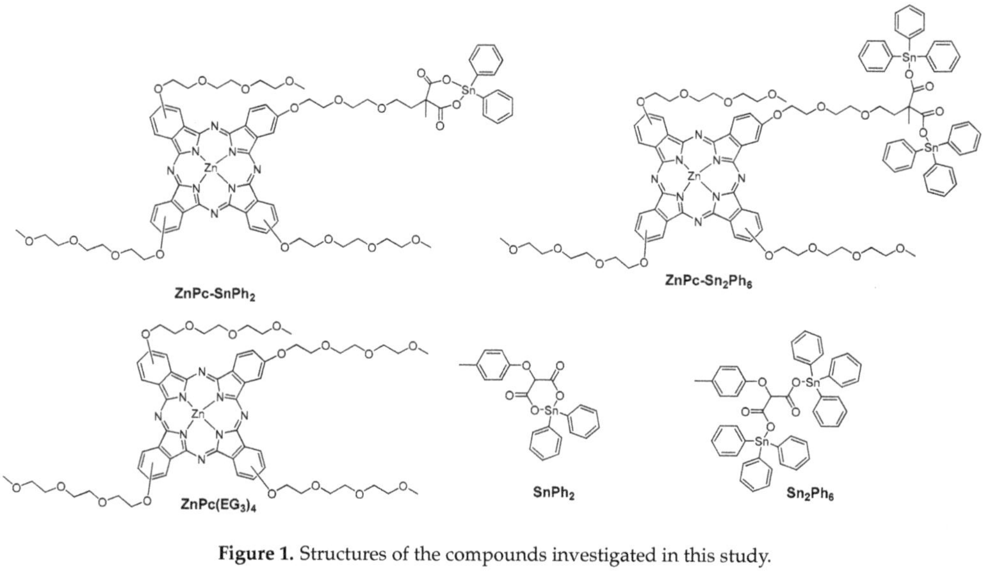

Figure 1. Structures of the compounds investigated in this study.

2. Results and Discussion.

2.1. Synthesis of the Compounds

Scheme 1 illustrates the synthetic route used to prepare the two dyads **ZnPc-SnPh$_2$** and **ZnPc-Sn$_2$Ph$_6$**. An excess of phthalonitrile **1** underwent mixed cyclisation with phthalonitrile **2**[7] in the presence of Zn(OAc)$_2$.2H$_2$O and 1,8-diazabicyclo [5.4.0]undec-7-ene (DBU) in n-pentanol to give the «3 + 1» unsymmetrical phthalocyanine **3** in 18% yield (Scheme 1). The ethyl esters of the malonate were subsequently hydrolyzed with sodium hydroxide in acetone and then acidified to afford the phthalocyanine **4** with 85% yield. The introduction of tin complex was accomplished by activating the carboxylic acid group by deprotonation with triethylamine (TEA) before being reacted with the tin precursor containing two or one chloro ligands [24].

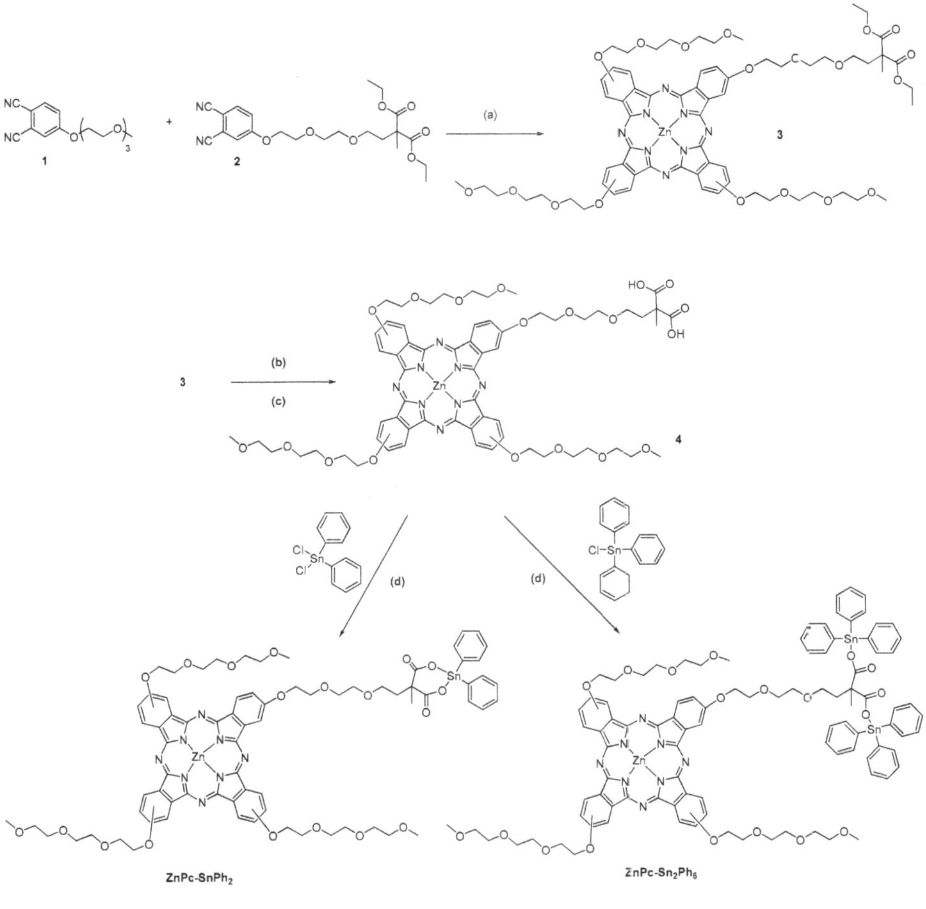

Scheme 1. Synthetic route for the preparation of the dyads composed of a zinc phthalocyanine linked to a tin complex. Reagents and conditions: (**a**) n-pentanol, DBU, Zn(OAc)$_2$.2H$_2$O, 140–150 °C, 12 h, 18%; (**b**) acetone, NaOH, reflux, 2 h; (**c**) HCl, 85%; (**d**) THF, TEA, room temperature, 12 h, 100%.

The complexes were characterized by proton NMR and elemental analysis as well as by high resolution mass spectrometry. The dyads **ZnPc-SnPh$_2$**, **ZnPc-Sn$_2$Ph$_6$** are soluble in CHCl$_3$, DMSO, THF and DMF.

2.2. UV-Vis Electronic Absorption Properties

The absorption spectra of the dyads are compared with those of their components, which are displayed in Figure 2 and the spectroscopic data are gathered in Table 1. In chloroform, **ZnPc(EG$_3$)$_4$** and the dyads **ZnPc-SnPh$_2$** and **ZnPc-Sn$_2$Ph$_6$** give typical UV-Vis spectra of non-aggregated phthalocyanines showing intense and sharp Q-bands in the red visible region around 680 nm along with the less intense vibronic overtone at 616 nm (Figure 2A). They also displayed a Soret band at 360 nm. The presence of the appended tin complex does not alter the spectrum of the dyads because: (i) it does not absorb beyond 300 nm (Supplementary Figure S1) and (ii) there is no electronic communication between the malonate group and the zinc phthalocyanine on the ground state.

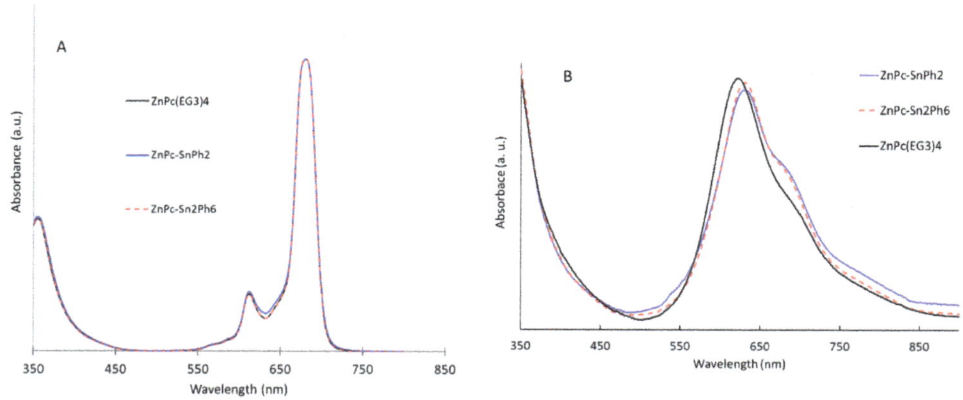

Figure 2. (**A**) Normalized absorption spectra of the phthalocyanine derivatives recorded in chloroform and (**B**) in water with 1% of DMSO at a concentration of about 10^{-6} M.

Table 1. Maximum absorption wavelengths and molar extinction coefficients of the compounds.

Compounds	λ_{abs}/nm (log ϵ/M^{-1} cm^{-1})
3 [a]	360 (4.83); 616 (4.46); 683 (5.19)
4 [a]	360 (4.86); 616 (4.48); 683 (5.21)
ZnPc(EG$_3$)$_4$ [b]	351 (5.45); 615 (5.04); 681 (5.71)
ZnPc-SnPh$_2$ [a]	360 (4.82); 616 (4.45); 683 (5.17)
ZnPc-Sn$_2$Ph$_6$ [a]	360 (4.84); 616 (4.47); 683 (5.2)

[a] recorded in DMF at room temperature. [b] recorded in $CHCl_3$ at room temperature.

The spectra were also recorded in water with 1% of DMSO to approach the conditions used in biological tests (Figure 2B). The shape and the relative intensity of the Q-bands are clearly different in aqueous environment with a broadening and an inversion of the intensity of the Q(0,0) and Q(0,1) bands, which are located respectively around 620 nm and 690 nm. This change witnesses the formation of H-aggregates. Although the solubility of the compounds is maintained in aqueous medium owing to the ethylene glycol chains, the hydrophobicity of phthalocyanine core induces the formation of π-stacked H-aggregates in aqueous environment.

2.3. Infra-Red (IR) Spectroscopic Study

Carboxylate ligand can chelate tin(IV) cation according to several potential binding modes (Figure 3) [25–27]. In solid state, the tin(IV) complexes are usually polymeric compounds with cis-trigonal bipyramidal geometry having five-coordinate tin sites under the bridging mode (type A).

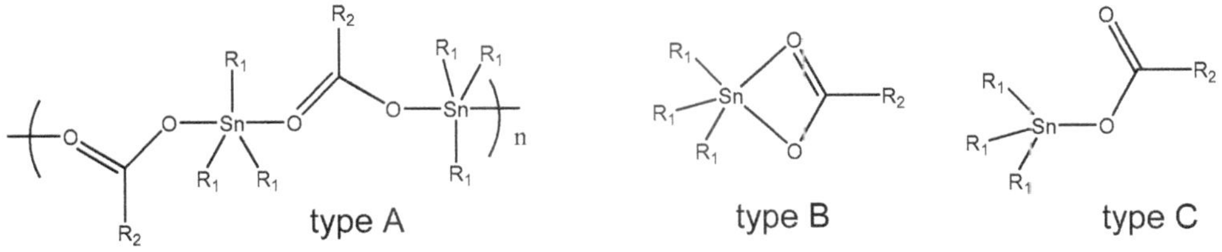

Figure 3. Illustration of three main binding modes of carboxylate ligand with tin(IV).

Particularly in solution, the complexes can exist as monomers either under a five-coordinate chelating bridged mode (type B) or as four-coordinated tin with carboxylate acting as a monodentate ligand (type C). ATR-IR spectra can be used as tools to diagnose the mode of coordination of tin complexes [28]. More specifically, the mode of binding can be deduced by the wavenumber difference ($\nu = \nu_{as} - \nu_s$) between the carboxylate antisymmetric (ν_{as}) and symmetric (ν_s) stretches. If this difference is larger than 250 cm^{-1}, tin(IV) is involved in a tetrahedral geometry (type C binding mode), while values between 250 and 150 cm^{-1} indicate a bridging polymeric structure (type A). When ν is lower than 150 cm^{-1}, then carboxylate is bidentate and binds to Sn(IV) under the chelating bridged mode (type B). The strong asymmetric stretching band of the carbonyl group in the acid form of compound **4**, situated at 1725 cm^{-1} has shrunk and is shifted to lower energy ($\approx$1550 cm^{-1}) after the reaction with chloro-phenyl tin derivatives, which is a clear indication that complexation of tin by the carboxylate has occurred. The wavenumber values of the stretching bands of the carboxylate before and after complexation of tin(IV) are gathered in Table 2.

Table 2. Wavenumber values of the stretching bands of the compounds recorded in the solid state by IR-ATR spectroscopy.

Compound	ν(Sn-O) cm^{-1}	ν_{as}(OCO) cm^{-1}	ν_s (OCO) cm^{-1}	ν
4	-	1725	1396	329
ZnPc-SnPh$_2$	452	1557	1430sh	127
ZnPc-Sn$_2$Ph$_6$	454	1547	1429	118

For the dyads **ZnPc-SnPh$_2$** and **ZnPc-Sn$_2$Ph$_6$**, the ν values are less than 150 cm^{-1}, which indicates that the carboxylate ligand is bidentate and binds to Sn(IV) under the chelating bridged mode.

2.4. Fluorescence Spectroscopy

Fluorescence spectra of the dyads **ZnPc-SnPh$_2$**, **ZnPc-Sn$_2$Ph$_6$** were compared to those of the reference zinc phthalocyanine **ZnPc(EG$_3$)$_4$** in order to study the possibility of a singlet excited state quenching by the tin complex. As for the electronic absorption spectra, the fluorescence emission measurements were conducted both in $CHCl_3$ and water with 1% of DMSO (Figure 4). Whatever the solvent conditions, the fluorescence intensity of the dyads **ZnPc-SnPh$_2$**, **ZnPc-Sn$_2$Ph$_6$** decreased compare to the reference phthalocyanine **ZnPc(EG$_3$)$_4$**.

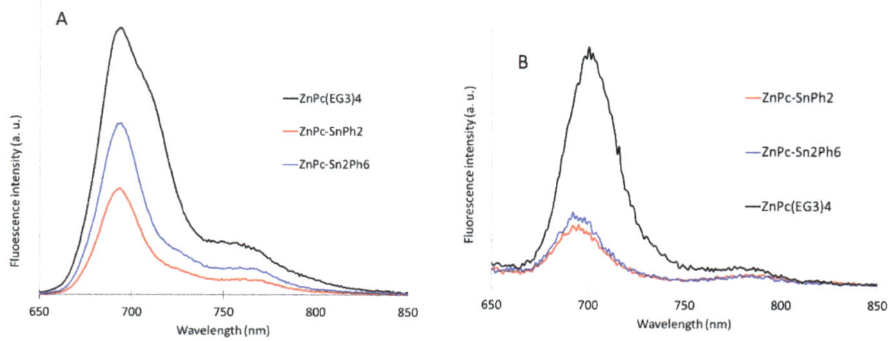

Figure 4. Fluorescence spectra of the dyads **ZnPc-SnPh$_2$**, **ZnPc-Sn$_2$Ph$_6$** recorded in (**A**) $CHCl_3$ or (**B**) in water with 1% of DMSO and compared to the spectra of the reference phthalocyanine **ZnPc(EG$_3$)$_4$** all excited at the same wavelength (λ_{exc} = 615 nm) with an iso-absorbing solution.

Based on the relative fluorescence intensities, the quenching of **ZnPc-SnPh$_2$** and **ZnPc-Sn$_2$Ph$_6$** is estimated at 60% and 36% in $CHCl_3$ and 70% and 75% in water + 1% DMSO respectively. In aqueous medium, the fluorescence intensity is lower than in $CHCl_3$, most probably because the H-aggregates impart a supplementary quenching process. In addition to the formation of H-aggregates, another fluorescence quenching process in the dyads could be attributed to photoinduced electron transfer to the tin complex (formation of «ZnP^{+}-Sn^{III}»). Indeed, energy transfer is unlikely from thermodynamic point of view because the organo-tin complex absorbs in the deep UV, therefore its lowest energy excited state is necessarily higher than that of the singlet excited state of the zinc phthalocyanine (Supplementary Figure S1).

Overall, the fluorescence of the zinc phthalocyanine in these dyads is significantly quenched relative to that of the reference **ZnPc(EG$_3$)$_4$**. The fluorescence inhibition is due to the presence of tin complex and the formation of H-aggregates, particularly in aqueous environment for the latter. The singlet excited state quenching could affect the efficiency of PDT, due to the reduction of the quantum yield of triplet excited state formation by the intervention of a competitive deactivation processes, which depopulate the singlet excited state (see below).

2.5. Singlet Oxygen Measurements

In order to directly estimate the impact of the tin complex on the production of singlet oxygen, the singlet oxygen quantum yield of the compounds was determined in $CHCl_3$ and deuterated water + 1% of DMSO by recording the fluorescence intensity of singlet oxygen emission at 1270 nm. D_2O was chosen instead of H_2O since the singlet oxygen lifetime value in D_2O is higher than in H_2O. This is known that solvents with high vibrational frequencies are more able to quench $^1O_2$ [29]. The data are collected in Table 3.

Table 3. Singlet oxygen quantum yields of the compounds determined in $CHCl_3$ and D_2O + 1% of DMSO upon excitation at 615 nm.

Compound	In $CHCl_3$	In D_2O + 1% of DMSO
ZnPc(EG$_3$)$_4$	0.21 ± 0.05	0 ± 0.05
ZnPc-SnPh$_2$	0.09 ± 0.05	0 ± 0.05
ZnPc-Sn$_2$Ph$_6$	0.12 ± 0.05	0 ± 0.05

In $CHCl_3$, the singlet oxygen quantum yield measurements reflect, within experimental errors, the degree of fluorescence quenching. Moreover, the singlet oxygen quantum yield of **ZnPc(EG$_3$)$_4$** is in agreement with previously reported in DMSO [30]. In D_2O, however, the singlet oxygen emission is importantly reduced and could not be detected because the quantum yields is too low for the sensitivity of this technique. However, we cannot exclude that the singlet oxygen quantum yield of **ZnPc(EG$_3$)$_4$** is not null in biological medium as the latter is quite active in PDT (see below).

2.6. Biological Effect Analysis of ZnPc(EG$_3$)$_4$:Cytotoxicity, PDT and ROS Production

The cytotoxic effect of **ZnPc(EG$_3$)$_4$** on human breast cancer cells (MCF-7) was determined and compared with that on healthy fibroblasts. For this, a dose-response experiment was performed by incubating cells in the darkness with increasing doses of **ZnPc(EG$_3$)$_4$** for 72 h. Then, the viability of cells was quantified and reported in Figure 5a. **ZnPc(EG$_3$)$_4$** exhibited a LC_{50} of 16.55 µM on MCF-7 cells and of 63.80 µM on fibroblasts. This difference (4 fold) between these two cell lines demonstrated that this compound was more cytotoxic on cancer cells than healthy cells.

The photodynamic therapeutic potential was also investigated on breast cancer and healthy cells. Several concentrations of compound (from 0.01 to 0.50 µM) were tested. Cells were incubated 24 h with **ZnPc(EG$_3$)$_4$** and submitted or not to laser excitation (650 nm, 20 min, 39 J.cm^{-2}). Two days later, the viability was quantified MCF-7 cells demonstrated a significant photosensitivity from 0.05 µM of **ZnPc(EG$_3$)$_4$** (Figure 5b). Indeed, cells submitted to laser excitation exhibited 31% cell death. This PDT effect increased in accordance with increasing concentration of **ZnPc(EG$_3$)$_4$** to reach 92% cell death when cells were incubated with 0.50 µM of **ZnPc(EG$_3$)$_4$** and submitted to photoexcitation. This effect is very strong and indicates that the reference compound **ZnPc(EG$_3$)$_4$** is quite active in PDT, most reasonably upon production of singlet oxygen. We also investigated this PDT efficiency on healthy cells in order to verify if the discriminatory effect already observed on dark cytotoxicity (Figure 5c) could be reproduced under laser excitation. Unfortunately, this effect is about the same on cancer or healthy cells, with a significant effect from 0.05 µM and a high PDT efficiency at 0.50 µM. However, we could note a difference between the two cell lines at this concentration; indeed, only 8% of cancer cells survived when 28% of fibroblasts survived after irradiation.

To identify the phototoxicity mechanism involved here, ROS quantification was performed and reported in Figure 5d. Cells were incubated with DCFH$_2$-DA (2′,7′-dichlorodihydrofluorescein diacetate), which is a non-fluorescent molecule. In the presence of ROS this molecule is oxidized into the fluorescent 2′,7′-dichlorodihydrofluorescein (DCF), whose green luminescence can be detected using a fluorescence microscope. The results showed that in the presence of 0.50 µM of **ZnPc(EG$_3$)$_4$**, light excitation at a wavelength of 650 nm induced green fluorescence inside the cells, mainly on cancer cells but also on healthy cells, thus demonstrating ROS production and confirming cell death induced by PDT effect.

Altogether these data demonstrated that this compound is very efficient for dark cytotoxicity and overall for PDT action, but it is not really discriminant between cancer and healthy cells. Although often, the main limitation of cancer therapy is the poor selectivity of cancer cells inducing a large panel of secondary effects on healthy organs and cells. So we decided to graft organotin complex on **ZnPc(EG$_3$)$_4$** to improve the effect and the selectivity on cancer cells.

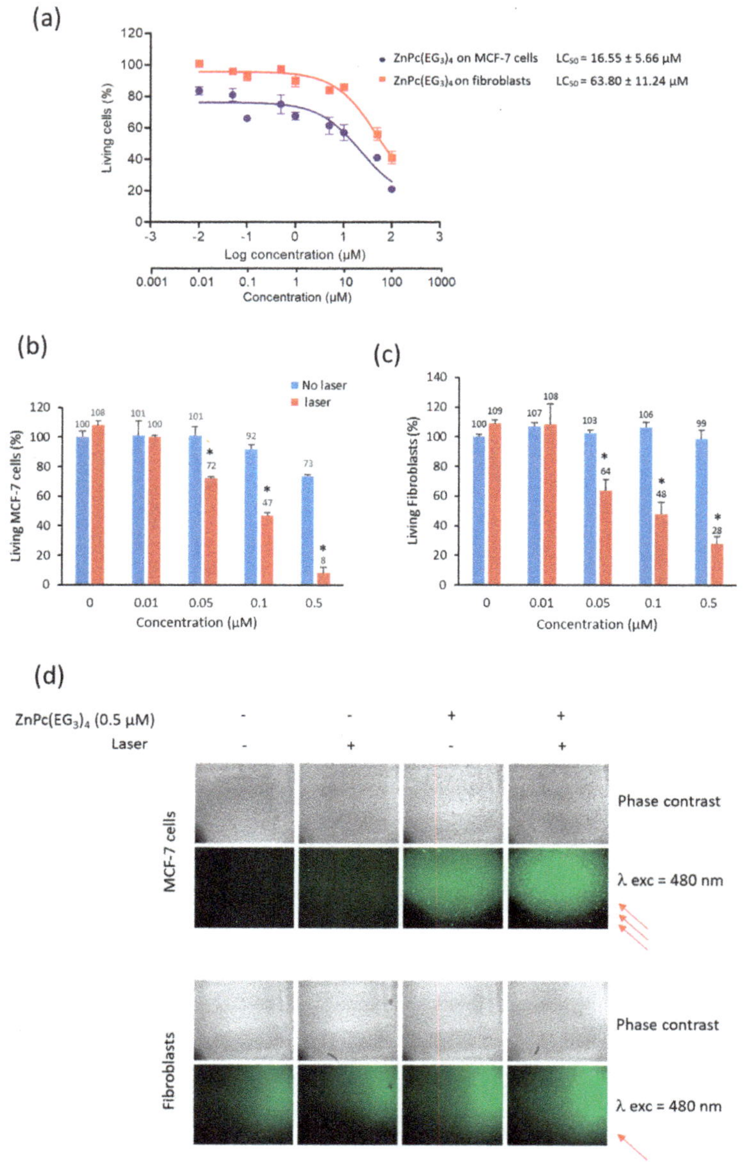

Figure 5. (**a**) Cytotoxicity measurement of increasing concentrations of **ZnPc(EG$_3$)$_4$**. Dose-responses on cancer MCF-7 cells and healthy fibroblasts incubated 72 h with **ZnPc(EG$_3$)$_4$**. (**b**) Phototoxicity of the compound **ZnPc(EG$_3$)$_4$** incubated at increasing concentrations (from 0.01 to 0.50 µM) with cancer cells MCF-7 or (**c**) healthy fibroblasts for 24 h and then irradiated (20 min, λ_{exc} = 650 nm, 39 J.cm^{-2}). * $p < 0.05$ statistically different from "No laser". The bar graph corresponds to the mean ± standard deviation of the values of 2 independent experiments in triplicate. (**d**) Demonstration of ROS production in cells (highlighted by red arrows) incubated for 24 h with 0.50 µM of **ZnPc(EG$_3$)$_4$**, then incubated with DCFH$_2$-DA (20 µM, 45 min) and irradiated for 10 min at 650 nm. The green fluorescence inside the cells reflects ROS production (imaged at 480 nm).

2.7. Cytotoxic Potential Analysis of the Dyads **ZnPc-SnPh$_2$** and **ZnPc-Sn$_2$Ph$_6$** on Cancer and Healthy Cells

In order to determine the therapeutic potential of these new compounds, the cytotoxic activity on MCF-7 cells and also on healthy fibroblasts was studied. Towards this goal, the cells were incubated 72 h in darkness, with increasing concentrations of each compound (from 0.01 to 100 μM). Drug concentrations leading to 50% cell mortality (LC_{50}) were determined using the classical sigmoidal dose-response curves of cytotoxicity obtained when plotted as a logarithmic function of the concentration (μM).

The analysis of the curves in Figure 6, shows that the compound **ZnPc-Sn$_2$Ph$_6$** is about 30 times more toxic on MCF-7 cells (LC_{50} = 0.016 μM) than the compound **ZnPc-SnPh$_2$** (LC_{50} = 0.453 μM). It is also interesting to note that both tin dyads present a higher toxicity on cancer cells compared to healthy cells (Figure 6). Indeed, the LC_{50} are lower on cancer cells than on fibroblasts with LC_{50} of 0.016 vs. 0.032 μM for the compound **ZnPc-Sn$_2$Ph$_6$**, and LC_{50} of 0.453 and 0.513 for the compound **ZnPc-SnPh$_2$**. These results suggest that the **ZnPc-SnPh$_2$** dyad may be a potentially interesting basis to develop targeting cancer cells.

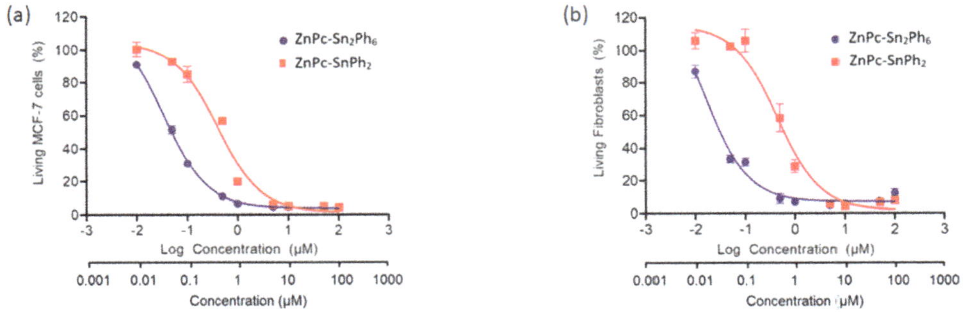

(c)

Compounds	LC_{50} (μM) MCF-7	LC_{50} (μM) Fibroblast
ZnPc-SnPh$_2$	0.453 ± 0.028	0.513 ± 0.133
ZnPc-Sn$_2$Ph$_6$	0.016 ± 0.002	0.032 ± 0.001

Figure 6. Cytotoxic study of **ZnPc-SnPh$_2$** and **ZnPc-Sn$_2$Ph$_6$**. (**a**) MCF-7 cells and (**b**) healthy fibroblasts were incubated 72 h with increasing concentrations of **ZnPc-SnPh$_2$** and **ZnPc-Sn$_2$Ph$_6$**, and maintained in darkness. (**c**) LC_{50} values are reported for each cell line. Values are means ± standard deviation of 2 experiments in triplicate.

In order to confirm the therapeutic interest of using a dyad rather than its constituents, we compared the cytotoxicity of tin dyads to the cytotoxic effect of equimolar mixtures of the phthalocyanine **ZnPc(EG$_3$)$_4$** and **SnPh$_2$** or **Sn$_2$Ph$_6$** (Figure 7). For all conditions tested here, and as reported in the Figure 7e, the cytotoxicity induced by dyads is always higher than the equimolar mixture of their constituents.

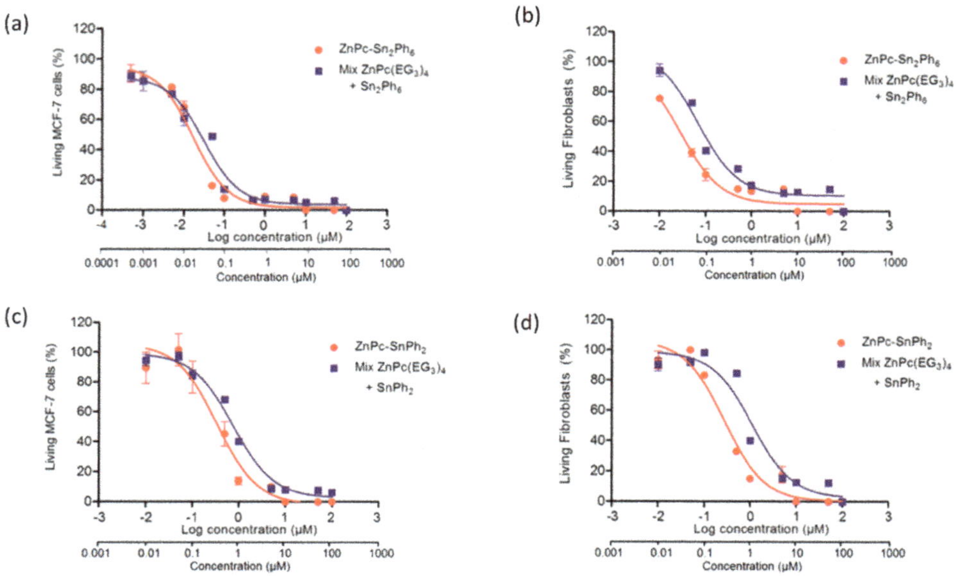

(e)

Compounds	LC_{50} (μM) MCF-7	LC_{50} (μM) Fibroblast
ZnPc-Sn_2Ph_6	0.016 ± 0.002	0.032 ± 0.001
Mix ZnPc$(EG_3)_4$+ Sn_2Ph_6	0.025 ± 0.002	0.097 ± 0.008
ZnPc-$SnPh_2$	0.453 ± 0.028	0.513 ± 0.133
Mix ZnPc$(EG_3)_4$+ $SnPh_2$	0.755 ± 0.109	1.224 ± 0.036

Figure 7. Study of the comparative cytotoxicity of the tin dyads with the equimolar mixture of their constituents, tin and zinc phthalocyanine. Comparison of the cytotoxic efficiency of the compound **ZnPc-Sn_2Ph_6** and the mixture **ZnPc$(EG_3)_4$ + Sn_2Ph_6** (a) on MCF-7 cells and (**b**) on healthy fibroblasts. Comparison of the cytotoxic efficiency of the compound **ZnPc-$SnPh_2$** and the mixture **ZnPc$(EG_3)_4$** + **$SnPh_2$** (c) on MCF-7 cells and (**d**) on healthy fibroblasts. Cells were incubated during 72 h with increasing concentrations of the dyads or the mixture (Mix) and maintained in darkness. (**e**) LC_{50} values are reported for each cell line. Values are means ± standard deviation of 2 experiments in triplicate.

2.8. Photodynamic Therapy Investigations of Dyads

The PDT efficiency was first studied on cancer cells with the dyads **ZnPc-Sn_2Ph_6** and **ZnPc-$SnPh_2$** under laser excitation of the zinc phthalocyanine in its Q-band at 650 nm. Experiments were performed at the concentration of LC_{50} and below up to a factor of 10. Several concentrations were tested, as reported in the tables below, (Tables 4 and 5) but unfortunately no specific mortality due to the laser was obtained under these conditions. This result can be explained by fluorescence quenching of the zinc phthalocyanine in these dyads, which most probably inhibits the energy transfer process to oxygen.

Table 4. Values of cellular survival in % corresponding to the mean ± standard deviation of 2 independent experiments conducted in duplicate.

MCF-7 with ZnPc-Sn_2Ph_6 (in μM)	0	0.005	0.01	0.05
% cell survival, in dark	100 ± 2	99 ± 3	77 ± 2	38 ± 3
% cell survival, λ_{exc} = 650 nm	98 ± 1	98 ± 4	75 ± 3	41 ± 3

Table 5. Values of cellular survival in % corresponding to the mean ± standard deviation of 2 independent experiments conducted in duplicate.

MCF-7 with ZnPc-$SnPh_2$ (in μM)	0	0.05	0.1	0.5
% cell survival, in dark	100 ± 5	92 ± 3	88 ± 4	38 ± 5
% cell survival, λ_{exc} = 650 nm	90 ± 9	90 ± 4	80 ± 2	36 ± 2

Another series of experiments was carried out under light excitation 390/420 nm, on MCF-7 cells and also on healthy fibroblasts. Again, no effect could be observed.

Taken together, these results show that these two dyads have a cytotoxic effect on cells and would allow cancer cells to be targeted more effectively than healthy cells. However, they did not show efficacy in PDT under all of the conditions tested most probably because the singlet excited state is quenched by the appended tin complex, that inhibits singlet oxygen production.

3. Materials and Methods

3.1. Synthesis of the Compounds

Generalities

^{1}H and ^{13}C NMR spectra were recorded on an *AVANCE 300 UltraShield BRUKER* and *AVANCE 400 BRUKER*. Chemical shifts for ^{1}H NMR spectra are referenced relative to residual proton in the deuterated solvent ($CDCl_3$ δ = 7.26 ppm for ^{1}H, DMSO-d_6 δ = 2.50 ppm for ^{1}H, THF-d_8 δ = 3.57, 1.72 ppm for ^{1}H) or to an internal reference (TMS, δ = 0 ppm for both ^{1}H and ^{13}C). NMR spectra were recorded at room temperature, chemical shifts are written in ppm and coupling constants in Hz. High-resolution mass (HR-MS) spectra were obtained by electrospray ionization coupled with high resolution ion trap orbitrap (LTQ-Orbitrap, ThermoFisher Scientific), working in ion-positive or ion-negative mode. UV-visible absorption spectra were recorded on a Variant Cary 300, using 1 cm path length cells. Emission spectra were recorded on a Fluoromax-4 Horiba Jobin Yvon spectrofluorometer (1 cm quartz cells).

Chemicals were purchased from Sigma-Aldrich or Alfa Aesar and used as received. $CHCl_3$ is stabilized with EtOH. Thin-layer chromatography (TLC) was performed on silica sheets precoated with Merck 5735 Kieselgel $60F_{254}$. Column chromatography was carried out with Merck 5735 Kieselgel 60F (0.040–0.063 mm mesh). Reference **ZnPc(EG_3)$_4$** was synthesized according to literature [30].

Phthalocyanine 3. A mixture of **2** 66 mg (1.52 mmol), phthalonitrile **1** 3.1 g (10.68 mmol) and $Zn(OAc)_2.2H_2O$ 83 mg (3.81 mmol) in n-pentanol (25 mL) was heated to 100°C, and a small amount of DBU (0.9 mL) was subsequently added. The mixture was heated at 150 °C for 12 h. After cooling the volatiles were removed under reduced pressure. The residue was dissolved in $CHCl_3$ and a part of the symmetrical phthalocyanine was separated from the desired compound by flash chromatography using $CHCl_3$/MeOH (98:2 *v*/*v*) as the eluent. The product was further purified by preparative TLC using the same eluent as before to give a blue green oily compound. Yield: (360 mg, 18%). UV-Vis (DMF, nm) λ_{max} (log ϵ/M^{-1} cm^{-1}): 360 (4.83); 616 (4.46); 683 (5.19). ^{1}H NMR ($CDCl_3$, 300 MHz) δ (ppm): 9.12 (br, 4H), 8.68 (s, 4H), 7.61 (br, 4H), 4.66 (br, 8H), 4.16–4.06 (m, 12H), 3.91–3.90 (m, 6H), 3.84–3.83 (t, 2H, *J* = 4.95 Hz), 3.81–3.78 (m, 6H), 3.73–3.66 (m, 8H), 3.62–3.56 (m, 8H), 3.37 (s, 9H), 2.26–2.22 (t, 2H, *J* = 6.98 Hz), 1.45 (s, 3H), 1.24–1.18 (tt, 6H, *J* = 1.56, 7.34 Hz). HRMS

(ES+) $[M + H]^+$ m/z 1367.5042 found, 1367.5066 calc. Elem. Anal. Exp C, 58.97; H, 6.47; N, 7.74. calc C, 58.59; H, 6.1; N, 8.11.

Phthalocyanine 4. A mixture of **3** 70 mg (5×10^{-2} mmol), NaOH 18 mg (45×10^{-2} mmol), and acetone (8 mL) was heated under reflux for 2 h. The volatiles were removed under reduced pressure. The green residue was washed thoroughly with acetone, and then redissolved in water and acidified with 3 M HCl until pH = 4. The green precipitate was washed thoroughly with water and ethanol, and then dried under vacuum. Yield: (57 mg, 85%). UV-Vis (DMF, nm) λ_{max} (log ϵ/M^{-1} cm^{-1}): 360 (4.86); 616 (4.48); 683 (5.21). ^{1}H NMR (DMSO-d_6 with a trace amount of pyridine, 300 MHz) δ (ppm): 9.29–9.24 (m, 4H), 8.93–8.88 (m, 4H), 7.79–7.75 (m, 4H), 4.68–4.63 (br, 8H), 4.07–4.04 (br, 8H), 3.64–3.66 (m, 10H), 3.57–3.52 (m, 12H), 3.42–3.40 (t, 8H, J = 6.06 Hz), 3.23 (s, 9H), 2.23–2.20 (2H, m), 1.36 (s, 3H). HRMS (ES-) [M-H] m/z 1309.4309 found, 1309.4238 calc. Elem. Anal. Exp C, 54.4; H, 5.52; N, 7.95. calc C, 54.82; H, 5.95; N, 8.12.

ZnPc-SnPh$_2$. A solution of phthalocyanine **4** 35 mg (2.6×10^{-2} mmol) in THF (1 mL) was added dropwise to a solution of $SnPh_2Cl_2$ 9 mg (2.6×10^{-2} mmol) in THF (1 mL) at room temperature. The reaction was stirred for 20 min, subsequently NEt_3 8 µL (5.3×10^{-2} mmol) was added dropwise. The reaction was then stirred overnight. The solvent was evaporated and the residue was washed with water and dried in vacuum. Yield: (42 mg, 100%). UV-Vis (DMF, nm) λ_{max} (log ϵ/M^{-1} cm^{-1}): 360 (4.82); 616 (4.45); 683 (5.17). ^{1}H NMR (THF-d_8, 400 MHz) δ (ppm): 9.26–9.24 (br, 4H), 8.90 (br, 4H), 7.77–7.72 (m, 8H), 7.45–7.37 (m, 6H), 4.68–4.66 (t, 8H, J = 4.71 Hz), 4.12–4.10 (t, 8H, J = 4.48 Hz), 3.84–3.82 (t, 8H, J = 5.17 Hz), 3.75–3.71 (m, 8H), 3.67–3.64 (t, 8H, J = 4.82 Hz), 3.52–3.50 (m, 8H), 3.31 (s, 9H), 1.28 (s, 3H). HRMS (ES+) $[M + H]^+$ m/z 1583.4023 found, 1583.4088 calc. Elem. Anal. Exp C, 51.43; H, 5.82; N, 6.63. calc C, 51.63; H, 5.77; N, 6.42.

ZnPc-Sn$_2$Ph$_6$. A solution of phthalocyanine **4** 33 mg (2.5×10^{-2} mmol) in THF (1 mL) was added dropwise to a solution of $SnPh_3Cl$ 19 mg (5×10^{-2} mmol) in THF (1 mL) at room temperature. The reaction was stirred for 20 min, subsequently NEt_3 7 µL (5×10^{-2} mmol) was added dropwise. The reaction was then stirred overnight. The solvent was evaporated and the residue was washed with water and dried in vacuum. Yield: (50 mg, 100%). UV-Vis (DMF, nm) λ_{max} (log ϵ/M^{-1} cm^{-1}): 360 (4.84); 616 (4.47); 683 (5.2). ^{1}H NMR (DMSO-d_6 with a trace amount of pyridine, 300 MHz) δ (ppm): 9.05–9.04 (br, 4H), 8.66–8.61 (br, 4H), 7.85–7.82 (br, 6H), 7.71–7.64 (br, 6H), 7.43–7.41 (br, 9H), 7.28 (br, 13H), 4.64–4.62 (br, 8H), 4.08 (br, 8H), 3.82–3.79 (br, 6H), 3.71–3.69 (br, 8H), 3.63–3.62 (br, 8H), 3.51–3.48 (t, 8H, J = 4.88 Hz), 3.26 (s, 9H), 1.19 (s, 3H). HRMS (ES+) $[M + H]^+$ m/z 2011.4662 found, 2011.4675 calc. Elem. Anal. Exp C, 57.28; H, 6.12; N, 5.54. calc C, 57.37; H, 5.88; N, 5.78.

3.2. *Biological Experiments*

3.2.1. Cell Culture

Human breast adenocarcinoma cells (MCF-7) (purchased from ATCC) were cultured in Dulbecco Eagle's Minimal Essential Medium (DMEM) supplemented with 10% fetal bovine serum (FBS) and 1% penicillin/streptomycin (P/S). Adult Human Dermal Fibroblast cells (FS 20-68) were maintained in Roswell Park Memorial Institute (RPMI) medium supplemented with 10% FBS and 1% P/S. Both cell lines were allowed to grow in humidified atmosphere at 37 °C under 5% CO_2. For cytotoxic studies, the compounds (powder) were first diluted in DMSO at the concentration of 10 mM. Then, they were sonicated during 30 s and diluted at the required concentrations in culture medium of each cell line.

3.2.2. Cytotoxicity Study

MCF-7 and fibroblasts (FS 20-68) cells were seeded into 96-well plates at a density of 1000 cells cm^{-2}. One day after cell growth, cells were incubated with or without different concentrations of compounds (from 0.01 to 100 µM) for 3 days. To quantify the percentage of living cells in each condition, cells were incubated 4 h with 0.5 mg.mL^{-1} MTT (3-(4, 5-dimethylthiazol-2-yl)-2,5-diphenyltetrazoliumbromide) in order to determine mitochondrial enzyme activity. Then, MTT precipitates were dissolved in ethanol/DMSO (1:1) solution and absorbance was measured at 540 nm.

3.2.3. Phototoxicity Assay at 650 nm

MCF-7 cancer cells were seeded into 96-well plates at a concentration of 1000 cells per well in 100 µL of culture medium and allowed to grow for 24 h. Then, cells were incubated 24 h, with or without various concentrations of compound solutions. After incubation, cells were submitted or not to laser irradiation (650 nm; 39 $J \cdot cm^{-2}$) during 20 min. Two days after irradiation, MTT assay was performed to measure the level of living cells.

3.3. *Statistical Analysis*

Statistical analysis was performed using the Student's *t*-test to compare paired groups of data. A p value < 0.05 was considered as statistically significant.

3.4. *Singlet Oxygen Quantum Yield*

Absorption spectra were recorded on a UV-3600 UV-visible double beam spectrophotometer (SHIMADZU, Marne La Vallee, France). Fluorescence and singlet oxygen spectra were recorded on a Fluorolog FL3-222 spectrofluorimeter (HORIBA Jobin Yvon, LONGJUMEAU, Paris, France) equipped with 450 W Xenon lamp, a thermo-stated cell compartment (25 °C), a UV-visible photomultiplier R928 (Hamamatsu, Japan) and an InGaAs infrared detector (DSS-16A020L Electro-Optical System Inc., Phoenixville, PA, USA)

Excitation beam is diffracted by a double ruled grating SPEX monochromator (1200 grooves/mm blazed at 330 nm). Emission beam is diffracted by a double ruled grating SPEX monochromator (1200 grooves/mm blazed at 500 nm). Singlet oxygen emission was detected through a double ruled grating SPEX monochromator (600 grooves/mm blazed at 1 µm) and a long-wave pass (780 nm). All spectra were measured in 4 faces quartz cuves. All the emission spectra (fluorescence and singlet oxygen luminescence) have been displayed with the same absorbance (less than 0.2) with the lamp and photomultiplier correction. The standard used for the singlet oxygen quantum yield was the Zinc(II) 2,9,16,23-(tetra-t-butyl)phthalocyanine with a standard quantum yield of 0.25.

4. Conclusions

Two new dyads **ZnPc-SnPh$_2$**, **ZnPc-Sn$_2$Ph$_6$** were prepared, by combining a zinc phthalocyanine sensitizer with a cytotoxic entity such as a tin-based complex. Cytotoxicity studies of the **ZnPc-SnPh$_2$** and **ZnPc-Sn$_2$Ph$_6$** dyads on cancer cells (MCF-7) and on healthy fibroblasts show that they exhibit high cytotoxicity. In particular, the dyad **ZnPc-Sn$_2$Ph$_6$** containing two tin atoms is highly potent towards cancer cells (LC_{50} = 0.16 µM) and is twice less toxic towards healthy human fibroblasts. In contrast, no PDT activity was detected for these two dyads at concentrations close to and below LC_{50}, most probably due to quenching of the singlet excited state of phthalocyanine by the tin complex, which decreased the singlet oxygen quantum yield. This result highlights that for the development of conjugates consisting of a sensitizer connected to a chemotherapeutic agent, one must pay attention that the photophysical properties of the sensitizer and in particular its singlet oxygen quantum yield is maintained in the conjugate in order to combine the PDT activity with the cytotoxicity of the anticancer drug. This can be estimated from Stern-Volmer quenching experiments before synthesizing the conjugates. This aspect has been mostly overlooked in many previous studies dealing with similar approach.

Supplementary Materials: The following are available online at https://www.mdpi.com/article/10.3390/ph14050413/s1, Figure S1: UV-Vis. absorption spectra of compounds **SnPh$_2$** and **Sn$_2$Ph$_6$** recorded in dichloromethane.; Figure S2: Mass spectrum of phthalocyanine **3**; Figure S3: Mass spectrum of phthalocyanine **4**; Figure S4: Mass spectrum of compound **ZnPcSnPh$_2$**; Figure S5: Mass spectrum of compound **ZnPcSn$_2$Ph$_6$**; Figure S6: ^{1}H NMR spectrum of phthalocyanine **3** recorded in $CDCl_3$; Figure S7: ^{1}H NMR spectrum of phthalocyanine **4** recorded in DMSO-d_6 with a trace of pyridine; Figure S8: ^{1}H NMR spectrum of compound **ZnPc-SnPh$_2$** recorded in THF-d$_8$; Figure S9: ^{1}H NMR spectrum of compound **ZnPc-Sn$_2$Ph$_6$** recorded in DMSO-d$_6$ with a trace of pyridine.

Author Contributions: Synthesis of the compounds: I.T. and supervision S.D.; Biological studies: C.N., M.P., E.M. and supervisison M.G.-B.; singlet oxygen measurements: P.A. and supervision C.F.; Conceptualization: F.O., M.G.-B. and M.K. All authors have read and agreed to the published version of the manuscript.

Funding: This research was funded by Campus France (Eiffel grant), Région Pays de la Loire (LUMOMAT project), CNRS-L and the Lebanese University.

Institutional Review Board Statement: Not applicable.

Informed Consent Statement: Not applicable.

Data Availability Statement: Raw data is available from the corresponding author upon request.

Acknowledgments: We thank Campus France (Eiffel grant), Région Pays de la Loire (LUMOMAT project), CNRS-L and the Lebanese University for the financial support. The authors greatly acknowledge J. Hémez and L. Arzel (AMaCC platform, CEISAM UMR CNRS 6230, University of Nantes) for the mass spectrometry analyses. We thank Mélanie Onofre for technical assistance.

Conflicts of Interest: The authors declare no conflict of interest.

References

1. Karges, J.; Yempala, T.; Tharaud, M.; Gibson, D.; Gasser, G. A Multi-action and Multi-target RuII–PtIV Conjugate Combining Cancer-Activated Chemotherapy and Photodynamic Therapy to Overcome Drug Resistant Cancers. *Angew. Chem. Int. Ed.* **2020**, *59*, 7069–7075. [CrossRef]
2. Chen, Y.; Gao, Y.; Li, Y.; Wang, K.; Zhu, J. Synergistic chemo-photodynamic therapy mediated by light-activated ROS-degradable nanocarriers. *J. Mater. Chem. B* **2019**, *7*, 460–468. [CrossRef]
3. Zuluaga, F.M.; Lange, N. Combination of Photodynamic Therapy With Anti-Cancer Agents. *Curr. Med. Chem.* **2008**, *15*, 1655–1673. [CrossRef]
4. Li, X.; Lovell, J.F.; Yoon, J.; Chen, X. Clinical development and potential of photothermal and photodynamic therapies for cancer. *Nat. Rev. Clin. Oncol.* **2020**, *17*, 657–674. [CrossRef]
5. Li, X.; Lee, S.; Yoon, J. Supramolecular photosensitizers rejuvenate photodynamic therapy. *Chem. Soc. Rev.* **2018**, *47*, 1174–1188. [CrossRef]
6. Li, X.; Yu, S.; Lee, D.; Kim, G.; Lee, B.; Cho, Y.; Zheng, B.-Y.; Ke, M.-R.; Huang, J.-D.; Nam, K.T.; et al. Facile Supramolecular Approach to Nucleic-Acid-Driven Activatable Nanotheranostics That Overcome Drawbacks of Photodynamic Therapy. *ACS Nano* **2018**, *12*, 681–688. [CrossRef] [PubMed]
7. Lau, J.T.; Lo, P.C.; Fong, W.P.; Ng, D.K. A zinc(II) phthalocyanine conjugated with an oxaliplatin derivative for dual chemo- and photodynamic therapy. *J. Med. Chem.* **2012**, *55*, 5446–5454. [CrossRef] [PubMed]
8. Ryu, J.H.; Koo, H.; Sun, I.C.; Yuk, S.H.; Choi, K.; Kim, K.; Kwon, I.C. Tumor-targeting multi-functional nanoparticles for theragnosis: New paradigm for cancer therapy. *Adv. Drug Deliv. Rev.* **2012**, *64*, 1447–1458. [CrossRef] [PubMed]
9. Lottner, C.; Bart, K.C.; Bernhardt, G.; Brunner, H. Hematoporphyrin-Derived Soluble Porphyrin-Platinum Conjugates with Combined Cytotoxic and Phototoxic Antitumor Activity. *J. Med. Chem.* **2002**, *45*, 2064–2078. [CrossRef] [PubMed]
10. Lottner, C.; Bart, K.C.; Bernhardt, G.; Brunner, H. Soluble Tetraarylporphyrin-Platinum Conjugates as Cytotoxic and Phototoxic Antitumor Agents. *J. Med. Chem.* **2002**, *45*, 2079–2089. [CrossRef]
11. Chung Enhanced apoptotic effect of combined modality of 9-hydroxypheophorbide α-mediated photodynamic therapy and carboplatin on AMC-HN-3 human head and neck cancer cells. *Oncol. Rep.* **1994**, *21*, 329–334.
12. Rizvi, I.; Celli, J.P.; Evans, C.L.; Abu-Yousif, A.O.; Muzikansky, A.; Pogue, B.W.; Finkelstein, D.; Hasan, T. Synergistic enhancement of carboplatin efficacy with photodynamic therapy in a three-dimensional model for micrometastatic ovarian cancer. *Cancer Res.* **2010**, *70*, 9319–9328. [CrossRef] [PubMed]
13. Kirveliene, V.; Grazeliene, G.; Dabkeviciene, D.; Micke, I.; Kirvelis, D.; Juodka, B.; Didziapetriene, J. Schedule-dependent interaction between Doxorubicin and mTHPC-mediated photodynamic therapy in murine hepatoma in vitro and in vivo. *Cancer Chemother. Pharmacol.* **2006**, *57*, 65–72. [CrossRef]
14. Aniogo, E.C.; George, B.P.A.; Abrahamse, H. In vitro combined effect of Doxorubicin and sulfonated zinc Phthalocyanine-mediated photodynamic therapy on MCF-7 breast cancer cells. *Tumour Biol.* **2017**, *39*, 1010428317727278. [CrossRef] [PubMed]
15. Ali, S.; Muhammad, S.; Khurshid, A.; Ikram, M.; Maqsood, M.; Fisher, C.; Cathcart, J.; Lilge, L. Effective phthalocyanines mediated photodynamic therapy with doxorubicin or methotrexate combination therapy at sub-micromolar concentrations in vitro. *Photodiagnosis Photodyn. Ther.* **2018**, *22*, 51–64. [CrossRef]
16. Schmitt, F.; Govindaswamy, P.; Zava, O.; Suss-Fink, G.; Juillerat-Jeanneret, L.; Therrien, B. Combined arene ruthenium porphyrins as chemotherapeutics and photosensitizers for cancer therapy. *J. Biol. Inorg. Chem.* **2009**, *14*, 101–109. [CrossRef]

17. Gianferrara, T.; Bergamo, A.; Bratsos, I.; Milani, B.; Spagnul, C.; Sava, G.; Alessio, E. Ruthenium-porphyrin conjugates with cytotoxic and phototoxic antitumor activity. *J. Med. Chem.* **2010**, *53*, 4678–4690. [CrossRef]
18. Ruiz-Gonzalez, R.; Milan, P.; Bresoli-Obach, R.; Stockert, J.C.; Villanueva, A.; Cañete, M.; Nonell, S. Photodynamic Synergistic Effect of Pheophorbide a and Doxorubicin in Combined Treatment against Tumoral Cells. *Cancers* **2017**, *9*, 18. [CrossRef]
19. van Leeuwen, M.; Beeby, A.; Fernandes, I.; Ashworth, S.H. The photochemistry and photophysics of a series of alpha octa(alkyl-substituted) silicon, zinc and palladium phthalocyanines. *Photochem. Photobiol. Sci.* **2014**, *13*, 62–69. [CrossRef]
20. Liu, J.Y.; Lo, P.C.; Fong, W.P.; Ng, D.K. Effects of the number and position of the substituents on the in vitro photodynamic activities of glucosylated zinc(II) phthalocyanines. *Org. Biomol. Chem.* **2009**, *7*, 1583–1591. [CrossRef] [PubMed]
21. Zheng, B.-Y.; Zhang, H.-P.; Ke, M.-R.; Huang, J.-D. Synthesis and antifungal photodynamic activities of a series of novel zinc(II) phthalocyanines substituted with piperazinyl moieties. *Dyes Pigment.* **2013**, *99*, 185–191. [CrossRef]
22. Durmus, M.; Nyokong, T. Synthesis, photophysical and photochemical studies of new water-soluble indium(III) phthalocyanines. *Photochem. Photobiol. Sci.* **2007**, *6*, 659–668. [CrossRef] [PubMed]
23. Li, X.; Zheng, B.-D.; Peng, X.-H.; Li, S.-Z.; Ying, J.-W.; Zhao, Y.; Huang, J.-D.; Yoon, J Phthalocyanines as medicinal photosensitizers: Developments in the last five years. *Coord. Chem. Rev.* **2019**, *379*, 147–160. [CrossRef]
24. Gómez-Ruiz, S.; Žižak, Ž.; Kaluđerović, G.N. Structural studies and cytotoxic activity against human cancer cell lines of mono and dinuclear tin(IV) complexes with the α,α'-dimercapto-o-xylene ligand. *Inorg. Chim. Acta* **2014**, *423*, 117–122. [CrossRef]
25. Anjorin, C.; Song, X.; Kumar, D.; Pike, R.; De Dios, A.; Eng, G. Crystal structures of two polymeric bis(triphenyltin) malonates. *J. Coord. Chem.* **2010**, *63*, 4031–4040. [CrossRef]
26. Thorpe, D.; Callejas, A.; Royzman, D.; Pike, R.D.; Eng, G.; Song, X. Synthesis and crystal structures of ionic triphenyltin complexes with oxalic and malonic acid. *J. Coord. Chem.* **2013**, *66*, 3647–3659. [CrossRef]
27. Song, X.; Zapata, A.; Hoerner, J.; de Dios, A.C.; Casabianca, L.; Eng, G. Synthesis, larvicidal, QSAR and structural studies of some triorganotin 2,2,3,3-tetramethylcyclopropanecarboxylates. *Appl. Organomet. Chem.* **2007**, *21*, 545–550. [CrossRef]
28. Davies, A.G. 6-Tin. In *Comprehensive Organometallic Chemistry II*; Abel, E.W., Stone, F.G.A., Wilkinson, G., Eds.; Elsevier: Oxford, UK, 1995; pp. 217–303.
29. Foote, C.S.; Krasnovsky, A.A.; Fu, Y.; Selke, M.; Karney, W.L. Singlet Oxygen Dimol-Sensitized Luminescence and Reactions of Singlet Oxygen with Organometallics. In *The Activation of Dioxygen and Homogeneous Catalytic Oxidation*; Barton, D.H.R., Martell, A.E., Sawyer, D.T., Eds.; Springer: Boston, MA, USA, 1993; pp. 411–422.
30. Li, Y.; Wang, J.; Zhang, X.; Guo, W.; Li, F.; Yu, M.; Kong, X.; Wu, W.; Hong, Z. Highly water-soluble and tumor-targeted photosensitizers for photodynamic therapy. *Org. Biomol. Chem.* **2015**, *13*, 7681–7694. [CrossRef]

pharmaceuticals

MDPI

Article

Novel Photosensitizer β-Mannose-Conjugated Chlorin e6 as a Potent Anticancer Agent for Human Glioblastoma U251 Cells

Yo Shinoda [1,*], Kohei Kujirai [1], Kohei Aoki [1], Mai Morita [1], Masato Masuda [2], Lihao Zhang [2], Zhou Kaixin [2], Akihiro Nomoto [2], Tsutomu Takahashi [1], Yayoi Tsuneoka [1], Jiro Akimoto [3], Hiromi Kataoka [4], Rioko Rachi [5], Atsushi Narumi [5], Tomokazu Yoshimura [6], Shigenobu Yano [6] and Yasuyuki Fujiwara [1,*]

1 Department of Environmental Health, School of Pharmacy, Tokyo University of Pharmacy and Life Sciences, 1432-1 Horinouchi, Hachioji, Tokyo 192-0392, Japan; y151061@toyaku.ac.jp (K.K.); kohe1170928@gmail.com (K.A.); ma131.news2@gmail.com (M.M.); tsutomu@toyaku.ac.jp (T.T.); tsuneoka@toyaku.ac.jp (Y.T.)

2 Department of Applied Chemistry, Graduate School of Engineering, Osaka Prefecture University, 1-1 Gakuen-cho, Naka-ku, Sakai, Osaka 599-8531, Japan; szb02120@edu.osakafu-u.ac.jp (M.M.); syb02151@edu.osakafu-u.ac.jp (L.Z.); zhx245259630@gmail.com (Z.K.); nomoto@chem.osakafu-u.ac.jp (A.N.)

3 Department of Neurosurgery, Tokyo Medical University, 6-7-1 Nishi-Shinjuku, Shinjuku, Tokyo 160-0023, Japan; jiroaki@gmail.com

4 Department of Gastroenterology and Metabolism, Nagoya City University Graduate School of Medical Sciences, 1 Kawasumi, Mizuho-cho, Mizuho-ku, Nagoya 467-8601, Japan; hkataoka@med.nagoya-cu.ac.jp

5 Department of Organic Materials Science, Graduate School of Organic Materials Science, Yamagata University, Jonan 4-3-16, Yonezawa 992-8510, Japan; tcf00071@st.yamagata-u.ac.jp (R.R.); narumi@yz.yamagata-u.ac.jp (A.N.)

6 KYOUSEI Science Center for Life and Nature, Nara Women's University, Kitauoya-Higashimachi, Nara 630-8506, Japan; yoshimura@cc.nara-wu.ac.jp (T.Y.); yano-s@cc.nara-wu.ac.jp (S.Y.)

* Correspondence: shinoda@toyaku.ac.jp (Y.S.); yasuyuki@toyaku.ac.jp (Y.F.)

Received: 1 October 2020; Accepted: 13 October 2020; Published: 16 October 2020

Abstract: A photosensitizer is a molecular drug for photodynamic diagnosis and photodynamic therapy (PDT) against cancer. Many studies have developed photosensitizers, but improvements in their cost, efficacy, and side effects are needed for better PDT of patients. In the present study, we developed a novel photosensitizer β-mannose-conjugated chlorin e6 (β-M-Ce6) and investigated its PDT effects in human glioblastoma U251 cells. U251 cells were incubated with β-M-Ce6, followed by laser irradiation. Cell viability was determined using the Cell Counting Kit-8 assay. The PDT effects of β-M-Ce6 were compared with those of talaporfin sodium (TS) and our previously reported photosensitizer β-glucose-conjugated chlorin e6 (β-G-Ce6). Cellular uptake of each photosensitizer and subcellular distribution were analyzed by fluorescence microscopy. β-M-Ce6 showed 1000× more potent PDT effects than those of TS, and these were similar to those of β-G-Ce6. β-M-Ce6 accumulation in U251 cells was much faster than TS accumulation and distributed to several organelles such as the Golgi apparatus, mitochondria, and lysosomes. This rapid cellular uptake was inhibited by low temperature, which suggested that β-M-Ce6 uptake uses biological machinery. β-M-Ce6 showed potent PDT anti-cancer effects compared with clinically approved TS, which is a possible candidate as a next generation photosensitizer in cancer therapy.

Keywords: β-mannose-conjugated chlorin e6; talaporfin sodium; glioblastoma; U251; photodynamic therapy; PDT

---

1. Introduction

Photodynamic therapy (PDT) is a less invasive treatment for cancer [1–4]. PDT uses a photosensitizer that accumulates in cancer cells and a harmless laser to elicit the photosensitizer to produce reactive oxygen species (ROS) to kill cancer cells [1,5,6]. Many photosensitizers for PDT have been developed, but effective photosensitizers with low side effects and costs are needed for better therapeutic effects in patients.

Talaporfin sodium (TS, NPe6, mono-L-aspartyl chlorin e6, Laserphyrin®) is a second generation photosensitizer approved by the Japanese Ministry of Health, Labor, and Welfare and clinically used in PDT for early-stage lung cancer, primary malignant brain tumors, and locally remnant recurrent esophageal cancer in Japan [7–12]. TS shows fast clearance from the body and relatively short-term side effects, such as skin photosensitivity, compared with the first generation clinically approved photosensitizer porfimer sodium (Photofrin) [13].

Recently, we developed β-glucose-conjugated chlorin e6 (β-G-Ce6) as a next generation photosensitizer [14]. It shows drastic anti-tumor effects in normal immortalized esophageal epithelial cells and esophageal cancer cells both in vitro and in vivo. Glucose conjugation is based on Warburg effects whereby cancer cells in general consume more glucose than normal cells [15,16]. Another strategy to enhance the cancer specificity of a photosensitizer is targeting tumor-associated macrophages (TAMs), a class of immune cells that exist in high numbers in the solid tumor microenvironment [17–19]. TAMs with high expression of mannose receptor are pro-tumorigenic cells that negatively affect therapy responsiveness [17–20]. In this context, we also developed a mannose-conjugated chlorin derivative, H2TFPC-SMan, 5,10,15,20-tetrakis (4-(α-d-mannopyranosylthio)-2,3,5,6-tetrafluorophenyl)-2,3-(methano (*N*-methyl) iminomethano) chlorin, which shows significant anti-cancer effects and accumulation in M2-polarized macrophages [21,22]. In the present study, we synthesized novel β-mannose-conjugated chlorin e6 (β-M-Ce6) as a photosensitizer derived from the core structure of chlorin e6 (Figure 1). We compared the anti-cancer effects of β-M-Ce6 with those of TS and our previously reported photosensitizer β-G-Ce6 [14] in human glioblastoma cell line U251.

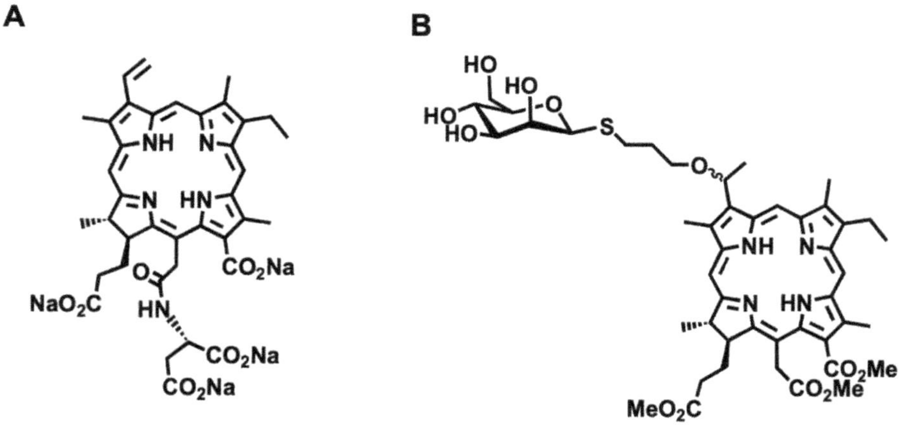

Figure 1. Chemical structures of talaporfin sodium (TS) and β-mannose-conjugated chlorin e6 (β-M-Ce6). (**A**) TS. (**B**) β-M-Ce6.

2. Results

2.1. Potent Anti-Cancer Effects of β-M-Ce6 Compared with TS

We first compared the anti-cancer effects of β-M-Ce6 with those of TS and β-G-Ce6 as PDT photosensitizers on the viability of human glioblastoma U251 cells. The cells were treated with several concentrations of each photosensitizer for 1 h and then irradiated by a 664 nm laser (1 J/cm^2)

(Figure 2). Photosensitizer-treated cells without irradiation did not show any cell death (data not shown). However, laser-irradiated cells showed significant cell death after photosensitizer treatment in a dose-dependent manner. Importantly, the median lethal dose (LD_{50}) of β-M-Ce6 (30 nM) was 1000× lower than that of TS (26 μM) (Figure 2D) and similar to that of β-G-Ce6 (21 nM), as reported previously [14]. These data suggest that β-M-Ce6 has potent anti-cancer effects compared with TS.

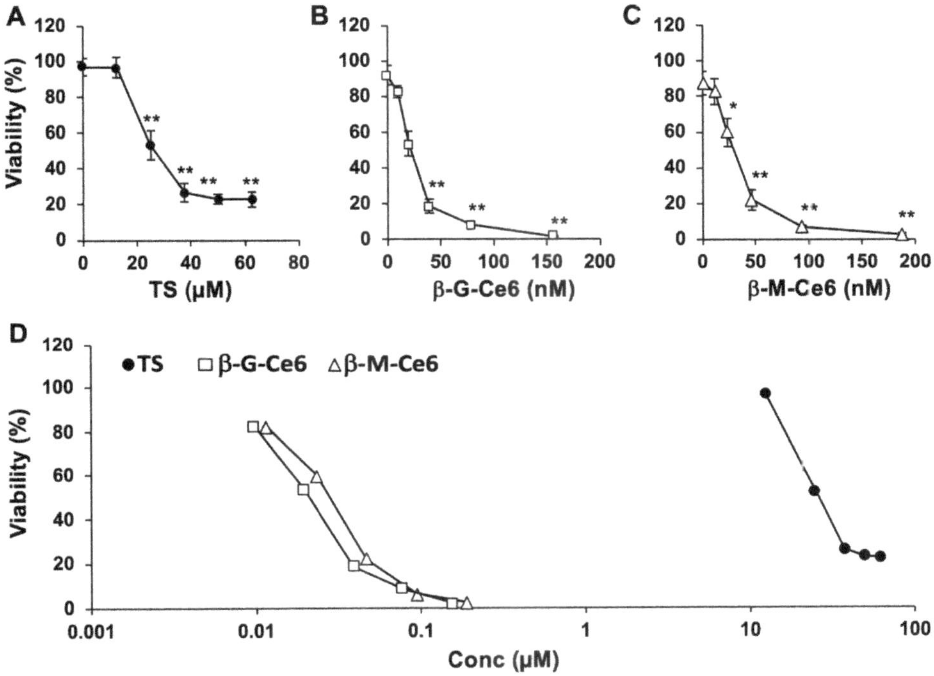

Figure 2. Dose dependency of anti-cancer effects for 1 h treatments. (**A**) Talaporfin sodium (TS), (**B**) β-glucose-chlorin e6 (β-G-Ce6), and (**C**) β-mannose-chlorin e6 (β-M-Ce6). (**D**) Semi-log graph of all tested compounds. $n = 6$, 6, and 8 for each compound and condition, respectively. * $p < 0.05$ and ** $p < 0.01$ vs. 0 μM of each photosensitizer, one-way analysis of variance (ANOVA) with the Tukey–Kramer post-hoc test.

2.2. Fast Cellular Accumulation Property of β-M-Ce6

Next, we investigated the cellular accumulation rate of each photosensitizer by cell viability assays (Figure 3). The concentrations of each photosensitizer we used in this experiment (62.5 μM for TS, 78.1 nM for β-G-Ce6, and 93.8 nM for β-M-Ce6) were determined by a preliminary experiment that showed around 5% cell viability at 180 min of treatment with each photosensitizer. Using these concentrations, we performed the cellular accumulation assay of each photosensitizer revealed by changing the incubation time from 5 to 180 min. TS treatment resulted in significant cell death (24.9% viability) at 30 min of treatment and the maximal effect (5.7% viability) at 180 min of treatment. However, the maximal effects (5–10% viabilities) of β-M-Ce6 and β-G-Ce6 were already obtained at 30 min of treatment and sustained until 180 min (Figure 3). These data suggest that β-M-Ce6 has faster cellular accumulation than TS and the rate is at least six times faster than that of TS.

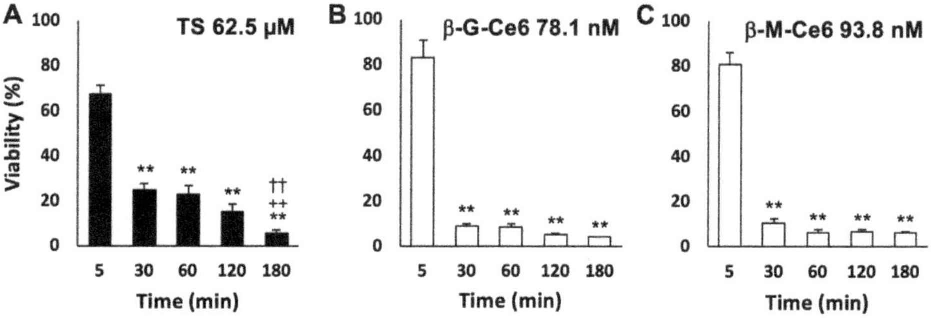

Figure 3. Time dependency of treatments. Doses that resulted in ~5% viability at 180 min of treatment were used. (**A**) Talaporfin sodium (TS, 62.5 μM), (**B**) β-glucose-chlorin e6 (β-G-Ce6, 78.1 nM), and (**C**) β-mannose-chlorin e6 (β-M-Ce6, 93.8 nM). n = 6, 6, and 8 for each compound and condition, respectively. ** $p < 0.01$ vs. 5 min, $^{++}$ $p < 0.01$ vs. 30 min, and †† $p < 0.01$ vs. 60 min of each photosensitizer, one-way analysis of variance (ANOVA) with the Tukey–Kramer post-hoc test.

2.3. *Fast Cellular Accumulation of β-M-Ce6 is Mainly Caused by Biological Machinery*

There are two major factors for cellular accumulation of a substrate. One is the physical property of the substrate, especially the partition coefficient. The other is biological machinery such as influx/efflux transporters, endo-/exo-cytosis, and several metabolic enzymes. To investigate whether the physical property of them or biological machinery were responsible for the fast cellular accumulation of β-M-Ce6, we first measured the partition coefficient (*log P*) defined by Equation (1):

$$\log P = \log \frac{[C_{1-octanol}]}{[C_{PBS}]} \tag{1}$$

where $[C_{1\text{-}octanol}]$ and $[C_{PBS}]$ denote the concentrations of the photosensitizers being portioned into the 1-octanol phase and the phosphate buffered saline (PBS) phase, respectively. The *log P*-values of β-M-Ce6 and β-G-Ce6 were 1.88 and 1.89, respectively. In contrast, the value of TS was less than −3.00 (pH 8–12), in accordance with its pharmaceutical interview form that supplements the package insert and is provided only in Japanese. These data suggested that β-M-Ce6 had the potential to more easily accumulate and penetrate lipid bilayer than TS. Biological machinery acts properly at the appropriate temperature. Therefore, if accumulation of β-M-Ce6 used biological machinery, its accumulation would be suppressed by a low temperature. We performed a photosensitizer transport assay at normal (37 °C) and cold (4 °C) temperatures and their accumulations were analyzed by fluorescence microscopy (Figure 4). TS showed almost no fluorescence under this experimental condition (2 μM, 30 min of exposure) at both temperatures. This result was consistent with the result of cell viability assays (Figure 2). β-M-Ce6 and β-G-Ce6 showed bright fluorescence at 37 °C, but the fluorescence of each photosensitizer was significantly lower at 4 °C. Taken together, these data suggested that the fast cellular accumulations of β-M-Ce6 and β-G-Ce6 were mainly mediated by biological machinery.

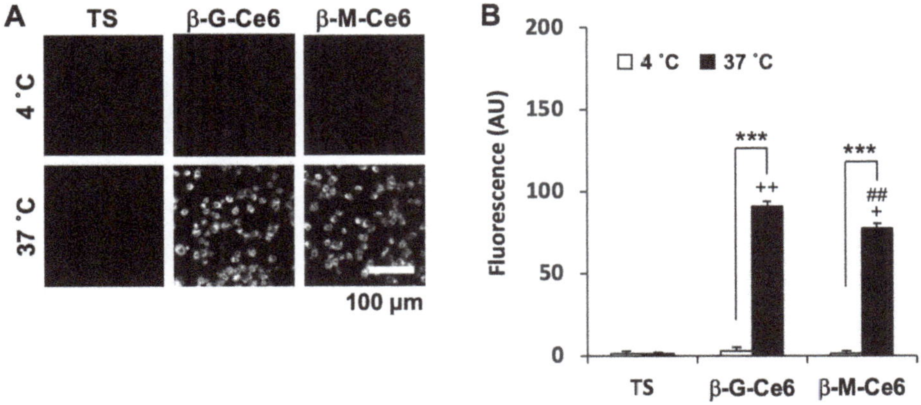

Figure 4. Accumulation of photosensitizers. (**A**) Representative fluorescence micrographs. (**B**) Mean fluorescent intensities of each photosensitizer (2 μM, 30 min) and temperature condition (4 and 37 °C), $n = 40$ for each photosensitizer and condition. *** $p < 0.001$, Student's t-test. $^{+}$ $p < 0.05$ and $^{++}$ $p < 0.01$ vs. 37 °C for TS and $^{\#\#}$ $p < 0.01$ vs. 37 °C for β-G-Ce6, one-way analysis of variance (ANOVA) with the Tukey–Kramer post-hoc test.

2.4. *β-M-Ce6 Accumulates in Lipid-Containing Organelles Such as the Golgi Apparatus, Mitochondria, and Lysosomes*

Subcellular distribution of a photosensitizer affects its PDT efficiency. Because of their photodynamic effect, ROS production and subsequent destruction of biological molecules, such as proteins, lipids, and nucleotides, are affected by the position of the photosensitizer. Therefore, we investigated the localization of photosensitizers by co-staining of several organelle markers and photosensitizers (Figure 5). Interestingly, all conjugated chlorins were distributed to all stained organelles such as the Golgi apparatus (Figure 5A), mitochondria (Figure 5B), and lysosomes (Figure 5C). Such phenomena were confirmed by simultaneous staining of all markers together with photosensitizers (Figure 5D). These data suggested that β-M-Ce6 and β-G-Ce6 were distributed to all lipid-containing organelles and this feature may be caused by their lipophilic properties.

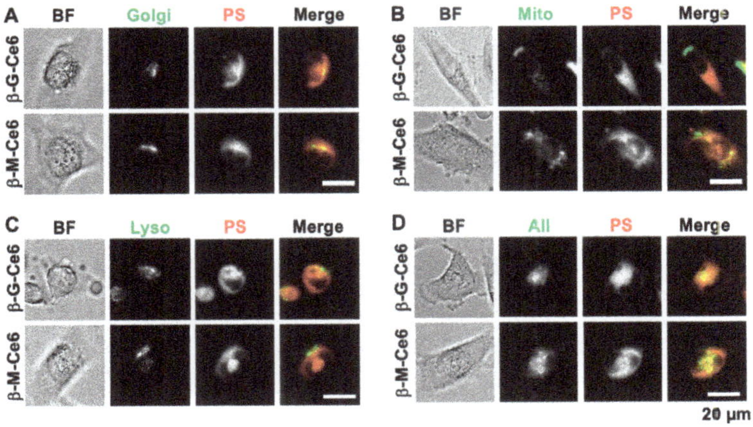

Figure 5. Representative images of the subcellular distribution of each photosensitizer. Brightfield (BF), organelle markers (Golgi apparatus (Golgi), Mitochondria (Mito), Lysosomes (Lyso), and co-staining of all markers (All)), 2 μM photosensitizer (PS), and merged images (green for organelle markers and red for PS) are shown. The distributions of photosensitizers compared with (**A**) the Golgi apparatus, (**B**) mitochondria, (**C**) lysosomes, and (**D**) all markers.

3. Discussion

In the present study, we demonstrated that novel β-M-Ce6 has a potent anticancer effect in human glioblastoma U251 cells. This anti-cancer effect of β-M-Ce6 was 1000× higher than that of TS, a clinically approved PDT drug in Japan. The cellular uptake rate of β-M-Ce6 was at least six times faster than that of TS and its uptake may employ biological machinery. Furthermore, the subcellular distribution of β-M-Ce6 may depend on its physical properties such as a high partition coefficient.

Compared with TS, β-M-Ce6 has two large different molecular regions. One molecular region of TS has four sodium carbonate regions, but β-M-Ce6 has three methyl ester regions. This difference makes β-M-Ce6 more lipophilic than TS. Consistently, the partition coefficient of β-M-Ce6 was much higher than that of TS. This *log P*-value of β-M-Ce6 (1.88) was reasonable compared with our previously reported chlorin derivative that has four glucosyl residues and a *log P*-value of 0.13 [23]. Lipophilicity is thought to simply increase cellular permeability and enhancement of lipophilicity is usually used in prodrug design [24]. In the present study, this physical feature of β-M-Ce6 may not largely affect intracellular uptake, but affect subcellular accumulation. A low temperature significantly decreased cellular uptake of β-M-Ce6, which indicated that uptake of β-M-Ce6 may employ biological machinery. However, β-M-Ce6 showed subcellular localization in not only the plasma membrane, but also several organelles such as the Golgi apparatus, mitochondria, and lysosomes. All these cellular structures have lipid bilayers. Therefore, β-M-Ce6 localization to these subcellular regions may be explained by its physical features, lipophilicity. Furthermore, this accumulative feature of β-M-Ce6 to several organelles may enhance PDT by simultaneous degeneration of organelle functions.

The other different molecular region of β-M-Ce6 compared with TS is a mannose residue. This approach was based on the high expression of the mannose receptor in TAMs [17,19,20,25]. TAMs influence progression, metastasis, and tumor recurrence, which originate mainly from circulating monocytes, but resident macrophages also develop in a tumor [26,27]. Therefore, mannose residues may enhance accumulation of a drug in a tumor by targeting the mannose receptor expressed on TAMs [28,29]. In the present study, we did not investigate whether β-M-Ce6 accumulates in TAMs or macrophages. However, it is noteworthy that the mannose conjugation enhanced the PDT effects in glioblastoma cells. One of the effects is thought to be associated with highly expressed sugar transporters in tumors [30,31]. Fourteen glucose transporters (GLUT) and 12 sodium-glucose cotransporters (SGLT) have been reported in humans, and at least GLUT1-3 and SGLT4/5 have some affinity for mannose [30,32,33]. Therefore, the expression of these glucose transporters may explain the potent anti-cancer effects of β-M-Ce6. In addition to glucose transporters, several proteins have the potential to bind mannose and affect cellular uptake [34,35]. Indeed, U251 cells express several genes encoding proteins that potentially bind to mannose, which include 16 genes that have mannose in their gene name (GSM723932; DNA microarray study deposited in Gene Expression Omnibus (http://www.ncbi.nlm.nih.gov/geo/)). In particular, mannose receptor C type 2 and mannose-6-phosphate receptor expressed in U251 cells are possible candidates to enhance PDT effects via mannose residues. In addition to these mannose-associated proteins, other candidates may exist. Therefore, the possible mechanism of enhanced PDT effects by mannose conjugation should be clarified in a future study.

Taken together, our results strongly suggest that β-M-Ce6 has potent PDT effects compared with clinically approved TS and potential as a next generation photosensitizer. Further investigation to reveal the mechanisms and possible side effects and an in vivo study should be required for future clinical trials.

4. Materials and Methods

4.1. Synthesis of β-M-Ce6

1-thio-β-d-tetraacetylmannose and 3-(3-bromopropoxy) chlorin-e6-TME were prepared by a literature method [36–38]. Et_3N (63.5 μL) was added to a stirred solution of 3-(3-bromopropoxy) chlorin-e6-TME (276 mg, 0.078 mmol) in CH_2Cl_2 under a N_2 atmosphere. To the mixture cooled at 0 °C, the solution of 1-thio-β-d-tetraacetylmannose (3.2 equiv.) in CH_2Cl_2 was added dropwise. After being stirred for 3 h at room temperature under a N_2 atmosphere, the mixture was transferred to a separating funnel, then CH_2Cl_2 and water were added, and the organic layer was separated. The aqueous layer was washed with brine and dried over sodium sulfate, then evaporated to dryness. The residue was purified by column chromatography (CH_2Cl_2/AcOEt, 1:2) to give acetylated β-M-Ce6. Deprotection was carried out in the ordinary way. NaOMe/MeOH suspension (10 equiv.) was added to acetylated β-M-Ce6 (0.152 mmol) in dried MeOH under the N_2 atmosphere. After being stirred for 0.5 h at room temperature, the solution was then quenched by AcOH (100 μL) and the mixture was evaporated to dryness. The residue was purified by column chromatography (CH_2Cl_2/AcOEt, 10:1) to give β-M-Ce6. ^{1}H NMR (400 MHz, $CDCl_3$): δ = 9.78 (d, *J* = 13.2 Hz, 1H), 9.67 (d, *J* = 5.2 Hz, 1H), 8.69 (d, *J* = 2.4 Hz, 1H), 5.85–5.86 (m, 1H), 5.20–5.40 (m, 4H), 4.35–4.50 (m, 2H), 4.20 (s, 3H), 3.85–3.65 (m, 6H), 3.54–3.60 (m, 11H), 3.40–3.50 (m, 2H), 3.42 (s, 3H), 3.29 (s, 3H), 2.90–3.20 (m, 2H), 2.30–2.80 (m, 7H), 2.20–2.05 (m, 3H), 1.65–1.80 (m, 8H), −1.55 (s, 2H). MALDI-TOF-MS: ($C_{46}H_{60}N_4O_{12}S$) calcd. 892.39; found. 892.37.

4.2. Cell Culture

Human glioblastoma U251 cells (Riken Cell Bank, Tsukuba, Japan) were cultured in 100 mm cell culture dishes (Thermo Scientific, Waltham, MA, USA) and 96-well plates (Thermo Scientific, Waltham, MA, USA) in Dulbecco's modified eagle medium (DMEM, Nissui, Tokyo, Japan) supplemented with 10% fetal bovine serum (Nichirei Bioscience, Tokyo, Japan) at 37 °C with 5% CO_2, as described previously [39]. Cells were seeded at 1×10^4 cells/well and cultured for 24 h before all experiments, except for the photosensitizer transport assay. Experimental protocols were approved by the Regulations for Biological Research at Tokyo University of Pharmacy and Life Sciences.

4.3. PDT

PDT was performed as described previously with small modifications [40]. The stock solutions of each photosensitizer were prepared with saline for TS (Meiji Seika Pharma, Tokyo, Japan) and dimethyl sulfoxide (DMSO) for other photosensitizers. For the dose dependency test, cultured cells were treated with 0–62.5 μM TS, 0–156.3 nM β-G-Ce6, or 0–187.5 nM β-M-Ce6 in fresh medium for 1 h. For the time dependency test, cells were treated with TS (62.5 μM), β-G-Ce6 (78.1 nM), or β-M-Ce6 (93.8 nM) in fresh medium for 5–180 min. Immediately after washing with PBS, cells were subjected to laser irradiation (wave length: 664 nm; laser power: 3.4 mW/cm^2; total dose of laser irradiation: 1 J/cm^2) using a semi-conductor laser irradiator, ZH-L5011HJP (Meiji Seika Pharma, Tokyo, Japan).

4.4. Cell Viability Assay

The cell viability assay was performed as described previously [41]. Twenty-four hours after laser irradiation, cell viability was measured using Cell Counting Kit-8 (Dojindo, Kumamoto, Japan). The Cell Count Kit-8 solution was mixed with culture medium at a final concentration of 10% and the mixture was applied to PDT-treated cells for 1 h at 37 °C. After incubation, absorbance was measured at 450 nm using a Varioskan Flash microplate reader (Thermo Scientific, Waltham, MA, USA).

4.5. Measurement of the Partition Coefficient

Both β-M-Ce6 and β-G-Ce6 were dissolved individually in DMSO to prepare stock solutions (5×10^{-3} M). The stock solution was diluted with a 1:1 1-octanol/acetone mixed solution, which was characterized by UV-vis spectroscopy (V-500 spectrophotometer, JASCO, Tokyo, Japan) to determine the molar absorbance coefficient in the 1:1 1-octanol/acetone-mixed solution. Similarly, the molar absorbance coefficient in a 1:1 PBS/acetone-mixed solution was determined. A 1:1 1-octanol/PBS-mixed solution was prepared. From the solution, the 1-octanol phase (600 μL) and PBS phase (600 μL) were collected and combined. A stock solution (3 μL) was added, mixed using a vortex mixer, and then centrifuged to obtain the partition solution. The $1 - octanol$ phase (500 μL) was collected from the partition solution and diluted with 1-octanol (500 μL) and acetone (1.00 mL), which was characterized by UV-vis spectroscopy to determine the concentration in the 1:1 1-octanol/acetone-mixed solution ($[C_{1-octanol}]$). Similarly, the PBS phase (500 μL) was collected from the partition solution and diluted with PBS (500 μL) and acetone (1.00 mL), which provided the concentration in the 1:1 1-PBS/acetone-mixed solution ($[C_{PBS}]$). The $log([C_{1-octanol}]/[C_{PBS}])$ values were calculated to provide the *log P*-values.

4.6. Photosensitizer Transport Assay

For fluorescence microscopy, U251 cells were cultured on 8-chamber glass-bottomed dishes coated with 0.04% Polyethyleneimine (Merck, Darmstadt, Germany) at 3×10^4 cells/well. Cells were cultured for 24 h before the transport assay. Precooled (4 °C) or prewarmed (37 °C) fresh HEPES containing DMEM without phenol red (FUJIFILM Wako Pure Chemical, Osaka, Japan) with 2 μM photosensitizer was applied to the culture for 30 min at 4 or 37 °C. Cells were washed twice with PBS and fixed with 4% paraformaldehyde (Merck, Darmstadt, Germany) in 0.1 M phosphate buffer for 30 min at room temperature. Fluorescence of the photosensitizer was observed under an Eclipse Ti-U inverted microscope (Nikon, Tokyo, Japan) equipped with a filter cube (Ex: 340–380 nm; DM: 400 nm; Em: 672–716 nm) and CMOS Zyla5.5 camera (Andor technology, Belfast, UK). Fluorescence data were collected and processed by NIS-elements (Nikon, Tokyo, Japan) and Photoshop CS6. Fluorescence intensities were measured by ImageJ 1.52k. Data were analyzed by Excel for Mac 2016.

4.7. Photosensitizer Distribution Assay

U251 cells were cultured on eight-chamber glass-bottomed dishes coated with 0.04% PEI at 3×10^4 cells/well. Cells were cultured for 24 h before the distribution assay. CytoPainter Golgi staining kit Green (200×, Abcam, Cambridge, UK), MitoBright LT Green (50 nM, Dojindo, Kumamoto, Japan), and Lysotracker Green (200 nM, Thermo Scientific, Waltham, MA, USA) were used in accordance with the manufacturers' protocols to stain the Golgi apparatus, mitochondria, and lysosomes, respectively. Briefly, fresh medium containing each dye and 10 μM photosensitizer was applied to the cells for 30 min at 37 °C with 5% CO_2. The cells were washed twice with PBS, and medium was replaced with HEPES containing DMEM without phenol red. Fluorescence of each organelle marker and the photosensitizer were obtained and processed by the same methods described above using a filter cube GFP-B for organelle markers (Ex: 460–500 nm, DM: 505 nm; Em: 510–560 nm) and photosensitizers (Ex: 340–380 nm; DM: 400 nm; Em: 672–716 nm).

4.8. Statistical Analysis

If not stated otherwise, data are expressed as the mean ± SEM. Differences between two datasets and multiple datasets were assessed using the Student's *t*-test and one-way analysis of variance (ANOVA) with the Tukey–Kramer post-hoc test, respectively. All data were collected and analyzed using a double-blinded approach.

Author Contributions: In the present investigation, the individual author contributions were made as follows: Conceptualization, Y.S., A.N. (Akihiro Nomoto), and S.Y.; investigation, Y.S., K.K., K.A., M.M. (Mai Morita), R.R., and A.N. (Atsushi Narumi); resources, M.M. (Masato Masuda), L.Z., Z.K., A.N. (Akihiro Nomoto), T.Y., and S.Y.; writing—original draft preparation, Y.S.; writing—review and editing, Y.S., A.N. (Akihiro Nomoto), T.T., Y.T., J.A., H.K., A.N. (Atsushi Narumi), S.Y., and Y.F.; supervision, Y.S. and Y.F.; project administration, Y.S. and Y.F.; funding acquisition, Y.S. and S.Y. All authors have read and agreed to the published version of the manuscript.

Funding: This work was partially supported by a grant from the Institute for Social Medicine at Tokyo University of Pharmacology and Life Science (2018 and 2019 to Y.S.), The Shimabara Science Promotion Foundation (2017 to Y.S.), GSK Japan (2015 to Y.S.), JSPS KAKENHI Grant Numbers 25288028 (2013–2015 to S.Y.) and 18K05161 (2018–2020 to S.Y.).

Acknowledgments: We thank Katsuhisa Inoue in the Tokyo University of Pharmacology and Life Science for technical advice using the transport assay. We also thank Mitchell Arico from Edanz Group (https://en-author-services.edanzgroup.com/ac) for editing a draft of this manuscript.

Conflicts of Interest: The authors declare no competing financial interest.

References

1. Dolmans, D.E.; Fukumura, D.; Jain, R.K. Photodynamic therapy for cancer. *Nat. Rev. Cancer* **2003**, *3*, 380–387. [CrossRef]
2. Juarranz, A.; Jaen, P.; Sanz-Rodriguez, F.; Cuevas, J.; Gonzalez, S. Photodynamic therapy of cancer. Basic principles and applications. *Clin. Transl. Oncol.* **2008**, *10*, 148–154. [CrossRef]
3. Colombeau, L.; Acherar, S.; Baros, F.; Arnoux, P.; Gazzali, A.M.; Zaghdoudi, K.; Toussaint, M.; Vanderesse, R.; Frochot, C. Inorganic Nanoparticles for Photodynamic Therapy. *Top. Curr. Chem.* **2016**, *370*, 113–134.
4. Savoia, P.; Deboli, T.; Previgliano, A.; Broganelli, P. Usefulness of Photodynamic Therapy as a Possible Therapeutic Alternative in the Treatment of Basal Cell Carcinoma. *Int. J. Mol. Sci.* **2015**, *16*, 23300–23317. [CrossRef] [PubMed]
5. Sibata, C.H.; Colussi, V.C.; Oleinick, N.L.; Kinsella, T.J. Photodynamic therapy: A new concept in medical treatment. *Braz. J. Med. Biol. Res.* **2000**, *33*, 869–880. [CrossRef]
6. Dougherty, T.J.; Gomer, C.J.; Henderson, B.W.; Jori, G.; Kessel, D.; Korbelik, M.; Moan, J.; Peng, Q. Photodynamic therapy. *J. Natl. Cancer Inst.* **1998**, *90*, 889–905. [CrossRef]
7. Wang, S.; Bromley, E.; Xu, L.; Chen, J.C.; Keltner, L. Talaporfin sodium. *Expert Opin. Pharmacother.* **2010**, *11*, 133–140. [CrossRef]
8. Akimoto, J. Photodynamic Therapy for Malignant Brain Tumors. *Neurol. Med. Chir.* **2016**, *56*, 151–157. [CrossRef]
9. Wu, H.; Minamide, T.; Yano, T. Role of photodynamic therapy in the treatment of esophageal cancer. *Dig. Endosc.* **2019**, *31*, 508–516. [CrossRef]
10. Usuda, J.; Tsutsui, H.; Honda, H.; Ichinose, S.; Ishizumi, T.; Hirata, T.; Inoue, T.; Ohtani, K.; Maehara, S.; Imai, K.; et al. Photodynamic therapy for lung cancers based on novel photodynamic diagnosis using talaporfin sodium (NPe6) and autofluorescence bronchoscopy. *Lung Cancer* **2007**, *58*, 317–323. [CrossRef]
11. Kato, H.; Furukawa, K.; Sato, M.; Okunaka, T.; Kusunoki, Y.; Kawahara, M.; Fukuoka, M.; Miyazawa, T.; Yana, T.; Matsui, K.; et al. Phase II clinical study of photodynamic therapy using mono-L-aspartyl chlorin e6 and diode laser for early superficial squamous cell carcinoma of the lung. *Lung Cancer* **2003**, *42*, 103–111. [CrossRef]
12. Spikes, J.D.; Bommer, J.C. Photosensitizing properties of mono-L-aspartyl chlorin e6 (NPe6): A candidate sensitizer for the photodynamic therapy of tumors. *J. Photochem. Photobiol. B* **1993**, *17*, 135–143. [CrossRef]
13. Gomer, C.J.; Ferrario, A. Tissue distribution and photosensitizing properties of mono-L-aspartyl chlorin e6 in a mouse tumor model. *Cancer Res.* **1990**, *50*, 3985–3990. [PubMed]
14. Nishie, H.; Kataoka, H.; Yano, S.; Yamaguchi, H.; Nomoto, A.; Tanaka, M.; Kato, A.; Shimura, T.; Mizoshita, T.; Kubota, E.; et al. Excellent antitumor effects for gastrointestinal cancers using photodynamic therapy with a novel glucose conjugated chlorin e6. *Biochem. Biophys. Res. Commun.* **2018**, *496*, 1204–1209. [CrossRef]
15. Warburg, O.; Wind, F.; Negelein, E. The Metabolism of Tumors in the Body. *J. Gen. Physiol.* **1927**, *8*, 519–530. [CrossRef]
16. Vander Heiden, M.G.; Cantley, L.C.; Thompson, C.B. Understanding the Warburg effect: The metabolic requirements of cell proliferation. *Science* **2009**, *324*, 1029–1033. [CrossRef]
17. Malfitano, A.M.; Pisanti, S.; Napolitano, F.; Di Somma, S.; Martinelli, R.; Portella, G. Tumor-Associated Macrophage Status in Cancer Treatment. *Cancers* **2020**, *12*, 1987. [CrossRef]
18. Laviron, M.; Boissonnas, A. Ontogeny of Tumor-Associated Macrophages. *Front. Immunol.* **2019**, *10*, 1799. [CrossRef]
19. Chanmee, T.; Ontong, P.; Konno, K.; Itano, N. Tumor-associated macrophages as major players in the tumor microenvironment. *Cancers* **2014**, *6*, 1670–1690. [CrossRef]
20. Katara, G.K.; Jaiswal, M.K.; Kulshrestha, A.; Kolli, B.; Gilman-Sachs, A.; Beaman, K.D. Tumor-associated vacuolar ATPase subunit promotes tumorigenic characteristics in macrophages. *Oncogene* **2014**, *33*, 5649–5654. [CrossRef]
21. Hayashi, N.; Kataoka, H.; Yano, S.; Tanaka, M.; Moriwaki, K.; Akashi, H.; Suzuki, S.; Mori, Y.; Kubota, E.; Tanida, S.; et al. A novel photodynamic therapy targeting cancer cells and tumor-associated macrophages. *Mol. Cancer Ther.* **2015**, *14*, 452–460. [CrossRef] [PubMed]
22. Shinoda, Y.; Takahashi, T.; Akimoto, J.; Ichikawa, M.; Yamazaki, H.; Narumi, A.; Yano, S.; Fujiwara, Y. Comparative photodynamic therapy cytotoxicity of mannose-conjugated chlorin and talaporfin sodium in cultured human and rat cells. *J. Toxicol. Sci.* **2017**, *42*, 111–119. [CrossRef] [PubMed]

23. Hirohara, S.; Obata, M.; Alitomo, H.; Sharyo, K.; Ando, T.; Tanihara, M.; Yano, S. Synthesis, photophysical properties and sugar-dependent in vitro photocytotoxicity of pyrrolidine-fused chlorins bearing S-glycosides. *J. Photochem. Photobiol. B* **2009**, *97*, 22–33. [CrossRef]
24. Rautio, J.; Kumpulainen, H.; Heimbach, T.; Oliyai, R.; Oh, D.; Jarvinen, T.; Savolainen, J. Prodrugs: Design and clinical applications. *Nat. Rev. Drug Discov.* **2008**, *7*, 255–270. [CrossRef] [PubMed]
25. Coffelt, S.B.; Tal, A.O.; Scholz, A.; De Palma, M.; Patel, S.; Urbich, C.; Biswas, S.K.; Murdoch, C.; Plate, K.H.; Reiss, Y.; et al. Angiopoietin-2 regulates gene expression in TIE2-expressing monocytes and augments their inherent proangiogenic functions. *Cancer Res.* **2010**, *70*, 5270–5280. [CrossRef]
26. Cortez-Retamozo, V.; Etzrodt, M.; Newton, A.; Rauch, P.J.; Chudnovskiy, A.; Berger, C.; Ryan, R.J.; Iwamoto, Y.; Marinelli, B.; Gorbatov, R.; et al. Origins of tumor-associated macrophages and neutrophils. *Proc. Natl. Acad. Sci. USA* **2012**, *109*, 2491–2496. [CrossRef]
27. Movahedi, K.; Van Ginderachter, J.A. The Ontogeny and Microenvironmental Regulation of Tumor-Associated Macrophages. *Antioxid. Redox Signal* **2016**, *25*, 775–791. [CrossRef]
28. Dalle Vedove, E.; Costabile, G.; Merkel, O.M. Mannose and Mannose-6-Phosphate Receptor-Targeted Drug Delivery Systems and Their Application in Cancer Therapy. *Adv. Healthc. Mater.* **2018**, *7*, e1701398. [CrossRef] [PubMed]
29. Irache, J.M.; Salman, H.H.; Gamazo, C.; Espuelas, S. Mannose-targeted systems for the delivery of therapeutics. *Expert Opin. Drug Deliv.* **2008**, *5*, 703–724. [CrossRef] [PubMed]
30. Calvo, M.B.; Figueroa, A.; Pulido, E.G.; Campelo, R.G.; Aparicio, L.A. Potential role of sugar transporters in cancer and their relationship with anticancer therapy. *Int. J. Endocrinol.* **2010**, *2010*, 205357. [CrossRef]
31. Szablewski, L. Expression of glucose transporters in cancers. *Biochim. Biophys. Acta* **2013**, *1835*, 164–169. [CrossRef] [PubMed]
32. Wright, E.M.; Loo, D.D.; Hirayama, B.A. Biology of human sodium glucose transporters. *Physiol. Rev.* **2011**, *91*, 733–794. [CrossRef] [PubMed]
33. Wright, E.M. Glucose transport families SLC5 and SLC50. *Mol. Aspects Med.* **2013**, *34*, 183–196. [CrossRef] [PubMed]
34. Taylor, M.E.; Summerfield, J.A. Mammalian mannose-binding proteins. *Clin. Sci.* **1986**, *70*, 539–546. [CrossRef]
35. Turner, M.W. Mannose binding protein. *Biochem. Soc. Trans.* **1994**, *22*, 88–94. [CrossRef]
36. Bensoussan, C.; Rival, N.; Hanquet, G.; Colobert, F.; Reymond, S. Iron-catalyzed cross-coupling between C-bromo mannopyranoside derivatives and a vinyl Grignard reagent: Toward the synthesis of the C31–C52 fragment of amphidinol 3. *Tetrahedron* **2013**, *69*, 7759–7770. [CrossRef]
37. Shu, P.; Zeng, J.; Tao, J.; Zhao, Y.; Yao, G. Selective S-deacetylation inspired by native chemical ligation: Practical syntheses of glycosyl thiols and drug mercapto-analogues. *Green Chem.* **2015**, *17*, 2545–2551. [CrossRef]
38. Hargus, J.A.; Fronczek, F.R.; Vicente, G.H.; Smith, K.M. Mono-(L)-aspartylchlorin-e6. *Photochem. Photobiol.* **2007**, *83*, 1006–1015. [CrossRef] [PubMed]
39. Shinoda, Y.; Aoki, K.; Shinkai, A.; Seki, K.; Takahashi, T.; Tsuneoka, Y.; Akimoto, J.; Fujiwara, Y. Synergistic effect of dichloroacetate on talaporfin sodium-based photodynamic therapy on U251 human astrocytoma cells. *Photodiagn. Photodyn. Ther.* **2020**, *31*, 101850. [CrossRef] [PubMed]
40. Takahashi, T.; Misawa, S.; Suzuki, S.; Saeki, N.; Shinoda, Y.; Tsuneoka, Y.; Akimoto, J.; Fujiwara, Y. Possible mechanism of heme oxygenase-1 expression in rat malignant meningioma KMY-J cells subjected to talaporfin sodium-mediated photodynamic therapy. *Photodiagn. Photodyn. Ther.* **2020**, *32*, 102009. [CrossRef]
41. Takahashi, T.; Suzuki, S.; Misawa, S.; Akimoto, J.; Shinoda, Y.; Fujiwara, Y. Photodynamic therapy using talaporfin sodium induces heme oxygenase-1 expression in rat malignant meningioma KMY-J cells. *J. Toxicol. Sci.* **2018**, *43*, 353–358. [CrossRef]

pharmaceuticals

MDPI

Review

Systematic Review and Meta-Analysis of In Vitro Anti-Human Cancer Experiments Investigating the Use of 5-Aminolevulinic Acid (5-ALA) for Photodynamic Therapy

Yo Shinoda *, Daitetsu Kato, Ryosuke Ando, Hikaru Endo, Tsutomu Takahashi, Yayoi Tsuneoka and Yasuyuki Fujiwara *

Department of Environmental Health, School of Pharmacy, Tokyo University of Pharmacy and Life Sciences, 1432-1 Horinouchi, Hachioji, Tokyo 192-0392, Japan; y151041@toyaku.ac.jp (D.K.); y151008@toyaku.ac.jp (R.A.); doendoenshning1403@gmail.com (H.E.); tsutomu@toyaku.ac.jp (T.T.); tsuneoka@toyaku.ac.jp (Y.T.)
* Correspondence: shinoda@toyaku.ac.jp (Y.S.); yasuyuki@toyaku.ac.jp (Y.F.)

Abstract: 5-Aminolevulinic acid (5-ALA) is an amino acid derivative and a precursor of protoporphyrin IX (PpIX). The photophysical feature of PpIX is clinically used in photodynamic diagnosis (PDD) and photodynamic therapy (PDT). These clinical applications are potentially based on in vitro cell culture experiments. Thus, conducting a systematic review and meta-analysis of in vitro 5-ALA PDT experiments is meaningful and may provide opportunities to consider future perspectives in this field. We conducted a systematic literature search in PubMed to summarize the in vitro 5-ALA PDT experiments and calculated the effectiveness of 5-ALA PDT for several cancer cell types. In total, 412 articles were identified, and 77 were extracted based on our inclusion criteria. The calculated effectiveness of 5-ALA PDT was statistically analyzed, which revealed a tendency of cancer-classification-dependent sensitivity to 5-ALA PDT, and stomach cancer was significantly more sensitive to 5-ALA PDT compared with cancers of different origins. Based on our analysis, we suggest a standardized in vitro experimental protocol for 5-ALA PDT.

Keywords: 5-aminolevulinic acid; 5-ALA; dALA; δALA; PpIX; protoporphyrin IX; photodynamic therapy; PDT

Citation: Shinoda, Y.; Kato, D.; Ando, R.; Endo, H.; Takahashi, T.; Tsuneoka, Y.; Fujiwara, Y. Systematic Review and Meta-Analysis of In Vitro Anti-Human Cancer Experiments Investigating the Use of 5-Aminolevulinic Acid (5-ALA) for Photodynamic Therapy. *Pharmaceuticals* **2021**, *14*, 229. https://doi.org/10.3390/ph14030229

Academic Editor: Serge Mordon

Received: 10 February 2021
Accepted: 27 February 2021
Published: 7 March 2021

1. Introduction

5-Aminolevulinic acid (5-ALA) is a naturally occurring amino acid derivative that acts as a precursor of protoporphyrin IX (PpIX) [1–3]. 5-ALA administration to animals, including humans, leads to the synthesis of PpIX, especially in tumors [4–7]. PpIX is activated by violet light (405 nm) or orange-red light (635 nm), subsequently emitting red fluorescence (620–710 nm) or generating reactive oxygen species (ROS) [8]. These features can potentially be used to visualize or kill cancer. Specifically, 5-ALA has been clinically tested for photodynamic diagnosis (PDD) during surgery to visualize cancer cells by fluorescence and photodynamic therapy (PDT) to target unfavorable neoplasms by increasing ROS production [9,10]. To date, 5-ALA has been clinically approved by the U.S. Food and Drug Administration (FDA) as GLEOLAN® (GLIOLAN® according to the European Medicines Agency (EMA)) for PDD for malignant glioma, and LEVULAN® and AMELUZ® have been approved for the PDT of patients with actinic keratoses. However, there are no FDA- or EMA-approved applications of 5-ALA-PDT for cancer.

Clinical anti-cancer applications of 5-ALA-PDT have been widely reported for several organs, such as the brain [11–16], skin [17–23], pharynx [24], blood and lymph [25], esophagus [26], urethra and prostate [27], and uterus [28]. In addition, 97 clinical trials of 5-ALA PDT for cancer treatment have been registered in the U.S. National Library of Medicine (ClinicalTrials.gov) as of 10 March 2021. Therefore, 5-ALA will hopefully be approved in the near future as a PDT drug for cancer patients. These clinical trials and applications are based on in vivo animal experiments, and these animal experiments are based on in vitro cell culture experiments. For the clinical application of 5-ALA PDT in cancer, a comprehensive review of in vitro 5-ALA PDT experiments and analysis of these results are meaningful and may provide important opportunities to consider for the future direction of 5-ALA experiments and clinical trials.

In this study, we systematically extracted and listed in vitro experiments that investigated 5-ALA PDT. We also performed a meta-analysis of these data by calculating and comparing the effectiveness of 5-ALA PDT in several cancer cell types from each article. Finally, we suggest a standard experimental protocol for the validation of future in vitro 5-ALA PDT experiments.

2. Methods

2.1. Literature Search and Selection

A systematic literature review using PubMed was performed according to the Preferred Reporting Items for Systematic Reviews and Meta-analyses Statement (PRISMA) guidelines [29]. For the title or abstract search, we used the term sets [5-aminolevulinic acid, aminolevulinic acid, dALA, δALA, 5-ALA, 5ALA], [in vitro, culture], and [photodynamic therapy, PDT] for OR searching in each term set, and each term set was used together for AND searching. In addition, the publication date was limited to the beginning of 1900 to the end of 2019 (available online). Together, the search query used was ((aminolevulinic acid [Title/Abstract] OR aminolevulinic acid [Title/Abstract] OR dALA [Title/Abstract] OR δALA [Title/Abstract] OR 5-ALA [Title/Abstract] OR 5ALA [Title/Abstract]) AND (in vitro [Title/Abstract] OR culture [Title/Abstract]) AND (photodynamic therapy [Title/Abstract] OR PDT [Title/Abstract])) AND (1900/01/01 [Date-Publication]: 2019/12/31 [Date-Publication]). The searched articles were further selected based on whether they included all of the following information: cell name, fluence, irradiation wavelength, time of incubation with 5-ALA, duration between 5-ALA treatment and irradiation, and duration between irradiation and viability assays. The selected articles that described the median lethal concentration (LC_{50}) in the text or those in which the LC_{50} could be estimated and/or calculated from the table or graph were included. Estimated LC_{50} and fluence from graphs were rounded.

2.2. Consistency of Terminology

Some papers used different terms to express the same thing. In addition, some standardizations of different terms were required to perform statistical analysis. Therefore, some terminologies were unified as follows: astrocytoma was used as glioblastoma, glioma stem-like cell was used as glioma stem cell, and glioblastoma stem-like cell was used as glioblastoma stem cell.

2.3. Data Collection, Processing, and Statistics

For data comparisons, the effectiveness of each application was calculated by the reciprocal of the fluence multiplied by the LC_{50} ($cm^2/(J \cdot \mu M)$). This effectiveness is thought to be proportional to the sensitivity of the cell to 5-ALA PDT. If there were more than three articles that used the same classified cells or cells from the same organ, similar wavelengths for irradiation (around 635 nm), the same duration of 5-ALA incubation (4 h), and the same duration between irradiation and viability assays (24 h), the values of effectiveness were averaged and assessed by one-way analysis of variance (ANOVA) with the post hoc Tukey–Kramer test using Statcel2 software (Seiunsha, Tokyo, Japan). If the sample size was greater than six, the data were assessed by the Wilcoxon rank-sum test using JMP Pro software (SAS Institute Japan, Tokyo, Japan).

3. Results

3.1. Collection of In Vitro 5-ALA PDT Experiments

The initial search resulted in 412 articles, of which 77 articles met the inclusion criteria mentioned in the Methods section. These articles included a total of 146 in vitro viability assays of cells treated with PDT under different conditions. They included 116, 12, 9, and 9 viability assays for cell lines derived from humans, mice, rats, and canines, respectively. The PDT experiments using human cell lines are listed in Table 1, and the total number of studies was 62. Eighty cell lines from different origins (21 organs) and classifications (16 classes) were tested. Brain cancer and adenocarcinoma were the most tested cancer origin and classification, respectively. Overall, several different experimental conditions were adapted in each study, including various durations of incubation with 5-ALA, irradiation wavelengths, fluences, and durations between irradiation and viability assays. Therefore, it was difficult to directly compare these experimental results. To ensure the comparability of the results, we extracted the data from studies that used similar experimental conditions (irradiation wavelength, duration of 5-ALA incubation, and duration between irradiation and viability assays). Then, we roughly estimated the effectiveness of 5-ALA PDT for different cells using the following equation. The effectiveness is the new parameter we introduced, which is thought to correlate to the sensitivity of the cells to 5-ALA PDT because both LC_{50} and fluence parameters are roughly inversely proportional to the sensitivity of the cells in PDT experiments [30,31].

$$Effectiveness = \frac{1}{LC_{50} \times Fluence}\left(\mathrm{cm}^2/[\mathrm{J}\cdot\mu\mathrm{M}]\right) \quad (1)$$

The effectiveness indicates the extent of 5-ALA effects on the treated cells under the individual experimental conditions. The larger the effectiveness value, the more effective 5-ALA PDT was against the cell. Although there were different effectiveness values estimated for the same cell types in different papers (such as 0.1 to 1,131.9 for A431 cells), most results showed a similar range of effectiveness values.

Table 1. In vitro 5-Aminolevulinic acid photodynamic therapy (5-ALA PDT) experiments for human cancer cell lines.

Organ	Classification	Name [s]: Sphere	Effectiveness ($\times 10^{-4}$ cm^2/(J·μM))	LC_{50} (μM)	Duration of Incubation (h)	Irradiation Wavelength (nm)	Fluence (J/cm^2)	Duration between Irradiation and Viability Assay (h)	Ref.
Bladder	Carcinoma	HCV-29	0.2	597	4	635	100	24	[32]
	Carcinoma	J82	1.1	597	4	635	15	24	[32]
	Carcinoma	J82	12.0	597	3	590–700	1.4	48	[33]
	Carcinoma	RT112	41.9	597	3	590–700	0.4	48	[33]
	Carcinoma	RT4	2.2	298	4	635	15	24	[32]
	Carcinoma	RT4	41.9	597	3	590–700	0.4	48	[33]
	Carcinoma	RT4 [s]	20.9	597	3	400–700	0.8	24	[34]
Bone	Chordoma	U-CH2	3.0	181	6	635	18.75	24	[35]
	Osteosarcoma	HOSM-1	2.5	200	6	580–740	20	24	[36]
	Osteosarcoma	HOSM-2	0.5	1000	12	600–1600	20	24	[37]
Brain	AT/RT	BT-16	1.1	370	4	635	25	12	[38]
	Glioblastoma	A172	3.3	1000	24	635	3	24	[39]
	Glioblastoma	ACBT [s]	0.6	597	4	635	30	24	[40]
	Glioblastoma	U251MG	3.3	1000	0.5	All (white)	3	20-24	[41]
	Glioblastoma	U251MG	5.0	1000	4	627	2	O/N	[42]
	Glioblastoma	U373	3.7	144	4	635	18.3	24	[43]
	Glioblastoma	U373	1.2	650	4	635	12.75	24	[44]
	Glioblastoma	U373	3.2	315	4	635 ± 20	10	24	[45]
	Glioblastoma	U373MG	3.5	500	2	635	5.7	48	[46]
	Glioblastoma	U373MG	7.1	1000	4	627	1.4	O/N	[47]
	Glioblastoma	U373MG	5.1	1000	4	635	1.95	48	[48]
	Glioblastoma	U373vIII	0.7	1100	4	635	12.75	24	[44]
	Glioblastoma	U373vIII	2.5	407	4	635 ± 20	10	24	[45]
	Glioblastoma	U87	1.5	510	4	635	12.75	24	[44]
	Glioblastoma	U87	1.1	931	4	635 ± 20	10	24	[45]

Table 1. *Cont.*

Organ	Classification	Name [s]: Sphere	Effectiveness ($\times 10^{-4}$ cm^2/(J·μM))	LC_{50} (μM)	Duration of Incubation (h)	Irradiation Wavelength (nm)	Fluence (J/cm^2)	Duration between Irradiation and Viability Assay (h)	Ref.
Brain (cont.)	Glioblastoma	U87MG	2.5	1000	0.5	All (white)	4	20–24	[41]
	Glioblastoma	U87MG	2.4	1000	6	634 ± 7	4.1	48	[31]
	Glioblastoma	U87MG	4.2	1000	4	627	2.4	O/N	[47]
	Glioblastoma	U87MG	3.3	1000	4	627	3	O/N	[42]
	Glioblastoma	U87vIII	0.3	2800	4	635	12.75	24	[44]
	Glioblastoma	U87vIII	0.9	1161	4	635 ± 20	10	24	[45]
	GSC	BT273 [s]	4.4	122	4	635	18.8	24	[43]
	GSC	BT275 [s]	10.7	49.5	4	635	18.8	24	[43]
	GSC	BT379 [s]	8.8	60.3	4	635	18.8	24	[43]
	GSC	GS3 [s]	4.3	124	4	635	18.8	24	[43]
	GSC	GS5 [s]	22.3	23.9	4	635	18.8	24	[43]
	Glioma stem cell	GS2	2.6	298	4	635	12.75	24	[44]
	Glioma stem cell	GS2	7.7	130	4	635 ± 20	10	24	[45]
	Glioma stem cell	GSC30 [s]	10.8	93	4	635 ± 20	10	24	[45]
	Medulloblastoma	D283 Med	5.3	500	2	635	3.8	48	[46]
	Medulloblastoma	Daoy	1.7	239	4	635	25	12	[38]
	Meningioma	KT21-MG1	1.2	448	24	635	18.75	1.5	[49]
	Neuroblastoma	SK-N-SH	1.4	1000	8	500–	7.2	48	[50]
	PNET	PFSK-1	1.7	239	4	635	25	12	[38]
Breast	Adenocarcinoma	MDA-MB-231	5.0	1000	0.5	633 ± 6	2	20	[51]
	Adenocarcinoma	MDA-MB-231	4.3	1000	6	634 ± 7	2.3	48	[31]
	Carcinoma	HB4a-Ras	166.7	1000	3	400-700	0.06	19	[52]
	Carcinoma	T47D	5.0	500	24	624 ± 5	4	24	[53]
	Carcinoma	T47D	16.7	1000	4	635	0.6	48	[48]
Colon	Adenocarcinoma	Caco-2	9.9	597	3	590–700	1.7	48	[54]
	Adenocarcinoma	HT-29	2.5	1000	3	635	4	24	[55]
	Adenocarcinoma	HT-29	4.3	597	3	590–700	3.9	48	[54]
	Adenocarcinoma	SW480	0.1	1500	4	600–720	50	24	[56]
	Adenocarcinoma	SW480	0.1	1500	4	600–720	53	24	[57]
	Adenocarcinoma	SW480	4.9	597	3	590–700	3.4	48	[54]
	Adenocarcinoma	SW620	0.4	1500	4	600–720	18	24	[56]
	Adenocarcinoma	SW620	0.4	1000	4	600–720	24	24	[57]
Esophagus	SCC	Eca-109	1.0	1000	24	630	10	24	[58]
	SCC	Eca-109	0.1	750	6	630	100	24	[59]
Gingiva	SCC	Ca9-22	20.8	1000	3	633	0.48	24	[60]
Hypopharynx	SCC	FADU	8.9	1000	24	635	1.12	24	[39]
Kidney	Carcinoma	A498	1.2	1000	6	634 ± 7	8.2	48	[31]
Larynx	SCC	AMC-HN3	7.0	239	24	632	6	24	[61]
Liver	Carcinoma	HepG2	10.0	1000	28	600–800	1	2	[62]
	Carcinoma	HepG2	2.2	185	3	632	25	24	[63]
Lung	Adenocarcinoma	LC-T	1.0	5000	9	600–700	2.1	0	[64]
	Carcinoma	H1299	5.0	1000	4	633	2	2.3	[65]
	Carcinoma	QU-DB	0.8	5000	9	600–700	2.5	0	[64]
Lymph	Lymphoma	HuT78	41.9	59.7	2	630	4	24	[66]
	Lymphoma	Ramos (RA1)	16.8	59.7	2	630	10	24	[66]
Nasopharynx	Carcinoma	HNE-1	3.0	328	4	630	10	24	[67]
	Carcinoma	KJ-1	3.6	1000	3	633	2.8	24	[60]
Oral Cavity	Dysplasia	DOK	1.2	810	4	635	10	24	[68]
Ovary	Adenocarcinoma	ES2	1.1	882	4	631	10.4	24	[69]
	Adenocarcinoma	KOC7C	1.1	857	4	631	10.4	24	[69]
	Adenocarcinoma	OV2774	1.3	1000	4	635	8	48	[48]
	Adenocarcinoma	OVMANA	9.9	97	4	631	10.4	24	[69]
	Adenocarcinoma	OVTOKO	3.9	244	4	631	10.4	24	[69]
	Adenocarcinoma	RMG1	17.1	56	4	631	10.4	24	[69]
	Adenocarcinoma	RMG2	17.1	56	4	631	10.4	24	[69]
	Adenocarcinoma	TOV21G	2.9	330	4	631	10.4	24	[69]
Prostate	Adenocarcinoma	LNCaP	11.2	298	4	631	3	24	[70]
Skin	Melanoma	A375	2.0	500	4	636	10	24	[71]

Table 1. *Cont.*

Organ	Classification	Name [s]: Sphere	Effectiveness ($\times 10^{-4}$ cm^2/(J·μM))	LC_{50} (μM)	Duration of Incubation (h)	Irradiation Wavelength (nm)	Fluence (J/cm^2)	Duration between Irradiation and Viability Assay (h)	Ref.
Skin (cont.)	Melanoma	A375	0.6	358	4	420–1400	45	24	[72]
	Melanoma	LOX	0.03	4000	4	635	100	20	[73]
	SCC	A431	0.6	6000	3	635 ± 9	3	0	[74]
	SCC	A431	17.0	393	20	630	1.5	24	[30]
	SCC	A431	12.5	100	24	632.8	8	24	[75]
	SCC	A431	0.1	2000	4	635	40	24	[76]
	SCC	A431	1131.9	1.77	48	630 ± 15	5	48	[77]
	SCC	HSC-5	1.0	200	2	545–700	50	3	[78]
	SCC	SCC-13	0.1	6000	1	635 ± 9	12.21	0	[74]
Stomach	Adenocarcinoma	KKLS	13.2	700	4	630	1.08	24	[79]
	Adenocarcinoma	MKN28	23.1	400	4	630	1.08	24	[79]
	Adenocarcinoma	MKN45	185.2	50	4	630	1.08	24	[79]
Tongue	SCC	CAL-27	1.6	620	4	635	10	24	[68]
	SCC	SCC-15	11.2	59.7	12	630	15	6	[80]
	SCC	SCC-4	5.3	187	4	640	10	24	[81]
	SCC	SCC-4	2.7	375	4	640	10	24	[82]
Uterus	Adenocarcinoma	BCC	16.7	500	4.5	532 ± 20	1.2	20	[83]
	Adenocarcinoma	HeLa	4.0	500	4	635	5	24	[84]
	Adenocarcinoma	HeLa	100.0	200	8	630	0.5	24	[85]
	Adenocarcinoma	HeLa	98.0	10.2	5	630	10	3	[86]
	Adenocarcinoma	HeLa	0.3	300	4	635	100	20	[87]
	Adenocarcinoma	HeLa	16.7	1000	24	635	0.6	24	[88]
	Adenocarcinoma	KB	1.3	200	6	580–740	40	24	[36]
	SCC	C-33A	588.2	1.7	6	630	10	3	[86]
	SCC	C-4 I	12.9	77.7	6	630	10	3	[86]
	SCC	Ca Ski	28.4	35.2	6	630	10	3	[86]
	SCC	HT-3	3.0	332	6	630	10	3	[86]
	SCC	Me-180	1373.6	0.728	6	630	10	3	[86]
	SCC	Me-180	660.1	0.505	4	632.8	30	4	[89]
	SCC	SiHa	3.0	332	6	630	10	3	[86]

AT/RT: atypical teratoid/rhabdoid tumor, GSC: glioblastoma stem cell, SCC: squamous cell carcinoma, PNET: primitive neuroectodermal tumor, O/N: overnight; LC_{50}, median lethal concentration.

3.2. 5-ALA PDT Effect on Cells of Different Cancer Classifications

The effectiveness against cells of the same cancer classifications was averaged and compared with each class (Figure 1). Several reports showed that the feature of the cells was altered by their microenvironments, such as 2D monolayer culture or 3D aggregation-forming spheres [90–93]. In the present review, because the inclusion and exclusion of the data from sphere cultures did not show any statistical differences, all of the data were included and averaged. Although there were no statistical significances, adenocarcinoma, glioblastoma stem cell (GSC), and glioma stem cell tended to show higher effectiveness values than squamous cell carcinoma (SCC), glioblastoma, and carcinoma, which may suggest that the effects of 5-ALA PDT are cancer-classification-dependent.

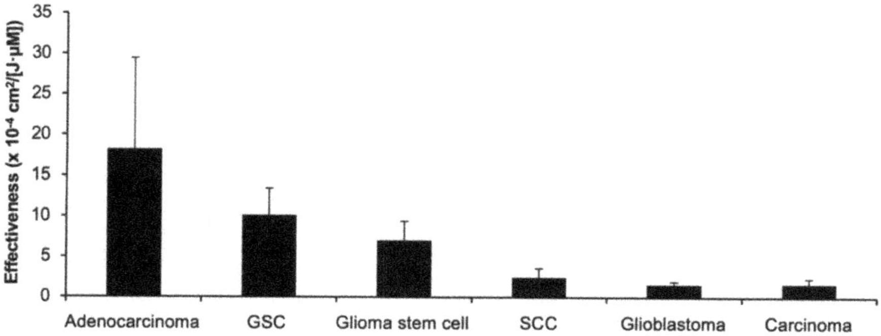

Figure 1. Effectiveness of 5-ALA PDT in cells of different cancer classifications. GSC: glioblastoma stem cell, SCC: squamous cell carcinoma. There were no statistical significances (one-way analysis of variance (ANOVA) with the post hoc Tukey–Kramer test). Adenocarcinoma ($n = 16$) and glioblastoma ($n = 10$) were statistically assessed using the Wilcoxon rank-sum test but showed no significant differences.

3.3. 5-ALA PDT Effect on Cells of Different Cancer Origins

Next, the effectiveness against cells of the same cancer origin was averaged and compared with each origin (Figure 2). Because the inclusion and exclusion of the data from sphere cultures and outliers did not show any statistical differences, all of the data were included and averaged. As a result, the stomach was identified as the organ most affected by 5-ALA PDT. Among the other organs, there were no statistically significant differences. However, the number of experiments using stomach-derived cells ($n = 3$) was small, and these experiments were performed in the same study. Therefore, this result should be considered carefully.

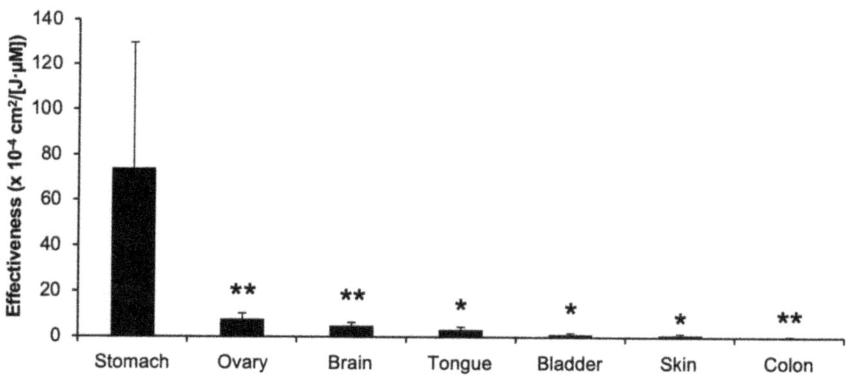

Figure 2. Effectiveness of 5-ALA PDT on cells of different cancer origins. The stomach was identified as the organ most affected by 5-ALA PDT (one-way analysis of variance (ANOVA) with the post hoc Tukey–Kramer test). * $p < 0.05$ and ** $p < 0.01$ compared with the stomach. The brain ($n = 18$) and ovary ($n = 7$) were statistically assessed using the Wilcoxon rank-sum test but showed no significant differences.

4. Discussion and Future Perspective

In the present review, we summarized past and recent (until the end of 2019) in vitro experiments investigating 5-ALA PDT for cancer cells and compared these data by calculating the effectiveness value. In total, 116 in vitro assays for human cancer cells were extracted, including cancer cells from 21 origins and 16 cancer classifications. Effectiveness values were calculated from the LC_{50} and fluence to compare the sensitivity of each cancer cell type to 5-ALA PDT. These data suggest that there are some tendencies of sensitivity to 5-ALA PDT in cells of different origins and classifications.

Several potential mechanisms may contribute to the differences in the sensitivity of each origin and classification to 5-ALA PDT. The most important candidate that influences the PDT sensitivity is the protein family associated with redox reactions. Oxidative stress-related proteins, such as superoxide dismutase [94], catalase [95], and NO synthase [96], are reported to affect PDT sensitivity. Glutathione and related proteins, such as glutathione peroxidase, glutathione-S-transferase, glutathione transferase omega-1, and glutathione synthase, are also thought to be associated with cell resistance to PDT [94,97,98]. Heme oxygenase-1 (HO-1) is an inducible cytoprotective enzyme that protects cells from oxidative stress, and its expression is induced by PDT [99]. Apurinic/apyrimidinic endonuclease 1/redox factor-1 (APE1/Ref-1) regulates cell responses to oxidative stress, which affects the PDT sensitivity [100]. The NAD(P)H/FAD redox status has been reported to affect PDT sensitivity [101]. The expression levels of these redox-related proteins, peptides, and compounds can be altered in cancer cells and may influence their sensitivity to 5-ALA PDT.

Several papers have reported the anti-cancer property of 5-ALA PDT; however, most articles do not describe all of the experimental protocols, preventing reproducibility. Although similar problems occur and should be considered in all manuscripts, it is still important for authors to describe their detailed experimental protocols to ensure reproducibility by other researchers and for reviewers to carefully assess the manuscripts. For the in vitro PDT experiments, complete chemical formulation of 5-ALA (such as 5-ALA hydrochloride) should be described, not 5-ALA alone; otherwise, the dimerization is reported [102]. Moreover, the parameters we mention in Table 1 (cell name, duration of incubation, irradiation wavelength, fluence, and duration between irradiation and viability assays) should be clearly mentioned in Section 2 because all of these parameters possibly affect the sensitivity of cells to 5-ALA PDT and the results of cell viability assays. Comparisons of the duration of 5-ALA incubation [35,68,103], irradiation wavelength [104,105], and fluence [30,31] showed that each parameter strongly affected cell viability. In addition, the duration between irradiation and viability assays can also affect the results because cell proliferation after light irradiation can be influenced by the number of viable cells with a sigmoid shape.

In this review, we compared the effectiveness calculated from the LC_{50} and fluence. The calculated effectiveness may be a useful parameter to compare the sensitivity of cells to 5-ALA PDT among individual manuscripts; however, it has some limitations to consider. Although both LC_{50} and fluence parameters are roughly inversely proportional to the sensitivity of the cells in PDT experiments, this inverse relationship (especially for fluence) does not show absolute linearity [31]. Based on our calculation, high fluence usually showed a relatively high effectiveness [30,57,106,107]. Therefore, the calculated effectiveness may be overestimated in the experiments using high fluence. The cellular microenvironments, such as 2D or 3D culture, are also possible candidates that may affect 5-ALA PDT sensitivity. Therefore, 2D/3D comparisons might be required. In this manuscript, the inclusion and exclusion of 3D culture data did not show any statistical changes, but this might be caused by the small number of datasets. We discarded several potential parameters that may affect in vitro PDT assays, such as the components of 5-ALA (hydrochloric acid, nitrate, and phosphate), initial cell density, light source (laser, lamp (halogen, mercury, or xenon), or LED), light irradiance, wash conditions, and completeness of light shielding except for light irradiation. These parameters can also affect the PDT results. For example, the value of light irradiance strongly influences cell viability [105,108,109]. Therefore, especially for the reproducibility, these parameters should be described in the individual manuscript together with the parameters described above.

For the meta-analysis and data comparison between each report, the experimental procedure for in vitro 5-ALA PDT experiments should be standardized as much as possible. We propose a standardized protocol developed with reference to papers listed in Table 1 (Scheme 1), and a similar protocol for investigating 5-ALA PDT in breast cancer cells has been recommended previously [110]. This protocol is for in vitro 5-ALA PDT experiments, but it can potentially be used for other in vitro PDT experiments using different photosensitizers with some modifications. The steps of this protocol are culturing cells to ~80% confluency, incubating cells with 5-ALA for 4 h, irradiating immediately after providing cells with fresh medium, and performing the viability assay 24 h after irradiation. A cell confluency of approximately 80% should be used for prevention of the effect of contact inhibition (including initiation of cell cycle arrest, downregulation of proliferation, and mitogen signaling pathways) [111]. It has been reported that a 4 h incubation time of 5-ALA is usually sufficient to induce maximal PDT effect for several cancer cells [60,65,112]. In addition, many studies including the references in this review used 4 h for their general experimental condition. However, some cells and experimental conditions require the incubation time of 5-ALA to be more than 4 h [62,85,107,113] and the time may also be affected by the 5-ALA concentration [35]. The timing of irradiation should be performed immediately after washout of 5-ALA because intracellular concentration of PpIX is gradually reduced after washout by cell metabolism [114]. We hope that this standardized protocol may help researchers conducting 5-ALA (or other photosensitizer) PDT experiments.

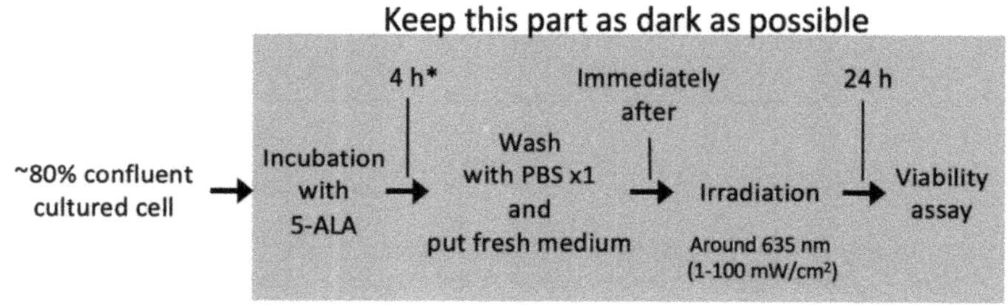

Scheme 1. Recommended experimental protocol for 5-ALA PDT. This protocol can be used as a standard protocol for in vitro 5-ALA PDT experiments. The duration of incubation (4 h may be a standard) can be changed if incubation time-dependency is investigated (*). Irradiance is particularly difficult to adjust because the light source is different in each lab, but the recommended irradiance is around 1 to 100 mW/cm^2. Note that the experimental procedure indicated inside the gray box should be performed in the dark as much as possible because undesirable irradiation from the fluorescent lights of laminar flow cabinets and/or experimental rooms can increase ROS production and subsequent cell death. PBS: phosphate-buffered saline.

More in vitro and in vivo 5-ALA PDT experiments for cancer should be performed to facilitate the clinical application of 5-ALA PDT in the future. In addition, 81 clinical trials have been registered in the U.S. National Library of Medicine. There are potential risks and side effects that should be considered during these trials and experiments; however, these efforts might advance the development of novel clinical approaches for the treatment of several cancers. To expand 5-ALA applications for several cancers, further in vitro 5-ALA PDT experiments are still required and should be continued.

Author Contributions: Conceptualization, Y.S.; methodology and validation, Y.S., T.T., and Y.T.; investigation, Y.S., D.K., R.A., and H.E.; writing—original draft preparation, Y.S.; writing—review and editing, Y.S., T.T., Y.T., and Y.F.; supervision, Y.S. and Y.F.; project administration, Y.S.; funding acquisition, Y.S. All authors have read and agreed to the published version of the manuscript.

Funding: This work was supported by a grant from the Institute for Social Medicine at Tokyo Univ. of Pharm. and Life Sci. (2018 and 2019 to Y.S.), the Shimabara Science Promotion Foundation (2017 to Y.S.), and GSK Japan (2015 to Y.S.).

Acknowledgments: We gratefully acknowledge Eisuke Enoki for his clinical advice.

Conflicts of Interest: The authors declare no competing financial interest.

References

1. Berlin, N.I.; Neuberger, A.; Scott, J.J. The metabolism of delta -aminolaevulic acid. 1. Normal pathways, studied with the aid of 15N. *Biochem. J.* **1956**, *64*, 80–90. [CrossRef] [PubMed]
2. Berlin, N.I.; Neuberger, A.; Scott, J.J. The metabolism of delta -aminolaevulic acid. 2. Normal pathways, studied with the aid of 14C. *Biochem. J.* **1956**, *64*, 90–100. [CrossRef]
3. Fukuda, H.; Casas, A.; Batlle, A. Aminolevulinic acid: From its unique biological function to its star role in photodynamic therapy. *Int. J. Biochem. Cell Biol.* **2005**, *37*, 272–276. [CrossRef]
4. Kennedy, J.C.; Pottier, R.H. Endogenous protoporphyrin IX, a clinically useful photosensitizer for photodynamic therapy. *J. Photochem. Photobiol. B* **1992**, *14*, 275–292. [CrossRef]
5. Kennedy, J.C.; Pottier, R.H.; Pross, D.C. Photodynamic therapy with endogenous protoporphyrin IX: Basic principles and present clinical experience. *J. Photochem. Photobiol. B* **1990**, *6*, 143–148. [CrossRef]
6. Krieg, R.C.; Messmann, H.; Rauch, J.; Seeger, S.; Knuechel, R. Metabolic characterization of tumor cell-specific protoporphyrin IX accumulation after exposure to 5-aminolevulinic acid in human colonic cells. *Photochem. Photobiol.* **2002**, *76*, 518–525. [CrossRef]
7. Zenzen, V.; Zankl, H. Protoporphyrin IX-accumulation in human tumor cells following topical ALA- and h-ALA-application in vivo. *Cancer Lett.* **2003**, *202*, 35–42. [CrossRef] [PubMed]
8. Berg, K.; Selbo, P.K.; Weyergang, A.; Dietze, A.; Prasmickaite, L.; Bonsted, A.; Engesaeter, B.O.; Angell-Petersen, E.; Warloe, T.; Frandsen, N.; et al. Porphyrin-related photosensitizers for cancer imaging and therapeutic applications. *J. Microsc.* **2005**, *218*, 133–147. [CrossRef] [PubMed]
9. Casas, A. Clinical uses of 5-aminolaevulinic acid in photodynamic treatment and photodetection of cancer: A review. *Cancer Lett.* **2020**, *490*, 165–173. [CrossRef] [PubMed]
10. Ishizuka, M.; Abe, F.; Sano, Y.; Takahashi, K.; Inoue, K.; Nakajima, M.; Kohda, T.; Komatsu, N.; Ogura, S.; Tanaka, T. Novel development of 5-aminolevurinic acid (ALA) in cancer diagnoses and therapy. *Int. Immunopharmacol.* **2011**, *11*, 358–365. [CrossRef]
11. Mahmoudi, K.; Garvey, K.L.; Bouras, A.; Cramer, G.; Stepp, H.; Jesu Raj, J.G.; Bozec, D.; Busch, T.M.; Hadjipanayis, C.G. 5-aminolevulinic acid photodynamic therapy for the treatment of high-grade gliomas. *J. Neurooncol.* **2019**, *141*, 595–607. [CrossRef] [PubMed]
12. Cramer, S.W.; Chen, C.C. Photodynamic Therapy for the Treatment of Glioblastoma. *Front. Surg.* **2020**, *6*, 81. [CrossRef] [PubMed]
13. Friesen, S.A.; Hjortland, G.O.; Madsen, S.J.; Hirschberg, H.; Engebraten, O.; Nesland, J.M.; Peng, Q. 5-Aminolevulinic acid-based photodynamic detection and therapy of brain tumors (review). *Int. J. Oncol.* **2002**, *21*, 577–582. [CrossRef]
14. Stepp, H.; Stummer, W. 5-ALA in the management of malignant glioma. *Lasers Surg. Med.* **2018**, *50*, 399–419. [CrossRef]
15. Nordmann, N.J.; Michael, A.P. 5-Aminolevulinic acid radiodynamic therapy for treatment of high-grade gliomas: A systematic review. *Clin. Neurol. Neurosurg.* **2020**, *201*, 106430. [CrossRef]
16. Tetard, M.C.; Vermandel, M.; Mordon, S.; Lejeune, J.P.; Reyns, N. Experimental use of photodynamic therapy in high grade gliomas: A review focused on 5-aminolevulinic acid. *Photodiagnosis Photodyn. Ther.* **2014**, *11*, 319–330. [CrossRef] [PubMed]
17. Champeau, M.; Vignoud, S.; Mortier, L.; Mordon, S. Photodynamic therapy for skin cancer: How to enhance drug penetration? *J. Photochem. Photobiol. B* **2019**, *197*, 111544. [CrossRef]
18. Blume, J.E.; Oseroff, A.R. Aminolevulinic acid photodynamic therapy for skin cancers. *Dermatol. Clin.* **2007**, *25*, 5–14. [CrossRef]
19. Naidoo, C.; Kruger, C.A.; Abrahamse, H. Simultaneous Photodiagnosis and Photodynamic Treatment of Metastatic Melanoma. *Molecules* **2019**, *24*, 3153. [CrossRef] [PubMed]
20. Zou, Y.; Zhao, Y.; Yu, J.; Luo, X.; Han, J.; Ye, Z.; Li, J.; Lin, H. Photodynamic therapy versus surgical excision to basal cell carcinoma: Meta-analysis. *J. Cosmet. Dermatol.* **2016**, *15*, 374–382. [CrossRef]
21. Marmur, E.S.; Schmults, C.D.; Goldberg, D.J. A review of laser and photodynamic therapy for the treatment of nonmelanoma skin cancer. *Dermatol. Surg.* **2004**, *30*, 264–271. [CrossRef] [PubMed]
22. Zeitouni, N.C.; Oseroff, A.R.; Shieh, S. Photodynamic therapy for nonmelanoma skin cancers. Current review and update. *Mol. Immunol.* **2003**, *39*, 1133–1136. [CrossRef]
23. De Vijlder, H.C.; Middelburg, T.; De Bruijn, H.S.; Martino Neumann, H.A.; Sterenborg, H.C.; Robinson, D.J.; De Haas, E.R. Optimizing ALA-PDT in the management of non-melanoma skin cancer by fractionated illumination. *G. Ital. Dermatol. Venereol.* **2009**, *144*, 433–439. [PubMed]
24. Jin, X.; Xu, H.; Deng, J.; Dan, H.; Ji, P.; Chen, Q.; Zeng, X. Photodynamic therapy for oral potentially malignant disorders. *Photodiagnosis Photodyn. Ther.* **2019**, *28*, 146–152. [CrossRef]
25. Oka, T.; Matsuoka, K.I.; Utsunomiya, A. Sensitive Photodynamic Detection of Adult T-cell Leukemia/Lymphoma and Specific Leukemic Cell Death Induced by Photodynamic Therapy: Current Status in Hematopoietic Malignancies. *Cancers* **2020**, *12*, 335. [CrossRef]
26. Gross, S.A.; Wolfsen, H.C. The role of photodynamic therapy in the esophagus. *Gastrointest. Endosc. Clin. N. Am.* **2010**, *20*, 35–53. [CrossRef]
27. Fukuhara, H.; Yamamoto, S.; Karashima, T.; Inoue, K. Photodynamic diagnosis and therapy for urothelial carcinoma and prostate cancer: New imaging technology and therapy. *Int. J. Clin. Oncol.* **2021**, *26*, 18–25. [CrossRef]
28. Matoba, Y.; Banno, K.; Kisu, I.; Aoki, D. Clinical application of photodynamic diagnosis and photodynamic therapy for gynecologic malignant diseases: A review. *Photodiagnosis Photodyn. Ther.* **2018**, *24*, 52–57. [CrossRef]

29. Moher, D.; Liberati, A.; Tetzlaff, J.; Altman, D.G. Preferred reporting items for systematic reviews and meta-analyses: The PRISMA statement. *PLoS Med.* **2009**, *6*, e1000097. [CrossRef] [PubMed]
30. Berlanda, J.; Kiesslich, T.; Engelhardt, V.; Krammer, B.; Plaetzer, K. Comparative in vitro study on the characteristics of different photosensitizers employed in PDT. *J. Photochem. Photobiol. B* **2010**, *100*, 173–180. [CrossRef] [PubMed]
31. Hartl, B.A.; Hirschberg, H.; Marcu, L.; Cherry, S.R. Characterizing low fluence thresholds for in vitro photodynamic therapy. *Biomed. Opt. Express* **2015**, *6*, 770–779. [CrossRef]
32. Riesenberg, R.; Fuchs, C.; Kriegmair, M. Photodynamic effects of 5-aminolevulinic acid-induced porphyrin on human bladder carcinoma cells in vitro. *Eur. J. Cancer* **1996**, *32A*, 328–334. [CrossRef]
33. Krieg, R.C.; Herr, A.; Raupach, K.; Ren, Q.; Schwamborn, K.; Knuechel, R. Analyzing effects of photodynamic therapy with 5-aminolevulinic acid (ALA) induced protoporphyrin IX (PPIX) in urothelial cells using reverse phase protein arrays. *Photochem. Photobiol. Sci.* **2007**, *6*, 1296–1305. [CrossRef]
34. Wild, P.J.; Krieg, R.C.; Seidl, J.; Stoehr, R.; Reher, K.; Hofmann, C.; Louhelainen, J.; Rosenthal, A.; Hartmann, A.; Pilarsky, C.; et al. RNA expression profiling of normal and tumor cells following photodynamic therapy with 5-aminolevulinic acid-induced protoporphyrin IX in vitro. *Mol. Cancer Ther.* **2005**, *4*, 516–528. [CrossRef] [PubMed]
35. Cornelius, J.F.; Eismann, L.; Ebbert, L.; Senger, B.; Petridis, A.K.; Kamp, M.A.; Sorg, R.V.; Steiger, H.J. 5-Aminolevulinic acid-based photodynamic therapy of chordoma: In vitro experiments on a human tumor cell line. *Photodiagnosis Photodyn. Ther.* **2017**, *20*, 111–115. [CrossRef] [PubMed]
36. Yanase, S.; Nomura, J.; Matsumura, Y.; Nagai, K.; Kinoshita, M.; Nakanishi, H.; Ohnishi, Y.; Tokuda, T.; Tagawa, T. Enhancement of the effect of 5-aminolevulinic acid-based photodynamic therapy by simultaneous hyperthermia. *Int. J. Oncol.* **2005**, *27*, 193–201. [CrossRef]
37. Nomura, J.; Yanase, S.; Tokuda, T.; Matsumura, Y.; Sekida, M.; Tagawa, T. Griseofulvin enhances the effect of aminolevulinic acid-based photodynamic therapy in vitro. *Photomed. Laser Surg.* **2006**, *24*, 186–191. [CrossRef] [PubMed]
38. Schwake, M.; Nemes, A.; Dondrop, J.; Schroeteler, J.; Schipmann, S.; Senner, V.; Stummer, W.; Ewelt, C. In-Vitro Use of 5-ALA for Photodynamic Therapy in Pediatric Brain Tumors. *Neurosurgery* **2018**, *83*, 1328–1337. [CrossRef]
39. Shi, L.; Buchner, A.; Pohla, H.; Pongratz, T.; Ruhm, A.; Zimmermann, W.; Gederaas, O.A.; Zhang, L.; Wang, X.; Stepp, H.; et al. Methadone enhances the effectiveness of 5-aminolevulinic acid-based photodynamic therapy for squamous cell carcinoma and glioblastoma in vitro. *J. Biophotonics* **2019**, *12*, e201800468. [CrossRef] [PubMed]
40. Hirschberg, H.; Sun, C.H.; Tromberg, B.J.; Yeh, A.T.; Madsen, S.J. Enhanced cytotoxic effects of 5-aminolevulinic acid-mediated photodynamic therapy by concurrent hyperthermia in glioma spheroids. *J. Neurooncol.* **2004**, *70*, 289–299. [CrossRef]
41. Fahey, J.M.; Emmer, J.V.; Korytowski, W.; Hogg, N.; Girotti, A.W. Antagonistic Effects of Endogenous Nitric Oxide in a Glioblastoma Photodynamic Therapy Model. *Photochem. Photobiol.* **2016**, *92*, 842–853. [CrossRef]
42. Albert, I.; Hefti, M.; Luginbuehl, V. Physiological oxygen concentration alters glioma cell malignancy and responsiveness to photodynamic therapy in vitro. *Neurol. Res.* **2014**, *36*, 1001–1010. [CrossRef]
43. Schimanski, A.; Ebbert, L.; Sabel, M.C.; Finocchiaro, G.; Lamszus, K.; Ewelt, C.; Etminan, N.; Fischer, J.C.; Sorg, R.V. Human glioblastoma stem-like cells accumulate protoporphyrin IX when subjected to exogenous 5-aminolaevulinic acid, rendering them sensitive to photodynamic treatment. *J. Photochem. Photobiol. B* **2016**, *163*, 203–210. [CrossRef]
44. Fisher, C.J.; Niu, C.; Foltz, W.; Chen, Y.; Sidorova-Darmos, E.; Eubanks, J.H.; Lilge, L. ALA-PpIX mediated photodynamic therapy of malignant gliomas augmented by hypothermia. *PLoS ONE* **2017**, *12*, e0181654. [CrossRef]
45. Fisher, C.; Obaid, G.; Niu, C.; Foltz, W.; Goldstein, A.; Hasan, T.; Lilge, L. Liposomal Lapatinib in Combination with Low-Dose Photodynamic Therapy for the Treatment of Glioma. *J. Clin. Med.* **2019**, *8*, 2214. [CrossRef]
46. Ritz, R.; Scheidle, C.; Noell, S.; Roser, F.; Schenk, M.; Dietz, K.; Strauss, W.S. In vitro comparison of hypericin and 5-aminolevulinic acid-derived protoporphyrin IX for photodynamic inactivation of medulloblastoma cells. *PLoS ONE* **2012**, *7*, e51974. [CrossRef]
47. Hefti, M.; Albert, I.; Luginbuehl, V. Phenytoin reduces 5-aminolevulinic acid-induced protoporphyrin IX accumulation in malignant glioma cells. *J. Neurooncol.* **2012**, *108*, 443–450. [CrossRef] [PubMed]
48. Sailer, R.; Strauss, W.S.; Wagner, M.; Emmert, H.; Schneckenburger, H. Relation between intracellular location and photodynamic efficacy of 5-aminolevulinic acid-induced protoporphyrin IX in vitro. Comparison between human glioblastoma cells and other cancer cell lines. *Photochem. Photobiol. Sci.* **2007**, *6*, 145–151. [CrossRef] [PubMed]
49. Cornelius, J.F.; Slotty, P.J.; El Khatib, M.; Giannakis, A.; Senger, B.; Steiger, H.J. Enhancing the effect of 5-aminolevulinic acid based photodynamic therapy in human meningioma cells. *Photodiagnosis Photodyn. Ther.* **2014**, *11*, 1–6. [CrossRef] [PubMed]
50. Wu, S.M.; Ren, Q.G.; Zhou, M.O.; Peng, Q.; Chen, J.Y. Protoporphyrin IX production and its photodynamic effects on glioma cells, neuroblastoma cells and normal cerebellar granule cells in vitro with 5-aminolevulinic acid and its hexylester. *Cancer Lett.* **2003**, *200*, 123–131. [CrossRef]
51. Fahey, J.M.; Girotti, A.W. Nitric oxide-mediated resistance to photodynamic therapy in a human breast tumor xenograft model: Improved outcome with NOS2 inhibitors. *Nitric Oxide* **2017**, *62*, 52–61. [CrossRef]
52. Calvo, G.; Saenz, D.; Simian, M.; Sampayo, R.; Mamone, L.; Vallecorsa, P.; Batlle, A.; Casas, A.; Di Venosa, G. Reversal of the Migratory and Invasive Phenotype of Ras-Transfected Mammary Cells by Photodynamic Therapy Treatment. *J. Cell. Biochem.* **2017**, *118*, 464–477. [CrossRef]

53. Ziegler, V.G.; Knaup, J.; Stahl, D.; Krammer, B.; Plaetzer, K. Fluorescence detection and depletion of T47D breast cancer cells from human mononuclear cell-enriched blood preparations by photodynamic treatment: Basic in vitro experiments towards the removal of circulating tumor cells. *Lasers Surg. Med.* **2011**, *43*, 548–556. [CrossRef]
54. Krieg, R.C.; Messmann, H.; Schlottmann, K.; Endlicher, E.; Seeger, S.; Scholmerich, J.; Knuechel, R. Intracellular localization is a cofactor for the phototoxicity of protoporphyrin IX in the gastrointestinal tract: In vitro study. *Photochem. Photobiol.* **2003**, *78*, 393–399. [CrossRef]
55. Hatakeyama, T.; Murayama, Y.; Komatsu, S.; Shiozaki, A.; Kuriu, Y.; Ikoma, H.; Nakanishi, M.; Ichikawa, D.; Fujiwara, H.; Okamoto, K.; et al. Efficacy of 5-aminolevulinic acid-mediated photodynamic therapy using light-emitting diodes in human colon cancer cells. *Oncol. Rep.* **2013**, *29*, 911–916. [CrossRef]
56. Wawrzyniec, K.; Kawczyk-Krupka, A.; Czuba, Z.P.; Krol, W.; Sieron, A. The influence of ALA-mediated photodynamic therapy on secretion of selected growth factors by colon cancer cells in hypoxia-like environment in vitro. *Photodiagnosis Photodyn. Ther.* **2015**, *12*, 598–611. [CrossRef] [PubMed]
57. Kawczyk-Krupka, A.; Czuba, Z.P.; Kwiatek, B.; Kwiatek, S.; Krupka, M.; Sieron, K. The effect of ALA-PDT under normoxia and cobalt chloride (CoCl2)-induced hypoxia on adhesion molecules (ICAM-1, VCAM-1) secretion by colorectal cancer cells. *Photodiagnosis Photodyn. Ther.* **2017**, *19*, 103–115. [CrossRef] [PubMed]
58. Chen, X.; Zhao, P.; Chen, F.; Li, L.; Luo, R. Effect and mechanism of 5-aminolevulinic acid-mediated photodynamic therapy in esophageal cancer. *Lasers Med. Sci.* **2011**, *26*, 69–78. [CrossRef]
59. Zhang, X.; Cai, L.; He, J.; Li, X.; Li, L.; Chen, X.; Lan, P. Influence and mechanism of 5-aminolevulinic acid-photodynamic therapy on the metastasis of esophageal carcinoma. *Photodiagnosis Photodyn. Ther.* **2017**, *20*, 78–85. [CrossRef] [PubMed]
60. Yang, T.H.; Chen, C.T.; Wang, C.P.; Lou, P.J. Photodynamic therapy suppresses the migration and invasion of head and neck cancer cells in vitro. *Oral Oncol.* **2007**, *43*, 358–365. [CrossRef] [PubMed]
61. Ahn, J.C.; Biswas, R.; Mondal, A.; Lee, Y.K.; Chung, P.S. Cisplatin enhances the efficacy of 5-aminolevulinic acid mediated photodynamic therapy in human head and neck squamous cell carcinoma. *Gen. Physiol. Biophys.* **2014**, *33*, 53–62. [CrossRef]
62. Yow, C.M.; Wong, C.K.; Huang, Z.; Ho, R.J. Study of the efficacy and mechanism of ALA-mediated photodynamic therapy on human hepatocellular carcinoma cell. *Liver. Int.* **2007**, *27*, 201–208. [CrossRef]
63. Vonarx-Coinsman, V.; Foultier, M.T.; de Brito, L.X.; Morlet, L.; Gouyette, A.; Patrice, T. HepG2 human hepatocarcinoma cells: An experimental model for photosensitization by endogenous porphyrins. *J. Photochem. Photobiol. B* **1995**, *30*, 201–208. [CrossRef]
64. Campbell, D.L.; Gudgin-Dickson, E.F.; Forkert, P.G.; Pottier, R.H.; Kennedy, J.C. Detection of early stages of carcinogenesis in adenomas of murine lung by 5-aminolevulinic acid-induced protoporphyrin IX fluorescence. *Photochem. Photobiol.* **1996**, *64*, 676–682. [CrossRef] [PubMed]
65. Su, G.C.; Wei, Y.H.; Wang, H.W. NADH fluorescence as a photobiological metric in 5-aminolevlinic acid (ALA)-photodynamic therapy. *Opt. Express* **2011**, *19*, 21145–21154. [CrossRef]
66. Boehncke, W.H.; Ruck, A.; Naumann, J.; Sterry, W.; Kaufmann, R. Comparison of sensitivity towards photodynamic therapy of cutaneous resident and infiltrating cell types in vitro. *Lasers Surg. Med.* **1996**, *19*, 451–457. [CrossRef]
67. Betz, C.S.; Lai, J.P.; Xiang, W.; Janda, P.; Heinrich, P.; Stepp, H.; Baumgartner, R.; Leunig, A. In vitro photodynamic therapy of nasopharyngeal carcinoma using 5-aminolevulinic acid. *Photochem. Photobiol. Sci.* **2002**, *1*, 315–319. [CrossRef] [PubMed]
68. Wang, X.; Jin, J.; Li, W.; Wang, Q.; Han, Y.; Liu, H. Differential in vitro sensitivity of oral precancerous and squamous cell carcinoma cell lines to 5-aminolevulinic acid-mediated photodynamic therapy. *Photodiagnosis Photodyn. Ther.* **2020**, *29*, 101554. [CrossRef] [PubMed]
69. Teshigawara, T.; Mizuno, M.; Ishii, T.; Kitajima, Y.; Utsumi, F.; Sakata, J.; Kajiyama, H.; Shibata, K.; Ishizuka, M.; Kikkawa, F. Novel potential photodynamic therapy strategy using 5-Aminolevulinic acid for ovarian clear-cell carcinoma. *Photodiagnosis Photodyn. Ther.* **2018**, *21*, 121–127. [CrossRef] [PubMed]
70. Chakrabarti, P.; Orihuela, E.; Egger, N.; Neal, D.E., Jr.; Gangula, R.; Adesokun, A.; Motamedi, M. Delta-aminolevulinic acid-mediated photosensitization of prostate cell lines: Implication for photodynamic therapy of prostate cancer. *Prostate* **1998**, *36*, 211–218. [CrossRef]
71. Robertson, C.A.; Abrahamse, H.; Evans, D. The in vitro PDT efficacy of a novel metallophthalocyanine (MPc) derivative and established 5-ALA photosensitizing dyes against human metastatic melanoma cells. *Lasers Surg. Med.* **2010**, *42*, 766–776. [CrossRef] [PubMed]
72. Cordoba, F.; Braathen, L.R.; Weissenberger, J.; Vallan, C.; Kato, M.; Nakashima, I.; Weis, J.; von Felbert, V. 5-aminolaevulinic acid photodynamic therapy in a transgenic mouse model of skin melanoma. *Exp. Dermatol.* **2005**, *14*, 429–437. [CrossRef]
73. Gilmore, B.F.; McCarron, P.A.; Morrow, D.I.; Murphy, D.J.; Woolfson, A.D.; Donnelly, R.F. In vitro phototoxicity of 5-aminolevulinic acid and its methyl ester and the influence of barrier properties on their release from a bioadhesive patch. *Eur. J. Pharm. Biopharm.* **2006**, *63*, 295–309. [CrossRef]
74. Novak, B.; Heesen, L.; Schary, N.; Lubbert, H. The influence of different illumination parameters on protoporphyrin IX induced cell death in squamous cell carcinoma cells. *Photodiagnosis Photodyn. Ther.* **2018**, *21*, 385–392. [CrossRef]
75. Shi, L.; Wang, X.; Zhao, F.; Luan, H.; Tu, Q.; Huang, Z.; Wang, H.; Wang, H. In vitro evaluation of 5-aminolevulinic acid (ALA) loaded PLGA nanoparticles. *Int. J. Nanomed.* **2013**, *8*, 2669–2676. [CrossRef]

76. Liu, H.; Daly, L.; Rudd, G.; Khan, A.P.; Mallidi, S.; Liu, Y.; Cuckov, F.; Hasan, T.; Celli, J.P. Development and evaluation of a low-cost, portable, LED-based device for PDT treatment of early-stage oral cancer in resource-limited settings. *Lasers Surg. Med.* **2019**, *51*, 345–351. [CrossRef]
77. Mei, L.H.; Yang, G.; Fang, F. Hyperbaric Oxygen Combined with 5-Aminolevulinic Acid Photodynamic Therapy Inhibited Human Squamous Cell Proliferation. *Biol. Pharm. Bull.* **2019**, *42*, 394–400. [CrossRef] [PubMed]
78. Ishida, N.; Watanabe, D.; Akita, Y.; Nakano, A.; Yamashita, N.; Kuhara, T.; Yanagishita, T.; Takeo, T.; Tamada, Y.; Matsumoto, Y. Etretinate enhances the susceptibility of human skin squamous cell carcinoma cells to 5-aminolaevulic acid-based photodynamic therapy. *Clin. Exp. Dermatol.* **2009**, *34*, 385–389. [CrossRef] [PubMed]
79. Hagiya, Y.; Endo, Y.; Yonemura, Y.; Takahashi, K.; Ishizuka, M.; Abe, F.; Tanaka, T.; Okura, I.; Nakajima, M.; Ishikawa, T.; et al. Pivotal roles of peptide transporter PEPT1 and ATP-binding cassette (ABC) transporter ABCG2 in 5-aminolevulinic acid (ALA)-based photocytotoxicity of gastric cancer cells in vitro. *Photodiagnosis Photodyn. Ther.* **2012**, *9*, 204–214. [CrossRef] [PubMed]
80. Takahashi, H.; Nakajima, S.; Sakata, I.; Ishida-Yamamoto, A.; Iizuka, H. Photodynamic therapy using a novel photosensitizer, ATX-S10(Na): Comparative effect with 5-aminolevulinic acid on squamous cell carcinoma cell line, SCC15, ultraviolet B-induced skin tumor, and phorbol ester-induced hyperproliferative skin. *Arch. Dermatol. Res.* **2005**, *296*, 496–502. [CrossRef] [PubMed]
81. Yang, D.F.; Chen, J.H.; Chiang, C.P.; Huang, Z.; Lee, J.W.; Liu, C.J.; Chang, J.L.; Hsu, Y.C. Improve efficacy of topical ALA-PDT by calcipotriol through up-regulation of coproporphyrinogen oxidase. *Photodiagnosis Photodyn. Ther.* **2014**, *11*, 331–341. [CrossRef]
82. Yang, D.F.; Lee, J.W.; Chen, H.M.; Huang, Z.; Hsu, Y.C. Methotrexate enhances 5-aminolevulinic acid-mediated photodynamic therapy-induced killing of human SCC4 cells by upregulation of coproporphyrinogen oxidase. *J. Formos. Med. Assoc.* **2014**, *113*, 88–93. [CrossRef]
83. JalalKamali, M.; Nematollahi-Mahani, S.N.; Shojaei, M.; Shamsoddini, A.; Arabpour, N. Effect of light polarization on the efficiency of photodynamic therapy of basal cell carcinomas: An in vitro cellular study. *Lasers Med. Sci.* **2018**, *33*, 305–313. [CrossRef] [PubMed]
84. Wei, X.Q.; Ma, H.Q.; Liu, A.H.; Zhang, Y.Z. Synergistic anticancer activity of 5-aminolevulinic acid photodynamic therapy in combination with low-dose cisplatin on Hela cells. *Asian Pac. J. Cancer Prev.* **2013**, *14*, 3023–3028. [CrossRef] [PubMed]
85. Zhou, X.; Wang, Y.; Si, J.; Zhou, R.; Gan, L.; Di, C.; Xie, Y.; Zhang, H. Laser controlled singlet oxygen generation in mitochondria to promote mitochondrial DNA replication in vitro. *Sci. Rep.* **2015**, *5*, 16925. [CrossRef] [PubMed]
86. He, G.F.; Bian, M.L.; Zhao, Y.W.; Xiang, Q.; Li, H.Y.; Xiao, C. Apoptosis-inducing effect of 5-aminolevulinic acid-mediated photodynamic therapy (5-ALA-PDT) on cervical cancer cell lines. *Chin. J. Cancer* **2008**, *27*, 195–200. [PubMed]
87. McCarron, P.A.; Donnelly, R.F.; Gilmore, B.F.; Woolfson, A.D.; McClelland, R.; Zawislak, A.; Price, J.H. Phototoxicity of 5-aminolevulinic acid in the HeLa cell line as an indicative measure of photodynamic effect after topical administration to gynecological lesions of intraepithelial form. *Pharm. Res.* **2004**, *21*, 1871–1879. [CrossRef] [PubMed]
88. Xie, J.; Wang, S.; Li, Z.; Ao, C.; Wang, J.; Wang, L.; Peng, X.; Zeng, K. 5-aminolevulinic acid photodynamic therapy reduces HPV viral load via autophagy and apoptosis by modulating Ras/Raf/MEK/ERK and PI3K/AKT pathways in HeLa cells. *J. Photochem. Photobiol. B* **2019**, *194*, 46–55. [CrossRef]
89. He, G.F.; Bian, M.L.; Zhao, Y.W.; Xiang, Q.; Li, H.Y.; Xiao, C. A study on the mechanism of 5-aminolevulinic acid photodynamic therapy in vitro and in vivo in cervical cancer. *Oncol. Rep.* **2009**, *21*, 861–868. [CrossRef]
90. Layer, P.G.; Robitzki, A.; Rothermel, A.; Willbold, E. Of layers and spheres: The reaggregate approach in tissue engineering. *Trends Neurosci.* **2002**, *25*, 131–134. [CrossRef]
91. Jensen, C.; Teng, Y. Is It Time to Start Transitioning From 2D to 3D Cell Culture? *Front. Mol. Biosci.* **2020**, *7*, 33. [CrossRef]
92. Knight, E.; Przyborski, S. Advances in 3D cell culture technologies enabling tissue-like structures to be created in vitro. *J. Anat.* **2015**, *227*, 746–756. [CrossRef]
93. Edmondson, R.; Broglie, J.J.; Adcock, A.F.; Yang, L. Three-dimensional cell culture systems and their applications in drug discovery and cell-based biosensors. *Assay. Drug. Dev. Technol.* **2014**, *12*, 207–218. [CrossRef] [PubMed]
94. Wright, K.E.; MacRobert, A.J.; Phillips, J.B. Inhibition of specific cellular antioxidant pathways increases the sensitivity of neurons to meta-tetrahydroxyphenyl chlorin-mediated photodynamic therapy in a 3D co-culture model. *Photochem. Photobiol.* **2012**, *88*, 1539–1545. [CrossRef] [PubMed]
95. Orlandi, V.T.; Martegani, E.; Bolognese, F. Catalase A is involved in the response to photooxidative stress in Pseudomonas aeruginosa. *Photodiagnosis Photodyn. Ther.* **2018**, *22*, 233–240. [CrossRef] [PubMed]
96. Girotti, A.W.; Fahey, J.M.; Korytowski, W. Nitric oxide-elicited resistance to anti-glioblastoma photodynamic therapy. *Cancer. Drug. Resist.* **2020**, *3*, 401–414. [CrossRef] [PubMed]
97. Theodossiou, T.A.; Olsen, C.E.; Jonsson, M.; Kubin, A.; Hothersall, J.S.; Berg, K. The diverse roles of glutathione-associated cell resistance against hypericin photodynamic therapy. *Redox. Biol.* **2017**, *12*, 191–197. [CrossRef] [PubMed]
98. Yokoyama, Y.; Shigeto, T.; Miura, R.; Kobayashi, A.; Mizunuma, M.; Yamauchi, A.; Futagami, M.; Mizunuma, H. Differences in the sensitivity of ovarian cancer to photodynamic therapy and the mechanisms for those differences. *Oncol. Lett.* **2017**, *13*, 4933–4938. [CrossRef]
99. Takahashi, T.; Misawa, S.; Suzuki, S.; Saeki, N.; Shinoda, Y.; Tsuneoka, Y.; Akimoto, J.; Fujiwara, Y. Possible mechanism of heme oxygenase-1 expression in rat malignant meningioma KMY-J cells subjected to talaporfin sodium-mediated photodynamic therapy. *Photodiagnosis Photodyn. Ther.* **2020**, *32*, 102009. [CrossRef] [PubMed]

100. Franchi, L.P.; de Freitas Lima, J.E.B.; Piva, H.L.; Tedesco, A.C. The redox function of apurinic/apyrimidinic endonuclease 1 as key modulator in photodynamic therapy. *J. Photochem. Photobiol. B* **2020**, *211*, 111992. [CrossRef]
101. Mikesova, L.; Mikes, J.; Koval, J.; Gyuraszova, K.; Culka, L.; Vargova, J.; Valekova, B.; Fedorocko, P. Conjunction of glutathione level, NAD(P)H/FAD redox status and hypericin content as a potential factor affecting colon cancer cell resistance to photodynamic therapy with hypericin. *Photodiagnosis Photodyn. Ther.* **2013**, *10*, 470–483. [CrossRef] [PubMed]
102. Bunke, A.; Zerbe, O.; Schmid, H.; Burmeister, G.; Merkle, H.P.; Gander, B. Degradation mechanism and stability of 5-aminolevulinic acid. *J. Pharm. Sci.* **2000**, *89*, 1335–1341. [CrossRef]
103. Ma, Y.; Qu, S.; Xu, L.; Lu, H.; Li, B. An in vitro study of the effect of 5-ALA-mediated photodynamic therapy on oral squamous cell carcinoma. *BMC Oral Health* **2020**, *20*, 258. [CrossRef] [PubMed]
104. Liu, B.; Farrell, T.J.; Patterson, M.S. Comparison of photodynamic therapy with different excitation wavelengths using a dynamic model of aminolevulinic acid-photodynamic therapy of human skin. *J. Biomed. Opt.* **2012**, *17*, 088001. [CrossRef]
105. Helander, L.; Krokan, H.E.; Johnsson, A.; Gederaas, O.A.; Plaetzer, K. Red versus blue light illumination in hexyl 5-aminolevulinate photodynamic therapy: The influence of light color and irradiance on the treatment outcome in vitro. *J. Biomed. Opt.* **2014**, *19*, 088002. [CrossRef] [PubMed]
106. Di Venosa, G.; Vallecorsa, P.; Giuntini, F.; Mamone, L.; Batlle, A.; Vanzuli, S.; Juarranz, A.; MacRobert, A.J.; Eggleston, I.M.; Casas, A. The use of dipeptide derivatives of 5-aminolaevulinic acid promotes their entry to tumor cells and improves tumor selectivity of photodynamic therapy. *Mol. Cancer Ther.* **2015**, *14*, 440–451. [CrossRef]
107. Perotti, C.; Fukuda, H.; DiVenosa, G.; MacRobert, A.J.; Batlle, A.; Casas, A. Porphyrin synthesis from ALA derivatives for photodynamic therapy. In vitro and in vivo studies. *Br. J. Cancer* **2004**, *90*, 1660–1665. [CrossRef]
108. Yamamoto, J.; Yamamoto, S.; Hirano, T.; Li, S.; Koide, M.; Kohno, E.; Okada, M.; Inenaga, C.; Tokuyama, T.; Yokota, N.; et al. Monitoring of singlet oxygen is useful for predicting the photodynamic effects in the treatment for experimental glioma. *Clin. Cancer Res.* **2006**, *12*, 7132–7139. [CrossRef] [PubMed]
109. Angell-Petersen, E.; Spetalen, S.; Madsen, S.J.; Sun, C.H.; Peng, Q.; Carper, S.W.; Sioud, M.; Hirschberg, H. Influence of light fluence rate on the effects of photodynamic therapy in an orthotopic rat glioma model. *J. Neurosurg.* **2006**, *104*, 109–117. [CrossRef]
110. Guney Eskiler, G.; Deveci Ozkan, A.; Sozen Kucukkara, E.; Kamanli, A.F.; Gunoglu, B.; Yildiz, M.Z. Optimization of 5-aminolevulinic acid-based photodynamic therapy protocol for breast cancer cells. *Photodiagnosis Photodyn. Ther.* **2020**, *31*, 101854. [CrossRef] [PubMed]
111. Levine, E.M.; Becker, Y.; Boone, C.W.; Eagle, H. Contact Inhibition, Macromolecular Synthesis, and Polyribosomes in Cultured Human Diploid Fibroblasts. *Proc. Natl. Acad. Sci. USA* **1965**, *53*, 350–356. [CrossRef] [PubMed]
112. Fujishiro, T.; Nonoguchi, N.; Pavliukov, M.; Ohmura, N.; Kawabata, S.; Park, Y.; Kajimoto, Y.; Ishikawa, T.; Nakano, I.; Kuroiwa, T. 5-Aminolevulinic acid-mediated photodynamic therapy can target human glioma stem-like cells refractory to antineoplastic agents. *Photodiagnosis Photodyn. Ther.* **2018**, *24*, 58–68. [CrossRef] [PubMed]
113. Tsai, J.C.; Hsiao, Y.Y.; Teng, L.J.; Chen, C.T.; Kao, M.C. Comparative study on the ALA photodynamic effects of human glioma and meningioma cells. *Lasers Surg. Med.* **1999**, *24*, 296–305. [CrossRef]
114. Iinuma, S.; Farshi, S.S.; Ortel, B.; Hasan, T. A mechanistic study of cellular photodestruction with 5-aminolaevulinic acid-induced porphyrin. *Br. J. Cancer* **1994**, *70*, 21–28. [CrossRef] [PubMed]

pharmaceuticals MDPI

Article

Synthesis of Multifunctional Nanoparticles for the Combination of Photodynamic Therapy and Immunotherapy

Mei-Hwa Lee [1,*], James L. Thomas [2], Jin-An Li [3], Jyun-Ren Chen [3], Tzong-Liu Wang [3] and Hung-Yin Lin [3,*]

1 Department of Materials Science and Engineering, I-Shou University, Kaohsiung 84001, Taiwan
2 Department of Physics and Astronomy, University of New Mexico, Albuquerque, NM 87131, USA; jthomas@unm.edu
3 Department of Chemical and Materials Engineering, National University of Kaohsiung, Kaohsiung 81148, Taiwan; m1075612@mail.nuk.edu.tw (J.-A.L.); 4a140109@stust.edu.tw (J.-R.C.); tlwang@nuk.edu.tw (T.-L.W.)
* Correspondence: meihwalee@ntu.edu.tw (M.-H.L.); linhy@ntu.edu.tw (H.-Y.L.)

Abstract: Programmed death-ligand 1 protein (PD-L1) has been posited to have a major role in suppressing the immune system during pregnancy, tissue allografts, autoimmune disease and other diseases, such as hepatitis. Photodynamic therapy uses light and a photosensitizer to generate singlet oxygen, which causes cell death (phototoxicity). In this work, photosensitizers (such as merocyanine) were immobilized on the surface of magnetic nanoparticles. One peptide sequence from PD-L1 was used as the template and imprinted onto poly(ethylene-*co*-vinyl alcohol) to generate magnetic composite nanoparticles for the targeting of PD-L1 on tumor cells. These nanoparticles were characterized using dynamic light scattering, high-performance liquid chromatography, Brunauer-Emmett-Teller analysis and superconducting quantum interference magnetometry. Natural killer-92 cells were added to these composite nanoparticles, which were then incubated with human hepatoma (HepG2) cells and illuminated with visible light for various periods. The viability and apoptosis pathway of HepG2 were examined using a cell counting kit-8 and quantitative real-time polymerase chain reaction. Finally, treatment with composite nanoparticles and irradiation of light was performed using an animal xenograft model.

Keywords: immunotherapy; photodynamic therapy; programmed death-ligand 1 protein (PD-L1); magnetic nanoparticles; peptide-imprinted polymer

Citation: Lee, M.-H.; Thomas, J.L.; Li, J.-A.; Chen, J.-R.; Wang, T.-L.; Lin, H.-Y. Synthesis of Multifunctional Nanoparticles for the Combination of Photodynamic Therapy and Immunotherapy. *Pharmaceuticals* **2021**, *14*, 508. https://doi.org/10.3390/ph14060508

Academic Editor: Serge Mordon

Received: 30 April 2021
Accepted: 21 May 2021
Published: 26 May 2021

1. Introduction

Molecularly imprinted polymers (MIPs) have been synthesized for use as biomimetic antibodies for bioseparation [1,2], biosensing [3–5] or gene activation [6,7]. The targets chosen for imprinting have been small molecules [5], proteins [5,8], or even cells [9,10]. Templates used included entire molecules [3,11], or epitopes [5,12] of targets. Especially with proteins, epitope imprinting is often used owing to the high cost of proteins, or solubility issues [1,13]. The specific recognition capabilities of MIPs are comparable to that of natural antibodies, and their stability is generally much better [14].

The science of protein epitope selection [5] or rational MIP design is immature, but several groups have demonstrated notable successes. Shea's group employed peptides that contain nine amino acids from the C-terminus of two proteins (melittin and green fluorescent protein (GFP)) to prepare MIP nanoparticles [15,16]. Zhang's group also used the C-terminal peptide for the recognition of albumin [17,18]. Kunter's group selected epitopes from the crystal database of proteins that could be digested with various proteinases [19], and then the epitopes on the outside of the proteins were chosen for the synthesis of MIPs (used as sensing elements on an extended-gate field-effect transistor [20,21].) Our earlier investigations have also demonstrated epitope-based recognition of a protein using peptide-imprinted polymers [5,13,22]. MIPs may also be employed as artificial receptors for the recognition of surface ligands of cells [23,24].

Immune checkpoint pathways, including the programmed death receptor-1/programmed death ligand-1 (PD-1/PD-L1) signaling pathway [25], are important in mediating self-tolerance and controlling self-damage. This pathway can sometimes be manipulated by cancer cells to evade immune surveillance [26]. PD-L1 binds to its receptor, PD-1 on activated T cells, B cells and myeloid cells, to modulate activation or inhibition [27]. Recently, cancer immunotherapies [28–30] that train or stimulate immunological systems to recognize, attack, and eradicate tumor cells with minimal damage to healthy cells have yielded promising clinical responses; those that involve nanoparticles [29,31] for chemotherapy, radiotherapy-, photothermal therapy, photodynamic therapy and hyperthermia are especially effective [31]. The development of strategies to block PD-L1-mediated immune inhibition [32] could further enhance the effectiveness of immunotherapy.

Photodynamic therapy (PDT) uses light and a photosensitizer to elicit cell death (phototoxicity) [33]. Singlet oxygen ($^1O_2$), as the major reactive oxygen species (ROS), is produced in PDT to treat cancer [34]; mechanisms of action include direct cytotoxic effects exerted on tumor cells, destruction of the tumor and peritumoral vasculature, and induction of local acute inflammatory reaction [35]. The commonly used photosensitizer merocyanine 540 (MC 540) [36] is a member of the family of benzoxazol merocyanine dyes that contain heterocyclic aromatic groups linked by a polymethine chain. It has recently been used in studies of up-converting nanoparticles (UCNP) [37,38].

Hepatocellular carcinoma (HCC) is a primary malignancy of the liver, occurring mostly in patients with underlying chronic liver disease and cirrhosis [39]. Liver cancer is one of the leading causes of cancer deaths globally, and an annual death toll of 700,000 has been recorded in recent years [40]. Hepatocellular carcinoma treatments include surgery, liver transplantation, the destruction of cancer cells using heat or cold, the delivery of chemotherapy, radiation therapy, targeted drug therapy, immunotherapy, and new liver cancer treatments currently in clinical trials [41,42]. A combination of treatment modalities, rather than a single treatment, may be most effective in slowing the progression of the disease in humans.

In the present work, multifunctional MIP NPs were synthesized for cancer cell destruction. As shown in Figure 1, merocyanine (MC540) molecules were immobilized on the surface of magnetic nanoparticles (MNPs). These MC540/MNPs were then mixed with a peptide from PD-L1 and poly(ethylene-*co*-vinyl alcohol) solution to form multifunctional magnetic peptide-imprinted composite nanoparticles (MPIPs/MC540). These MPIPs were characterized for their size distribution, recognition capacity, specific surface area, and magnetization. Natural killer-92 (NK-92) cells were added to human hepatoma (HepG2) cells, which were then incubated with these composite nanoparticles and illuminated with visible light for varying durations. The viability and apoptosis pathway of HepG2 were investigated using cell counting kit-8 (CCK8) and quantitative real-time polymerase chain reaction (qRT-PCR). Finally, treatment with composite nanoparticles and irradiation of light was performed on an animal xenograft model.

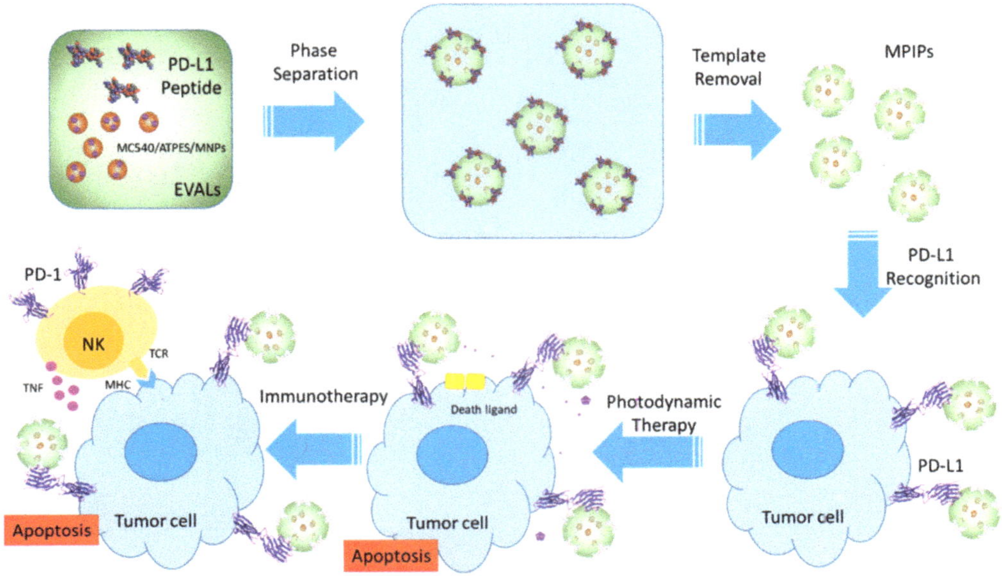

Figure 1. Preparation and administration of multifunctional magnetic PD-L1 peptide-imprinted composite nanoparticles.

2. Results and Discussion

Figure 2 presents the characteristics, including size distribution and mean size, of MIP NPs that were prepared with poly(ethylene-*co*-vinyl alcohol)s, EVALs, that contained various ethylene mole percents. As shown in Figure 2a, the magnetic nanoparticles (MNPs) had diameters of 61 ± 6 nm, while the freshly prepared magnetic peptide imprinted particles (MPIPs; made with 27 mol% EVAL and containing the MNPs) had diameters of 165 ± 28 nm. Interestingly, washing to remove the template actually increased the particle sizes (to 238 ± 30 nm) as measured by dynamic light scattering (DLS), perhaps owing to a swelling of the surface polymer layer. Rebinding of the template led to contraction of the MPIPs to 183 ± 47 nm. Figure 2b shows the mean sizes for particles prepared with different ethylene mole percentages in the EVAL. There is a trend to larger "before washing" sizes with increasing ethylene, but more interestingly, the intermediate ethylene percentages (32 and 38 mol%) gave larger particles after washing (~360 and ~490 nm) and yet the smallest particles on rebinding (~120 and 125 nm). N_2 adsorption-desorption isotherms (Figure 2c) were used to determine the specific surface area, using the multi-point Brunauer–Emmett–Teller (BET) method. The specific surface areas of MPIPs before and after template removal were 301.4 ± 23.9 m^2/g and 337.7 ± 35.4 m^2/g, respectively, for particles prepared using EVAL with ethylene at 27 mol%. The specific surface area in the MPIPs increased slightly upon template removal, as might be expected. The magnetization curves of the MNPs, and MPIPs before and after template removal, plotted in Figure 2d, reveal their superparamagnetic properties; their saturated magnetizations were found to be 58, 48 and 50 emu/g. The inset in Figure 2d shows the accumulation of MPIPs on the side of a vial nearest a magnet.

Figure S1a in the Supplementary Materials shows the adsorption of peptide on imprinted and non-imprinted nanoparticles made with EVALs with different ethylene contents. EVAL with the lowest ethylene mol% studied, 27 mol%, gave the largest imprinting effectiveness (defined as the ratio of adsorption of template molecules on MMIPs to that on the magnetic non-imprinted polymer composite nanoparticles, MNIPs) in Table S2. Figure S1b plots the isothermal adsorptions of peptide on MPIPs and on MNIPs. The maximum binding on MPIPs and NPIPs was 66.1 ± 12.8 and 39.1 ± 9.4 mg/g, respectively.

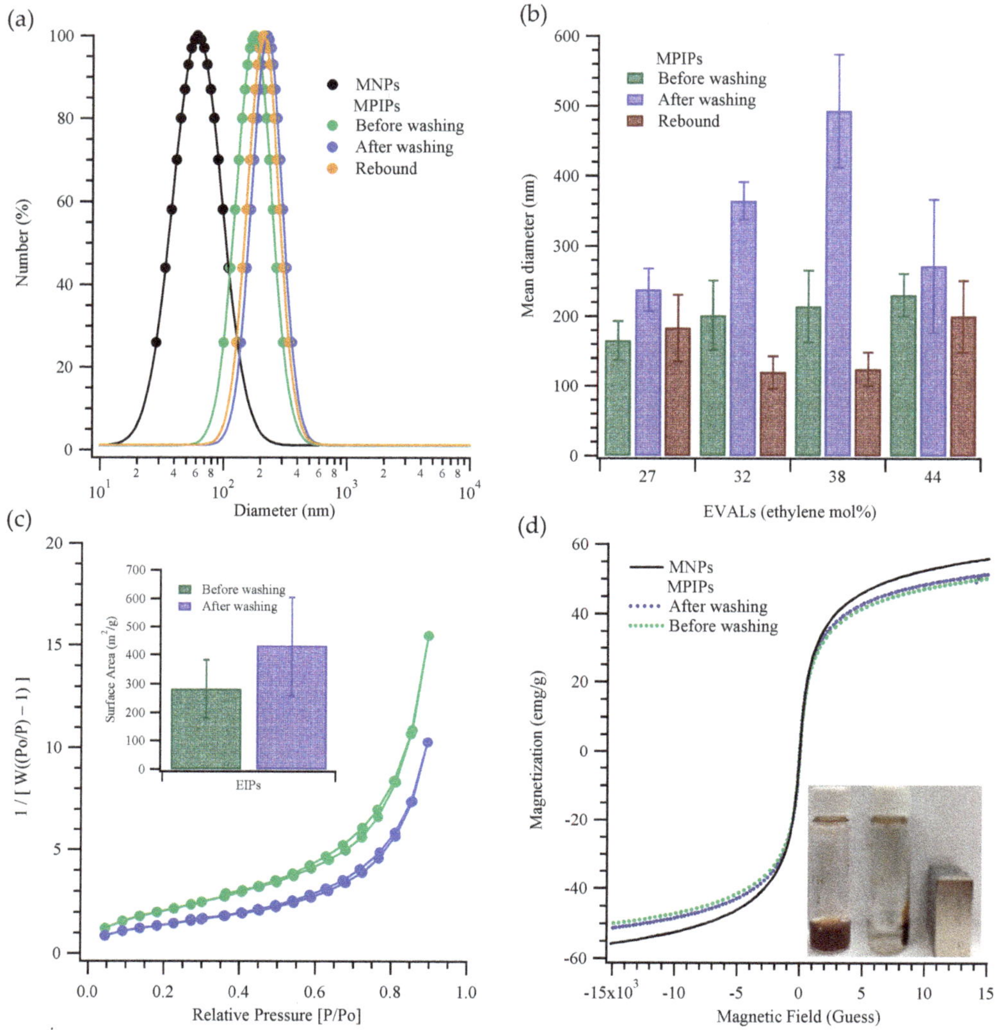

Figure 2. (**a**) Particles size distributions of MNPs (•); MPIPs before (•), after (•) template removal and rebound with template (•). (**b**) Mean diameters of MPIPs before, after template removal and rebound with template containing various ethylene mol% of EVAL. (**c**) Surface area of MPIPs before (•) and after (•) template removal measured by adsorption and desorption of nitrogen. (**d**) Magnetization of MNPs and MPIPs before and after template removal. Inset: MPIPs on the walls of a vial under magnetic field.

Figure S2 displays the viabilities of HepG2 cells under various conditions, but without immunotherapy (i.e. without incubation with NK-92 cells.) Figures S2a and S3 show the viabilities of cells (measured with the CCK-8 kit, normalized to the cell viability obtained with no treatment) incubated for 24 h with four different NPs: unmodified magnetic nanoparticles (MNP), magnetic nanoparticles with surface modified 3-triethoxysilylpropylamine (MNPs/APTES), magnetic nanoparticles with surface APTES and bound merocyanine photosensitizer (MNPs/APTES/MC540), and non-imprinted polymer-coated composite particles (MNIPs). In general, these controls show little effect on cellular viability, though there appears to be a small enhancement with particles containing photosensitizer.

Figure S2b shows HepG2 viability vs irradiation (illumination) duration, for cells in medium containing the free merocyanine photosensitizer MC540. 10 μg/mL MC540 did not suppress the growth of HepG2 cells even under irradiation for as long as 25 min; however, higher concentrations of MC540 dramatically reduced cellular viability when irradiated for longer than 10 min. In contrast, a short duration of irradiation (~5 min) and a large dosage of MC540 actually seemed to promote cell growth after 24 h. Figure S2c plots the viability of HepG2 cells with various concentrations of MEIP/MC540 NPs with or without 15 min of irradiation. The strong phototoxicity of MC540 is retained even when the photosensitizer is incorporated into the peptide-imprinted polymeric nanoparticles.

Figure 3 displays the viabilities of HepG2 cells when dosed with different concentrations of nanoparticles, with or without NK cells and with or without irradiation. Without NK-92 cells and without illumination, the viability of HepG2 cells with various concentrations of MNIPs was about 95%, increasing to 140 and 250% on the second and third days compared to the controls on day 0, as shown in Figure 3a (Viabilities >100% reflect continued cell growth and division.) In contrast, the viability of HepG2 cells with various concentrations of MPIP/MC540 was about 95%, increasing to 160 and 200% on the second and third days compared to the controls on day 0, as shown in Figure 3b. The phototoxicity of MPIP/MC540 up to 500 μg/mL after 1 day is shown in Figure S2c; the viability for every day administration of a lower concentration of MPIPs/MC540 after 1–3 days is shown in Figure 3c. Treatment with MPIPs/MC540, followed by NK-92 cells after 1 day ((but without irradiation), Figure 3d)), resulted in arrested growth of the cell population, essentially regardless of the MPIPs/MC540 concentration. High MPIPs/MC540 concentrations did result in greater prompt cell death, but even the lowest concentration (zero) showed negligible cell growth. The inhibition of cell growth thus appears to be caused by the NK-92 cells alone [43]. Figure 3e shows the combined effects of phototoxicity and NK-92 treatment. Cells were maintained in media containing different concentrations of MPIP/MC540; each day, cells were treated with photodynamic therapy for 15 min. (The medium was also changed every day, but replaced with fresh medium containing the specified concentration of MPIPs/MC540.) The viabilities are lower than with NK-92 killer cells alone. If MPIPs block PD-L1 and prevent it from binding to PD-1, they should enhance the ability of NK-92 cells to destroy HepG2 cancer cells. Figure 3f demonstrates that a combination of photodynamic therapy and immunotherapy can increase the in vitro suppression of HepG2 cells.

The viability of HepG2 cells treated with MNIPs with NK-92 cells was reduced by approximately 60%. Furthermore, photodynamic therapy can promote the suppression of HepG2 cells from 60% to about 30%; binding of MPIPs on HepG2 increased suppression to 20%. Similar results can be found in Figure S5 for the treatments of Hepa1-6 cells with MNIPs, MPIPs and MPIPs/MC540 with or without irradiation.

The pathway of photodynamic therapy with MPIPs/MC540 was analyzed by qRT-PCR, as presented in Figure 4. This analysis showed that apoptosis was induced using MPIPs/MC540 with irradiation by upregulating caspase-8, caspase-3 caspase-9 and Bax, and downregulation of Bcl-2 protein expression in Figure 4a. Figure 4b shows that gene expressions of caspase-8, caspase-3 caspase-9 and Bax were increased under irradiation of light, compared to no light control group. The apoptosis of HepG2 cells was caused by the activation of Casp9, but not Casp8, possibly indicating the endocytosis of MPIPs/MC540 both with and without the addition of NK-92 cells. $^1O_2$ was produced inside the HepG2 cells and increased the expression of proapoptotic Bcl family proteins (including Bax and Bid). However, the addition of NK-92 cells for immunotherapy increased the expression antiapoptotic proteins (such as Bcl2), possibly also inhibiting the expression of CYCS and then Casp9. Surprisingly, the combination of immunotherapy and photodynamic therapy dramatically promotes the expression of Casp8 and then Casp3, inducing the apoptosis of HepG2 cells.

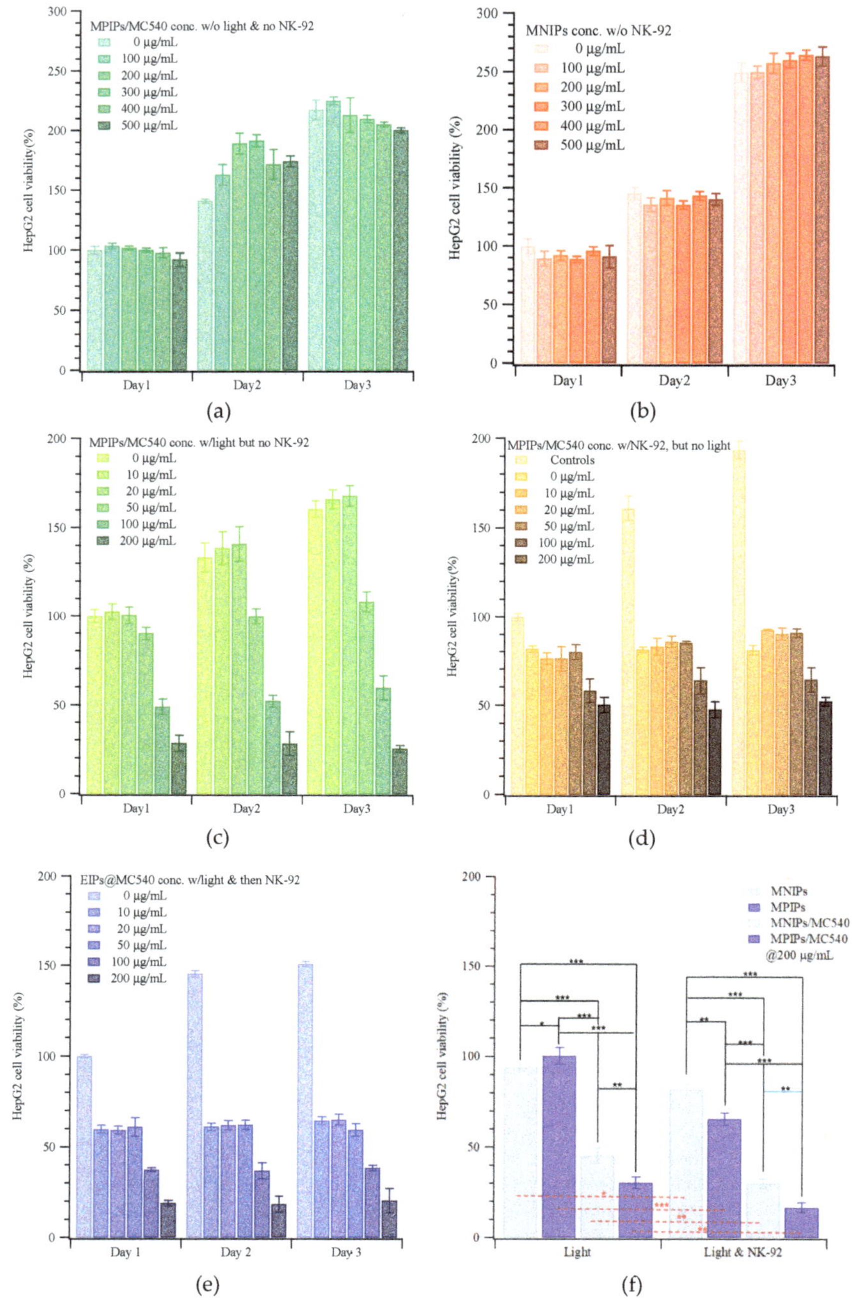

Figure 3. Continuous cellular viability measurements of HepG2 cells incubated with various concentrations of (**a**) MNIPs, (**b**) MPIPs/MC540 without irradiation or NK-92 treatment, (**c**) MPIPs/MC540 with irradiation. (**d**) MPIPs/MC540 with NK-92 treatment, (**e**) MPIPs/MC540 with irradiation or NK-92 treatment, (**f**) Comparison of cellular viability of HepG2 cells with NK-92 cells (2×10^4 cells/well), MNIPs/MC540 or MPIPs/MC540 (200 μg/mL) with or without addition of NK-92 cells NK-92 cells (2×10^4 cells/well) under irradiation (* $p < 0.05$; ** $p < 0.005$; *** $p < 0.0005$).

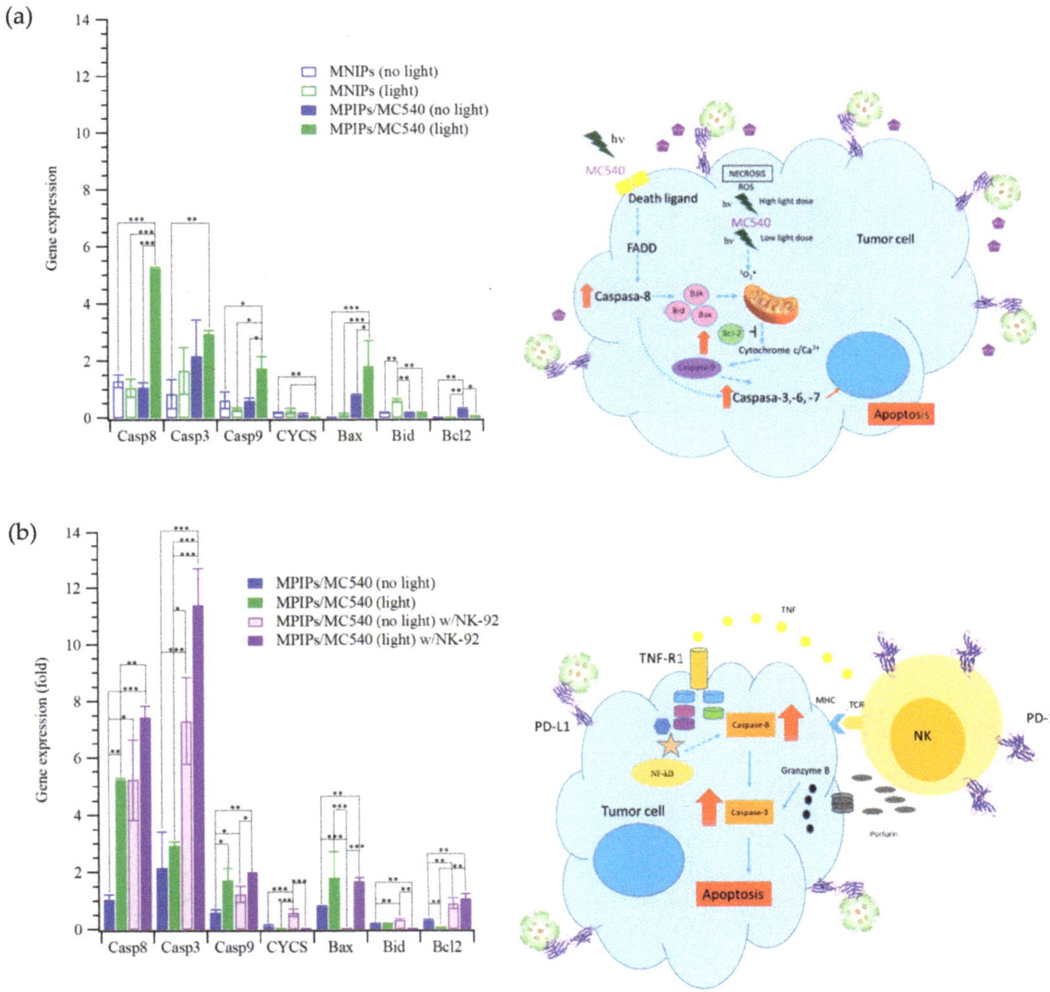

Figure 4. Relative gene expression of cellular apoptosis (Casp8, Casp3, Casp9, CYCS, Bax, Bid and Bcl-2) of HepG2 cells treated with (**a**) MNIPs, MPIPs/MC540 and (**b**) MPIPs/MC540 and additional NK-92 cells (2×10^4 cells/well) under irradiation or not (* $p < 0.05$; ** $p < 0.005$; *** $p < 0.0005$).

The treatment of an animal tumor model using composite nanoparticles is shown in Figure 5. In Figure 5a, photographs of mice treated with PBS (control), 10.0, 1.0 and 10.0 mg/kg of MPIPs/MC540; the latter two groups were irradiated with 15 min of 520 nm light. All mice had initially similar body mass of 24–29 g. The tumor volume for the control increased from 65 to 600 mm^3 over 11 days. Both the low and high dosages of MPIPs/MC540 accompanied by irradiation with light were effective at decreasing tumor size from 60–75 mm^3 to 33.1 $\pm$ 28.9 and 17.6 $\pm$ 4.8 mm^3, respectively. Surprisingly, the high dosage of MPIPs/MC540 (alone, without irradiation) still suppressed the growth of tumor, keeping it about 1/3 the size of the control. Nonetheless, the most effective treatment combined MPIPs/MC540 with irradiation.

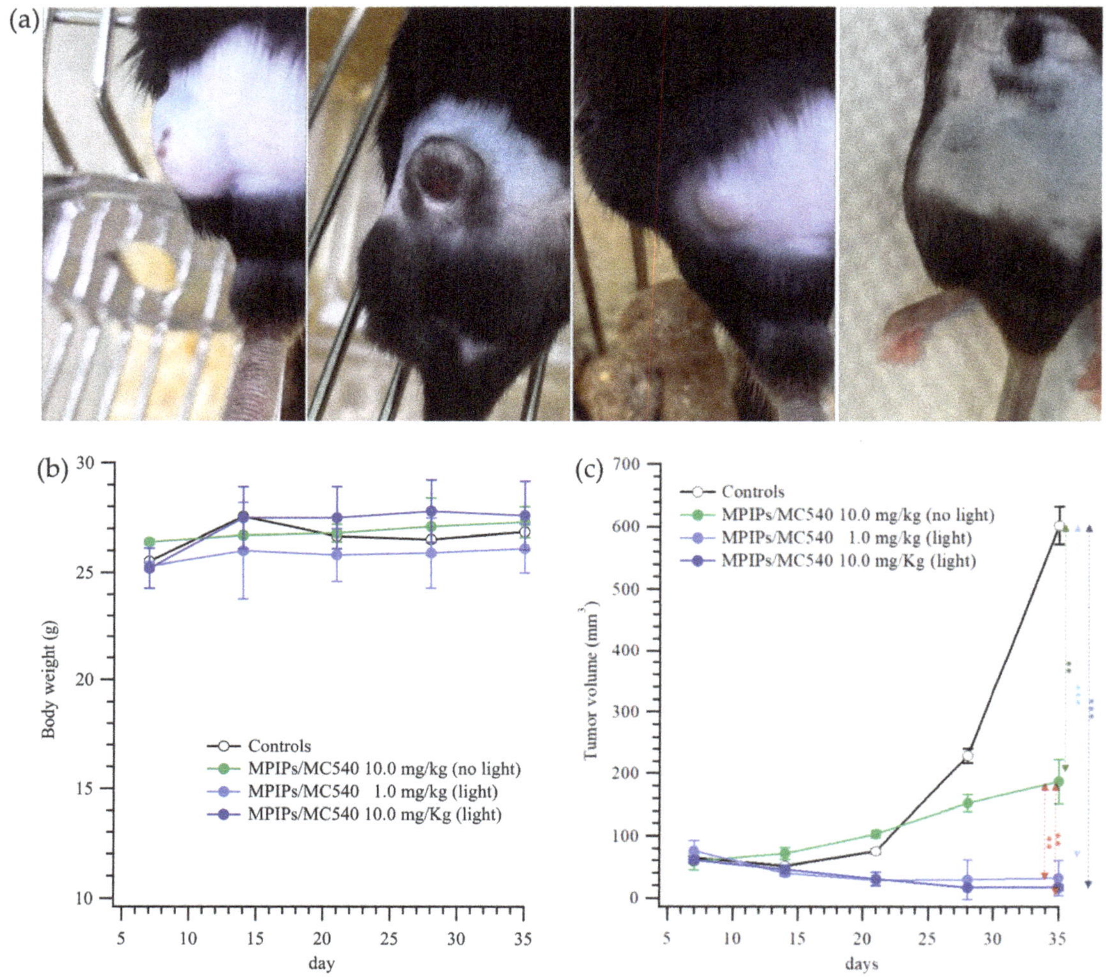

Figure 5. (**a**) Pictures of mice with treatment of PBS, 10.0, 1.0 and 10.0 mg/kg of MPIPs/MC540; the latter two groups were irradiated with 15 min of an LED mini dot light 1 h after injection. (**b**) Body weight of mice during the treatment. (**c**) The tumor size measured on the four groups of mice mentioned in 4 (**a**) (** $p < 0.005$; *** $p < 0.0005$).

The morphology and sizes of the tumors are shown in Figure 6, after sacrifice of the mice at 35 days, showing the control tumor, a high dose (10.0 mg/kg) of MPIPs/MC540, a low dose (1.0 mg/kg) with irradiation, and a high dose with irradiation. The tumor sizes are in agreement with the pictures and measurements in Figure 5a,c. The solid tumor had nearly disappeared using high dosages of MPIPs/MC540 combined with irradiation of light. The size of the solid tumor was bigger without irradiation at the same 1.0 mg/kg dose of MPIPs/MC540.

Hematoxylin and eosin (HE) stains and immunohistochemistry (IHC) stains (using proliferating cell nuclear antigen (PCNA) antibody) have been used for recognizing the growth and proliferation of cancer cells, respectively. The results of HE stains were shown for control, low/high dosages of MPIPs/MC540 with irradiation of light and low dosages of MPIPs/MC540 without irradiation in Figure 6b. Nuclei and cytoplasm are purple and pink, respectively, using HE stains for live cancer tissue. Comparing control and low/high dosages of MPIPs/MC540 with/without irradiation of light, the purple distribution was looser at high dosages of MPIPs/MC540 with irradiation of light. This supports that, at high dosages of MPIPs/MC540 with irradiation of light, tumor cell growth is suppressed. PCNA immunohistochemistry with antigen retrieval was used to measure the proliferation of cells. In Figure 6c, the brown and blue-purple were PCNA and hematoxylin stains, respectively; tumor proliferation was slightly reduced without irradiation of light at high dosages of MPIPs/MC540, but more effectively reduced under irradiation of light at low and high dosages of MPIPs/MC540.

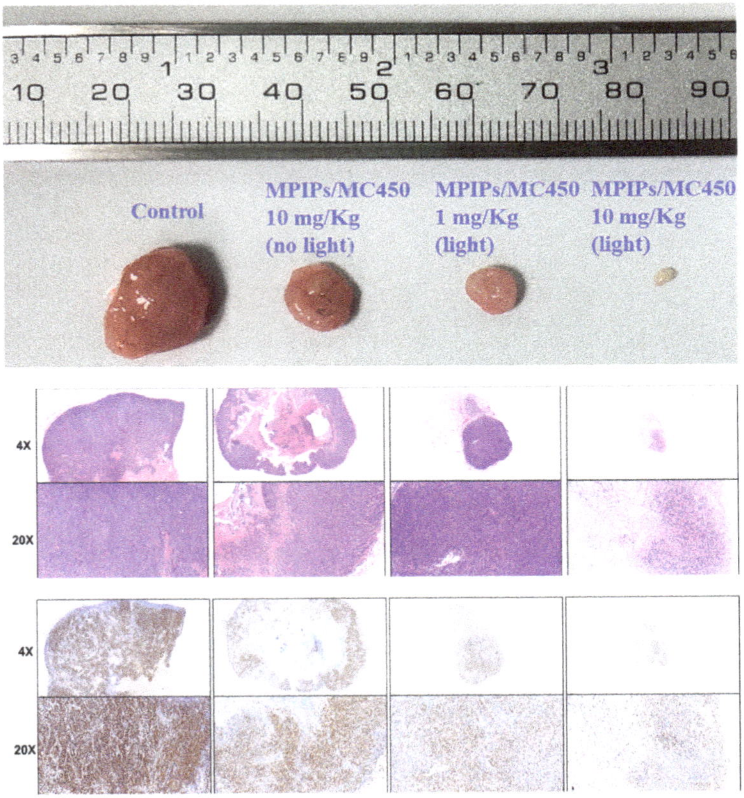

Figure 6. (**a**) Pictures of tumor specimens from the xenograft models with treatment of PBS (control), 10.0, 1.0 and 10.0 mg/kg of MPIPs/MC540; the latter two groups were irradiated with 15 min of an LED mini dot light 1 h after injection. (**b**) Hematoxylin & eosin (HE) staining of above tumor specimens. (**c**) Immunohistochemical (IHC) staining with anti-proliferating cell nuclear antigen (PCNA) primary antibodies.

3. Materials and Methods

3.1. Reagents and Chemicals

One peptide of PD-L1 (EDLKVQHSSYRQRA) was ordered from Yao-Hong Biotechnology Inc. (HPLC grade, New Taipei City, Taiwan). Poly(ethylene-*co*-vinyl alcohol), EVAL, with ethylene 27, 32, 38 and 44 mol%, 3-triethoxysilylpropylamine (APTES), merocyanine 540 (MC540) and RT-PCR primers, which were listed in Table S1, were from Sigma-Aldrich Co. (St. Louis, MO, USA). Iron (III) chloride 6-hydrate (97%), iron(II) sulphate 7-hydrate (99.0%) and dimethyl sulfoxide (DMSO) were from Panreac (Barcelona, Spain). DMSO was used as the solvent to dissolve EVAL polymer particles in the concentration of 1 wt%. Absolute ethyl alcohol was from J.T. Baker (ACS grade, Phillipsburg, NJ, USA). Both human hepatoblastoma (HepG2, #60364) and human natural killer cell (NK-92, #60414) were obtained from Bioresource Collection and Research Centre (BCRC), Taiwan.

3.2. *Formation of Multifunctional Magnetic Peptide-Imprinted Composite Nanoparticles*

The synthesis of multifunctional magnetic peptide-imprinted composite nanoparticles included the following steps: (1) magnetic nanoparticles, synthesized by co-precipitation of a mixture of iron (III) chloride 6-hydrate and iron (II) sulfate 7-hydrate by sodium hydroxide, were repeatedly washed while adsorbed on a magnetic plate [13]. (2) Two g of magnetic nanoparticles were mixed with 90 mL of ethanol to form a uniformly dispersed solution, and 180 µL APTES was then added. The mixture was stirred in a water bath at 90 °C for 90 min. Then, the APTES/MNPs were washed with 10 mL 95% alcohol for 5 min (3×), separated with a magnet after each washing. Three mL of MC540 (1.0 mg/mL) and 10 g of APTES/MNPs were mixed for 10 min and then washed with deionized water twice. (3) Peptide was dissolved in DMSO, at concentrations of 0.2, 2, 20, 100 and 200 µg/mL. 250 µL EVAL/DMSO solution was added into the same volume of peptide solution to form a clear EVAL solution, and 10 mg of the composite nanoparticles were then added. The EVAL was precipitated by dispersing 0.5 mL EVAL solution into 10 mL deionized water; then template was removed by washing in 10 mL deionized water 15 min (3×), separating the multifunctional magnetic peptide-imprinted polymer composite nanoparticles (MPIPs/MC540) magnetically after each washing. The magnetic non-imprinted polymer composite nanoparticles (MNIPs) were prepared identically, but without peptide addition.

3.3. *Cytotoxicity Test of HepG2 Cells with Magnetic Peptide-Imprinted Composite Nanoparticles and NK-92 Cells*

3.3.1. MTT Assay for Cell Viability

HepG2 cells were cultured in 90% of a 1:1 ratio mixture of Dulbecco's modified Eagle's medium (DMEM) and Ham's F12 medium with 10% heat-inactivated fetal bovine serum (FBS), supplemented with 0.4 mg/mL G418 (Geneticin) at 37 °C and 5% CO_2. For the cytotoxicity experiments 10 µL of 7.5.0x10^4 HepG2 cells and 190 µL culture medium per well (7.5 × 10^3 HepG2 cells per well) were seeded in 96 well culture plates and then incubated at 37 °C in 5% CO_2 for 24 h. Various concentrations of nanoparticles were added to each well at 37 °C for 24 h. 20 µL MTT (3-(4,5-dimethylthiazol-2-yl)-2,5 diphenyltetrazolium bromide, a yellow tetrazole) solution in phosphate buffered saline (PBS) was added to each well after 24 h, and then incubated in 5% CO_2 for 3 h at 37 °C. The solution was removed from each well. Then, 100 µL dimethyl sulfoxide (DMSO) was added to each well and incubated at 37 °C for 30 min in the dark, until cells have lysed and purple crystals have dissolved. The absorption intensities were measured by an ELISA reader (CLARIOstar, BMG Labtech, Offenburg, Germany) at 450 nm (I_{450}), and the reference absorption (I_{ref}, to account for turbidity and scattering) was obtained at 650 nm. The cellular viability (%) was then calculated from the ratio of effective absorption (I_{450}–I_{ref}) to controls.

3.3.2. Photodynamic Therapy (PDT)

The HepG2 cells at a density of 5 × 10^4 cells/mL were added to a 24-well culture plate and incubated for 24 h at 37 °C in 5% CO_2. Various concentrations of MNPs, APTES/MNPs, MNPs/MC540, and MNIP/MC540, MPIPs/MC540 nanoparticles were added to HepG2 cells, respectively. The well plate was exposed to photodynamic light irradiation using an LED mini dot light (HO HUA Electronic Components Company, Kaohsiung, Taiwan) for 15 min, with wavelengths from 460 to 580 nm and peak intensity at 520 nm. The distance was 1.5 cm between the LED device (50 W/m^2) and the cell plate. Then, the well culture plate was incubated for 24 h at 37 °C in 5% CO_2. The cellular viability of HepG2 cells was analyzed using MTT assay (see Section 3.3.1).

3.3.3. Immunotherapy with NK-92 Cells

NK-92 cells were cultured in RPMI1640 with L-glutamine and 10% FBS at 37 °C and 5% CO_2. After PDT, a CCK-8 cell counting kit (Sigma Aldrich, Kumamoto, Japan) was used to repeatedly assess cell viability. Particles were removed from wells and 500 µL of CCK-8 solution was added before the absorption intensity measurements by an ELISA reader, as described in Section 3.3.1. Only NK-92 (2×10^4 cells/well) were added on the second and third day.

3.4. Gene Expression of HepG2 Cells Treated with MPIPs, NK92 and PDT

The sequence (5′–3′) of primers for GAPDH, Caspase 8, Caspase 3, Caspase 9, Cytochrome C (CYCS), Bax, Bid, Bcl-2 was listed in Table S1. The total RNA extraction from the HepG2 cells cultured one day after UV irradiation was performed using the KingFisher Total RNA Kit and the KingFisher mL magnetic particle processors, both from Thermo Scientific (Vantaa, Finland). Complementary DNA was obtained following a Deoxy+ real-time 2× SYBR green RT-PCR kit (Yeastern Biotech Co., Ltd., Taipei, Taiwan) protocol. The RT-PCR was then performed in a PikoReal RT-PCR system (Thermo Scientific). Relative gene expression was determined using the $\Delta\Delta$Cq method [44] and normalized to a reference gene (GAPDH) and to a control (HepG2).

3.5. Animal Model and Immunohistochemical Staining

The animal experiments were approved by the Institutional Animal Care and Use Committees (IACUC) of National University of Kaohsiung (NUK) (protocol No. 10708, 10 November 2018) and performed in accordance with the Association for Assessment and Accreditation of Laboratory Animal Care guidelines (http://www.aaalac.org, accessed on 25 May 2021). To generate murine subcutaneous tumors, (1×10^6 cells/mouse) Hepa1-6 tumor cells were transplanted into the right flank of C57CL/6 mice. When the tumor volumes were around 50∼100 mm^3, the mice were randomly divided into four groups and were intravenously injected with PBS, MEIPs/MC540 (1.0 or 10.0 mg/kg). One hour after injection, the low (1.0 mg/kg) and (10.0 mg/kg) groups were then irradiated with light (wavelength 520 nm, intensity 1W) for 15 min. This treatment was performed twice a week for 5 weeks. On fifth week after the start of treatment, tumors were removed. Tumors were measured twice weekly, and the tumor volume (mm^3) was calculated as (long diameter × short diameter^2)/2. Once the mice exhibit signs of impaired health or when the volume of the tumor exceeded 1.5 cm^3, the mice were euthanized with CO_2.

Tumor specimens from the xenograft models were cut into 5 µm slices, fixed in 10% neutral buffered formalin, and embedded in paraffin. Slides were preincubated in 5% goat serum (Abcam, Cambridge, UK) in PBS and immunostained with anti-proliferating cell nuclear antigen (PCNA; GeneTex, Irvine, CA, USA) primary antibodies (both diluted 1:500 at 37 °C for 2 h. Slides were treated with hematoxylin for 30 s for visualization under a light microscope (UltraVision System; Thermo).

3.6. Data Analysis

All experiments were carried out in triplicate and data are expressed as means ± standard deviation. The cellular viability, gene expression data and tumour volume were analyzed with Student's t-test. Statistical significance was set at a p-value of less 0.05, significant as $p < 0.05$, highly significant as $p < 0.005$, extremely significant as $p < 0.0005$.

4. Conclusions

In this work, molecularly imprinted polymer (MIP) particles were synthesized to target the programmed death receptor (PD-L1) on HepG2 cells, and a photosensitizer (MC540) was incorporated to enhance the efficacy of photodynamic therapy. Additionally, blocking the PD-L1 protein with the MIP particles increased the functionality of natural killer (NK-92) cells. A biomolecular pathway investigation of HepG2 cells revealed that the singlet oxygen produced by irradiation promoted the expression of proapoptotic Bcl family proteins. Furthermore, the NK-92 cells further promoted the expression of Casp8 in the apoptosis of HepG2 cells. To summarize, this work demonstrated the efficacy of this combined therapy on a tumor model in mice and identified pathways and mechanisms of action.

Supplementary Materials: The following are available online at https://www.mdpi.com/article/10.3390/ph14060508/s1, Experimental Section S1.1 Characterization of magnetic peptide-imprinted polymer composite nanoparticles; Table S1: The sequence of primers used in this work, including GAPDH, Casp8, Casp3, Casp9, CYCS, Bax, Bid, and Bcl-2; Table S2: Prescreening of the binding of peptide molecules to magnetic peptide- and non-imprinted poly(ethylene-*co*-vinyl alcohol) nanoparticles; Figure S1. (a) Adsorption of peptide (50 μg/mL) on the MPIPs with imprinted onto various ethylene mol% of EVAL (b) Isothermal adsorption of MPIPs and MNIPs with various of peptide concentrations; Figure S2. Cellular viability of HepG2 cells with (a) various concentrations of MNPs, MNP/APTES, MNP/APTES/MC540, and MNIPs/MC540; (b) various concentrations of free MC540 and irradiation durations; (c) various concentrations of MPIPs/MC540 with and without irradiation; Figure S3. Optical, DAPI stained and merged images (from left to right columns) of (a) HepG2 cells; with MNPs (b); with MNPs/APTES (c), with MNPs/APTES/MC540 (d); and MNIPs/MC540 (e); Figure S4. (a) Immunostaining of CD-133 on HepG2 cells, (b) MPIPs/MC540 and (c) their merge image; Figure S5. Continuous cellular viability measurements of Hepa1-6 cells incubated with 100 μg/mL of MNIPs, MPIPs and MPIPs/MC540 (a) without or (b) with irradiation.

Author Contributions: Conceptualization, T.-L.W. and H.-Y.L.; methodology, M.-H.L. and H.-Y.L.; validation, J.L.T., M.-H.L. and H.-Y.L.; writing—original draft preparation, H.-Y.L.; writing—review and editing, J.L.T., M.-H.L. and H.-Y.L.; supervision, M.-H.L. and H.-Y.L.; project administration, M.-H.L. and H.-Y.L.; funding acquisition, M.-H.L., T.-L.W. and H.-Y.L.; experiments: J.-A.L. and J.-R.C. All authors have read and agreed to the published version of the manuscript.

Funding: This research was funded by Ministry of Science and Technology of ROC under Contact nos. MOST 107-2923-M-390-001-MY3, MOST 108-2314-B-214-005-, MOST 108-2221-E-390-004-, MOST 109-2221-E-390-024- and MOST 109-2314-B-390-001-MY3.

Institutional Review Board Statement: All the experimental procedures were approved by the Institutional Animal Care and Use Committee (IACUC) of National University of Kaohsiung (protocol No. 10708; 10 November 2018) and supported from Ministry of Science and Technology of ROC under Contract nos. MOST 108-2314-B-214-005-.

Informed Consent Statement: Not applicable.

Data Availability Statement: The authors confirm that the data supporting the findings of this study are available within the article and its supplementary materials.

Conflicts of Interest: The authors declare no conflict of interest.

References

1. Lee, M.-H.; Thomas, J.L.; Wang, H.-Y.; Chang, C.-C.; Lin, C.-C.; Lin, H.-Y. Extraction of resveratrol from polygonum cuspidatum with magnetic orcinol-imprinted poly(ethylene-co-vinyl alcohol) composite particles and their in vitro suppression of human osteogenic sarcoma (HOS) cell line. *J. Mater. Chem.* **2012**, *22*, 24644–24651. [CrossRef]
2. Lee, M.-H.; Thomas, J.L.; Chen, Y.-C.; Wang, H.-Y.; Lin, H.-Y. Hydrolysis of Magnetic Amylase-Imprinted Poly(ethylene-co-vinyl alcohol) Composite Nanoparticles. *ACS Appl. Mater. Interfaces* **2012**, *4*, 916–921. [CrossRef] [PubMed]
3. Lin, H.-Y.; Ho, M.-S.; Lee, M.-H. Instant formation of molecularly imprinted poly(ethylene-co-vinyl alcohol)/quantum dot composite nanoparticles and their use in one-pot urinalysis. *Biosens. Bioelectron.* **2009**, *25*, 579–586. [CrossRef] [PubMed]
4. Huang, C.-Y.; Tsai, T.-C.; Thomas, J.L.; Lee, M.-H.; Liu, B.-D.; Lin, H.-Y. Urinalysis with molecularly imprinted poly(ethylene-co-vinyl alcohol) potentiostat sensors. *Biosens. Bioelectron.* **2009**, *24*, 2611–2617. [CrossRef]
5. Lee, M.H.; Thomas, J.L.; Su, Z.L.; Yeh, W.K.; Monzel, A.S.; Bolognin, S.; Schwamborn, J.C.; Yang, C.H.; Lin, H.Y. Epitope imprinting of alpha-synuclein for sensing in Parkinson's brain organoid culture medium. *Biosens Bioelectron.* **2020**. [CrossRef]
6. Lee, M.-H.; Lin, C.-C.; Thomas, J.L.; Li, J.-A.; Lin, H.-Y. Cellular reprogramming with multigene activation by the delivery of CRISPR/dCas9 ribonucleoproteins via magnetic peptide-imprinted chitosan nanoparticles. *Mater. Today Bio* **2021**, *9*, 100091. [CrossRef]
7. Lee, M.-H.; Lin, C.-C.; Thomas, J.L.; Chan, C.-K.; Lin, H.-Y. Epitope recognition of magnetic peptide-imprinted chitosan composite nanoparticles for the extraction of CRISPR/dCas9a proteins from transfected cells. *Nanotechnology* **2021**, *32*, 18LT02. [CrossRef]

8. Lee, M.-H.; Thomas, J.L.; Chang, Y.-C.; Tsai, Y.-S.; Liu, B.-D.; Lin, H.-Y. Electrochemical sensing of nuclear matrix protein 22 in urine with molecularly imprinted poly(ethylene-co-vinyl alcohol) coated zinc oxide nanorod arrays for clinical studies of bladder cancer diagnosis. *Biosens. Bioelectron.* **2016**, *79*, 789–795. [CrossRef]
9. Chen, W.-J.; Lee, M.-H.; Thomas, J.L.; Lu, P.-H.; Li, M.-H.; Lin, H.-Y. Microcontact Imprinting of Algae on Poly (ethylene-co-vinyl alcohol) for Biofuel Cells. *ACS Appl. Mater. Interfaces* **2013**, *5*, 11123–11128. [CrossRef]
10. Lee, M.-H.; Thomas, J.L.; Chen, W.-J.; Li, M.-H.; Shih, C.-P.; Lin, H.-Y. Fabrication of bacteria-imprinted polymer coated electrodes for microbial fuel cells. *ACS Sustain. Chem. Eng.* **2015**, *3*, 1190–1196. [CrossRef]
11. Lin, H.-Y.; Hsu, C.-Y.; Thomas, J.L.; Wang, S.-E.; Chen, H.-C.; Chou, T.-C. The microcontact imprinting of proteins: The effect of cross-linking monomers for lysozyme, ribonuclease A and myoglobin. *Biosens. Bioelectron.* **2006**, *22*, 534–543. [CrossRef]
12. Lee, M.-H.; Liu, K.-T.; Thomas, J.L.; Su, Z.-L.; O'Hare, D.; van Wuellen, T.; Chamarro, J.M.; Bolognin, S.; Luo, S.-C.; Schwamborn, J.C.; et al. Peptide-Imprinted Poly(hydroxymethyl 3,4-ethylenedioxythiophene) Nanotubes for Detection of α Synuclein in Human Brain Organoids. *ACS Appl. Nano Mater.* **2020**, *3*, 8027–8036. [CrossRef]
13. Lee, M.-H.; Thomas, J.L.; Liao, C.-L.; Jurcevic, S.; Crnogorac-Jurcevic, T.; Lin, H.-Y. Epitope recognition of peptide-imprinted polymers for Regenerating protein 1 (REG1). *Sep. Purif. Technol.* **2018**, *192*, 213–219. [CrossRef]
14. Refaat, D.; Aggour, M.G.; Farghali, A.A.; Mahajan, R.; Wiklander, J.G.; Nicholls, I.A.; Piletsky, S.A. Strategies for molecular imprinting and the evolution of MIP nanoparticles as plastic antibodies—Synthesis and applications. *Int. J. Mol. Sci.* **2019**, *20*, 6304. [CrossRef]
15. Nishino, H.; Huang, C.S.; Shea, K.J. Selective protein capture by epitope imprinting. *Angew. Chem. Int. Ed.* **2006**, *45*, 2392–2396. [CrossRef] [PubMed]
16. Zeng, Z.; Hoshino, Y.; Rodriguez, A.; Yoo, H.; Shea, K.J. Synthetic polymer nanoparticles with antibody-like affinity for a hydrophilic peptide. *ACS Nano* **2009**, *4*, 199–204. [CrossRef] [PubMed]
17. Zhao, X.-L.; Li, D.-Y.; He, X.-W.; Li, W.-Y.; Zhang, Y.-K. An epitope imprinting method on the surface of magnetic nanoparticles for specific recognition of bovine serum album. *J. Mater. Chem. B* **2014**, *2*, 7575–7582. [CrossRef]
18. Qin, Y.-P.; Jia, C.; He, X.-W.; Li, W.-Y.; Zhang, Y.-K. Thermosensitive Metal Chelation Dual-Template Epitope Imprinting Polymer Using Distillation–Precipitation Polymerization for Simultaneous Recognition of Human Serum Albumin and Transferrin. *ACS Appl. Mater. Interfaces* **2018**, *10*, 9060–9068. [CrossRef]
19. Bossi, A.M.; Sharma, P.S.; Montana, L.; Zoccatelli, G.; Laub, O.; Levi, R. Fingerprint-Imprinted Polymer: Rational Selection of Peptide Epitope Templates for the Determination of Proteins by Molecularly Imprinted Polymers. *Anal. Chem.* **2012**, *84*, 4036–4041. [CrossRef] [PubMed]
20. Iskierko, Z.; Sharma, P.S.; Noworyta, K.R.; Borowicz, P.; Cieplak, M.; Kutner, W.; Bossi, A.M. Selective PQQPFPQQ Gluten Epitope Chemical Sensor with a Molecularly Imprinted Polymer Recognition Unit and an Extended-Gate Field-Effect Transistor Transduction Unit. *Anal. Chem.* **2019**, *91*, 4537–4543. [CrossRef] [PubMed]
21. Lach, P.; Cieplak, M.; Majewska, M.; Noworyta, K.R.; Sharma, P.S.; Kutner, W. "Gate Effect" in p-Synephrine Electrochemical Sensing with a Molecularly Imprinted Polymer and Redox Probes. *Anal. Chem.* **2019**, *91*, 7546–7553 [CrossRef]
22. Lee, M.-H.; Thomas, J.L.; Liao, C.-L.; Jurcevic, S.; Crnogorac-Jurcevic, T.; Lin, H.-Y. Polymers imprinted with three REG1B peptides for electrochemical determination of Regenerating Protein 1B, a urinary biomarker for pancreatic ductal adenocarcinoma. *Microchim. Acta* **2017**, *184*, 1773–1780. [CrossRef]
23. Pan, J.; Chen, W.; Ma, Y.; Pan, G. Molecularly imprinted polymers as receptor mimics for selective cell recognition. *Chem. Soc. Rev.* **2018**, *47*, 5574–5587. [CrossRef]
24. Piletsky, S.; Canfarotta, F.; Poma, A.; Bossi, A.M.; Piletsky, S. Molecularly Imprinted Polymers for Cell Recognition. *Trends Biotechnol.* **2020**, *38*, 368–387. [CrossRef]
25. Guan, J.; Lim, K.S.; Mekhail, T.; Chang, C.-C. Programmed death ligand-1 (PD-L1) expression in the programmed death receptor-1 (PD-1)/PD-L1 blockade: A key player against various cancers. *Arch. Pathol. Lab. Med.* **2017**, *141*, 851–861. [CrossRef] [PubMed]
26. Lau, J.; Cheung, J.; Navarro, A.; Lianoglou, S.; Haley, B.; Totpal, K.; Sanders, L.; Koeppen, H.; Caplazi, P.; McBride, J.; et al. Tumour and host cell PD-L1 is required to mediate suppression of anti-tumour immunity in mice. *Nat. Commun.* **2017**, *8*, 14572. [CrossRef] [PubMed]
27. Zou, W.; Wolchok, J.D.; Chen, L. PD-L1 (B7-H1) and PD-1 pathway blockade for cancer therapy: Mechanisms, response biomarkers, and combinations. *Sci. Transl. Med.* **2016**, *8*, rv324–rv328. [CrossRef]
28. Milling, L.; Zhang, Y.; Irvine, D.J. Delivering safer immunotherapies for cancer. *Adv. Drug Deliv. Rev.* **2017**, *114*, 79–101. [CrossRef] [PubMed]
29. Jo, S.D.; Nam, G.-H.; Kwak, G.; Yang, Y.; Kwon, I.C. Harnessing designed nanoparticles: Current strategies and future perspectives in cancer immunotherapy. *Nano Today* **2017**, *17*, 23–37. [CrossRef]
30. Qian, H.; Liu, B.; Jiang, X. Application of nanomaterials in cancer immunotherapy. *Mater. Today Chem.* **2018**, *7*, 53–64. [CrossRef]
31. Duan, X.; Chan, C.; Lin, W. Nanoparticle-Mediated Immunogenic Cell Death Enables and Potentiates Cancer Immunotherapy. *Angew. Chem. Int. Ed.* **2019**, *58*, 670–680. [CrossRef]
32. Liu, Q.; Tian, J.; Tian, Y.; Sun, Q.; Sun, D.; Wang, F.; Xu, H.; Ying, G.; Wang, J.; Yetisen, A.K.; et al. Near-Infrared-II Nanoparticles for Cancer Imaging of Immune Checkpoint Programmed Death-Ligand 1 and Photodynamic/Immune Therapy. *ACS Nano* **2021**, *15*, 515–525. [CrossRef]

33. Robertson, C.A.; Evans, D.H.; Abrahamse, H. Photodynamic therapy (PDT): A short review on cellular mechanisms and cancer research applications for PDT. *J. Photochem. Photobiol. B Biol.* **2009**, *96*, 1–8. [CrossRef]
34. Hoebeke, M.; Piette, J.; van de Vorst, A. Photosensitized production of singlet oxygen by merocyanine 540 bound to liposomes. *J. Photochem. Photobiol. B Biol.* **1991**, *9*, 281–294. [CrossRef]
35. Firczuk, M.; Nowis, D.; Gołąb, J. PDT-induced inflammatory and host responses. *Photochem. Photobiol. Sci.* **2011**, *10*, 653–663. [CrossRef] [PubMed]
36. Wang, H.; Liu, Z.; Wang, S.; Dong, C.; Gong, X.; Zhao, P.; Chang, J. MC540 and upconverting nanocrystal coloaded polymeric liposome for near-infrared light-triggered photodynamic therapy and cell fluorescent imaging. *ACS Appl. Mater. Interfaces* **2014**, *6*, 3219–3225. [CrossRef]
37. Idris, N.M.; Gnanasammandhan, M.K.; Zhang, J.; Ho, P.C.; Mahendran, R.; Zhang, Y. In vivo photodynamic therapy using upconversion nanoparticles as remote-controlled nanotransducers. *Nat. Med.* **2012**, *18*, 1580. [CrossRef]
38. Xu, J.; Yang, P.; Sun, M.; Bi, H.; Liu, B.; Yang, D.; Gai, S.; He, F.; Lin, J. Highly emissive dye-sensitized upconversion nanostructure for dual-photosensitizer photodynamic therapy and bioimaging. *ACS Nano* **2017**, *11*, 4133–4144. [CrossRef] [PubMed]
39. Trinchet, J.-C.; Beaugrand, M. Treatment of hepatocellular carcinoma in patients with cirrhosis. *J. Hepatol.* **1997**, *27*, 756–765. [CrossRef]
40. Singh, A.K.; Kumar, R.; Pandey, A.K. Hepatocellular carcinoma: Causes, mechanism of progression and biomarkers. *Curr. Chem. Genom. Transl. Med.* **2018**, *12*, 9. [CrossRef] [PubMed]
41. Bruix, J.; Reig, M.; Sherman, M. Evidence-based diagnosis, staging, and treatment of patients with hepatocellular carcinoma. *Gastroenterology* **2016**, *150*, 835–853. [CrossRef] [PubMed]
42. Heimbach, J.K.; Kulik, L.M.; Finn, R.S.; Sirlin, C.B.; Abecassis, M.M.; Roberts, L.R.; Zhu, A.X.; Murad, M.H.; Marrero, J.A. AASLD guidelines for the treatment of hepatocellular carcinoma. *Hepatology* **2018**, *67*, 358–380. [CrossRef]
43. Lee, M.-H.; Liu, K.-H.; Thomas, J.L.; Chen, J.-R.; Lin, H.-Y. Immunotherapy of Hepatocellular Carcinoma with Magnetic PD-1 Peptide-Imprinted Polymer Nanocomposite and Natural Killer Cells. *Biomolecules* **2019**, *9*, 651. [CrossRef] [PubMed]
44. Pfaffl, M.W. A new mathematical model for relative quantification in real-time RT-PCR. *Nucleic Acids Res.* **2001**, *29*, e45. [CrossRef] [PubMed]

pharmaceuticals

MDPI

Article

Terbium-Based AGuIX-Design Nanoparticle to Mediate X-ray-Induced Photodynamic Therapy

Joël Daouk [1], Mathilde Iltis [1], Batoul Dhaini [2], Denise Béchet [1], Philippe Arnoux [2], Paul Rocchi [3], Alain Delconte [1], Benoît Habermeyer [4], François Lux [3], Céline Frochot [2], Olivier Tillement [3], Muriel Barberi-Heyob [1,*] and Hervé Schohn [1]

1 Department of Biology, Signals and Systems in Cancer and Neuroscience, UMR 7039 Research Center for Automatic Control (CRAN), Université de Lorraine–French National Scientific Research Center (CNRS), F-54000 Nancy, France; joel.daouk@univ-lorraine.fr (J.D.); mathilde.iltis@yahoo.fr (M.I.); denisebechet@hotmail.com (D.B.); alain.delconte@univ-lorraine.fr (A.D.); herve.schohn@univ-lorraine.fr (H.S.)

2 Reactions and Chemical Engineering Laboratory (LRGP), UMR 7274, Université de Lorraine–French National Scientific Research Center (CNRS), F-54000 Nancy, France; batoul.dhaini@univ-lorraine.fr (B.D.); philippe.arnoux@univ-lorraine.fr (P.A.); celine.frochot@univ-lorraine.fr (C.F.)

3 Light Matter Institute, UMR-5306, Université de Lyon–French National Scientific Research Center (CNRS), F-69000 Lyon, France; paul.rocchi@univ-lyon1.fr (P.R.); francois.lux@univ-lyon1.fr (F.L.); olivier.tillement@univ-lyon1.fr (O.T.)

4 Porphychem SAS, F-21000 Dijon, France; b.habermeyer@porphychem.com

* Correspondence: muriel.barberi-heyob@univ-lorraine.fr; Tel.: +33-(0)3-72-74-61-14

Abstract: X-ray-induced photodynamic therapy is based on the energy transfer from a nanoscintillator to a photosensitizer molecule, whose activation leads to singlet oxygen and radical species generation, triggering cancer cells to cell death. Herein, we synthesized ultra-small nanoparticle chelated with Terbium (Tb) as a nanoscintillator and 5-(4-carboxyphenyl succinimide ester)-10,15,20-triphenyl porphyrin (P1) as a photosensitizer (AGuIX@Tb-P1). The synthesis was based on the AGuIX@ platform design. AGuIX@Tb-P1 was characterised for its photo-physical and physico-chemical properties. The effect of the nanoparticles was studied using human glioblastoma U-251 MG cells and was compared to treatment with AGuIX@ nanoparticles doped with Gadolinium (Gd) and P1 (AGuIX@Gd-P1). We demonstrated that the AGuIX@Tb-P1 design was consistent with X-ray photon energy transfer from Terbium to P1. Both nanoparticles had similar dark cytotoxicity and they were absorbed in a similar rate within the cells. Pre-treated cells exposure to X-rays was related to reactive species production. Using clonogenic assays, establishment of survival curves allowed discrimination of the impact of radiation treatment from X-ray-induced photodynamic effect. We showed that cell growth arrest was increased (35%-increase) when cells were treated with AGuIX@Tb-P1 compared to the nanoparticle doped with Gd.

Keywords: glioblastoma multiforme; AGuIX®; terbium; gadolinium; photodynamic therapy; X-ray-induced photodynamic therapy; singlet oxygen

Citation: Daouk, J.; Iltis, M.; Dhaini, B.; Béchet, D.; Arnoux, P.; Rocchi, P.; Delconte, A.; Habermeyer, B.; Lux, F.; Frochot, C.; et al. Terbium-Based AGuIX-Design Nanoparticle to Mediate X-ray-Induced Photodynamic Therapy. *Pharmaceuticals* **2021**, *14*, 396. https://doi.org/10.3390/ph14050396

Academic Editor: Dimitris Tsiourvas

Received: 24 March 2021
Accepted: 19 April 2021
Published: 22 April 2021

1. Introduction

Glioblastoma multiforme (GBM) is one of the main incurable brain tumors, mainly due to the presence of infiltrated cells within the parenchyma, responsible for GBM recurrence into the surrounding brain tissue [1]. The conventional treatment of GBM tumors consists of surgical resection followed by X-ray radiation and adjuvant temozolomide administration which improves modestly patient survival [2]. Brain exposure to X-ray involves the generation of oxidative stress, which are responsible for DNA alteration, lipid peroxidation, protein oxidation, and cell redox statue changes, triggering cells to cell death [3]. However, these effects are not limited to malignant cells, but also alter surrounding cells.

Alternative therapeutic strategies have been developed, notably photodynamic therapy (PDT). PDT appears as an innovative technology being investigated to fulfil the need for a targeted cancer treatment that may reduce recurrence and extend survival with minimal side effects [3,4]. It aims at selectively killing neoplastic cells by the combined action of a photosensitizer and visible light in the presence of oxygen, whose combined action mainly results in the formation of reactive species, especially singlet oxygen which is the main mediator of PDT reaction. To improve PDT efficiency, photosensitizer can be bound to ligands such as monoclonal antibodies or peptide moieties and be delivered by carrier systems such as nanoparticle [5,6]. Moreover, the nanoparticles can be modified by functional groups for additional biochemical properties. In addition, nanoparticles accumulation in the solid tumor site is improved by the enhanced permeability and retention effect (EPR) [7,8]. Numerous clinical studies, including phase III randomized prospective clinical trials, have been reported for PDT, using alternative methods such as interstitial PDT and intraoperative PDT [9–20]. Interstitial PDT offers a localized treatment approach in which improvements in local control of GBM may result in significant enhanced survival [11,12,20]. Several photosensitizers have been used, including porfimer sodium (Photofrin®), 5-aminolevulinic acid (5-ALA, Gliolan®), m-tetrahydroxyphenylchlorin (mTHPC, temoporfin, Foscan®) and benzoporphyrin derivative monoacids ring A (BPD-MA, verteporfin, Visudine®).

Compared to radiotherapy, the light irradiation used in PDT is less energetic and it cannot penetrate deeply enough into the tumor, as most tissue chromophores absorb visible light commonly used in clinical practice [4,6]. The penetration depth of 630 nm light in brain-adjacent-to-tumor is estimated at 2.5 mm. A breakthrough strategy to treat GBM via nanomedicine and X-ray has been suggested by combining the principles of radiotherapy and PDT, both clinically proven modalities, while maintaining their main benefits and decreasing their drawbacks. The principle of the so-named X-ray-induced PDT (X-PDT) is based on the conversion of X-ray photons into visible photons, known as X-ray excited optical luminescence, from the nanoscintillator embedded in the nanoparticle and linked to the photosensitizer, which, in turn, produces singlet oxygen and other oxygen reactive species [21,22]. X-PDT proof-of-concept with nanoparticles was first introduced by Cheng and Wang, who described simultaneous radiation and X-ray-induced photodynamic effects [23]. The strategy requires nanoparticles, exhibiting appropriated physical properties to establish energy transduction from the nanoscintillator to the photosensitizer, a high scintillation quantum yield and an optimal energy transfer from the scintillator onto the photosensitizer [24,25]. It must be pointed that only PDT can generate singlet oxygen which is highly cytotoxic to tumor tissue and to treat deeply lesions without invasive approach such as interstitial PDT. It is possible to use X-ray as an excitation source instead of light. Thus, the light penetration problem through the tumor tissue is overcome, and the activation of the photosensitizer within tumor tissue is performed by classical radiotherapy using X-ray. In addition, the cumulative effects between conventional radiotherapy and PDT should allow the use of conventional X-ray doses. In metal-hybrid system, the metal-based nanoparticle consists of a nanoscintillator coated with polyethylene glycol or a polysiloxane layers to ensure biocompatibility which allows covalently coupling of the photosensitizer [24,25]. Members of the lanthanide family have been used in nanoparticle synthesis, such as mesoporous lanthanum fluoride doped with Cerium or Terbium (Tb) and grafted with porphyrin derivative, Tb_2O_3 coated with a polysiloxane layer or silica-doped with lanthamide [26–31].

Among them, ultra-small Gadolinium (Gd) based nanoparticles, namely AGuIX®, were developed [32]. The nanoparticle design was first proposed for a non-toxic resonance magnetic agent and its imaging properties [33]. Moreover, in vitro and in vivo pre-clinic experiments demonstrated that AGuIX doped with Gd act as a theranostic agent, enhancing radiosensitization of tumor cells in diverse experimental conditions, notably, at different photon radiation energies (with a range from kiloelectron volts to million electron volts) and with different types of radiation [34]. The radiosensitizing effects are associated to diverse processes: Gd mediated generation of electrophotons and Auger photons amplifying the local production of reactive oxygen derived species, as demonstrated for Gold embedded nanoparticles [35,36]; or an impairment of DNA breaks reparation, as a consequence of irradiation, and reactive species production triggering cells to cell death [34,37]. In X-PDT, the main goal of the treatment consists of the transfer energy from irradiated nanoscintillator to the photosensitizer, limiting the delivery of high radiation energies and deposits to kill cancer cells without any alteration to adjoining normal cells. Recently, we demonstrated that AGuIX@ doped with Gd and 5-(4-carboxyphenyl succinimide ester)-10,15,20-triphenylporphyrin (P1) can be used to target Neuropilin-1, a transmembrane receptor localized in endothelial cells within mouse grafted human GBM tumors [38]. The synthesized nanoparticle, referred as AGuIX@Gd-P1, behaved with similar properties to the original AGuIX@Gd.

Therefore, in order to use the AGuIX platform for X-PDT, we suggested the replacement of Gd in the Gd-based AGuIX nanoparticle by Tb as a nanoscintillator and the grafting of P1 (Scheme 1). In these conditions, an external light source will not be required to simultaneously support a photodynamic effect. The nanoparticles (referred herein as AGuIX@Tb and AGuIX@Tb-P1) were characterised for their photophysical and chemical properties. We evaluated the effect of X-PDT on human GBM U-251 MG cell survival after cell exposure to nanoparticles. In parallel, we tested AGuIX@ doped with Gd and P1 (referred herein as AGuIX@Gd and AGuIX@Gd-P1), which has been characterised previously [38]. We highlighted that chelated Tb-P1 nanoparticles can react linearly to X-ray energy and flow, and were able to activate P1 to produce singlet oxygen. In vitro, using human U-251 MG glioblastoma cells, experiments confirmed the interest of these AGuIX design, notably at a 3.0 Gy·min^{-1} dose rate. Moreover, cell exposure to AGuIX@Tb-P1 improved the effect on cell growth arrest when it was compared to similar treatment with AGuIX@Gd or AGuIX@Gd-P1.

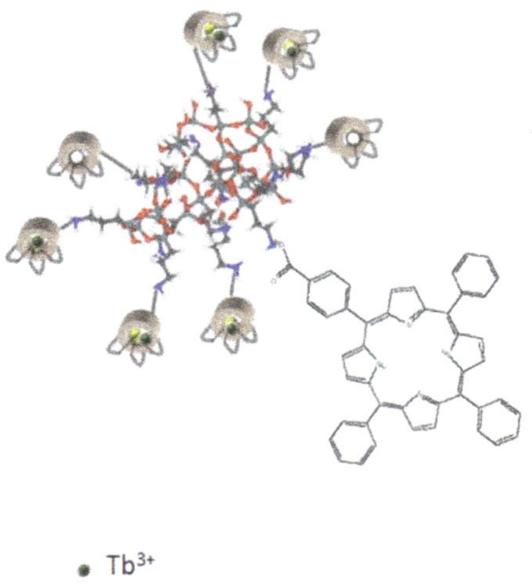

Scheme 1. Graphical design of the AguIX@Tb-P1. The AGuIX@ platform consists of a polysiloxane core (in red) surrounded by 1,4,7,10-tetraazacyclododecane-1,4,7,10-tetraacetic acid (DOTA)/Tb^{3+} complexes, covalently grafted to the inorganic matrix (in grey) by a maleimide arm (in blue). The same reaction was used for P1 grafting on the inorganic core.

2. Results

2.1. Characteristics of the AGuIX@Tb-P1

We already demonstrated that the grafting of P1 on AGuIX doped with Gd induced a hydrodynamic diameter at about 11.1 nm, twice as large as the original AGuIX@Gd nanoparticle, with an estimated diameter at 4.9 nm on average [38]. Replacing Gd of the original AGuIX-designed nanoparticle by Tb did not induce any size modification. The hydrodynamic diameters of AGuIX@Tb and AGuIX@Tb-P1 were estimated, respectively, at 3.8 ± 1.0 nm and 11 ± 0.8 nm. Moreover, the ζ potential raised from -11.8 to -45.6 mV, when measurements were achieved with AGuIX@Tb and AGuIX@Tb-P1, respectively, supporting a high stability of the latter conjugate (Figure S1). AGuIX@Tb and AGuIX@Gd emission spectra after UV light excitation are presented in Figure 1 as well as P1 absorption spectrum, in water. An overlay between P1 absorption spectrum and Tb emission (Figure 1a,b) could be observed. In contrast, Gd emission did not overlay P1 emission spectrum. Calculation of the spectral overlap between Tb emission and P1 absorption was estimated at $J = 1.15 \times 10^{14}$ M^{-1} nm^4 cm^{-1}. The corresponding Förster radius was found to be 2.5 nm (from 1 up to 10 nm). Moreover, Tb luminescence lifetime was 390 µs, long enough to allow energy transfer to P1. Thus, Tb luminescence in presence of P1 showed an exponential decay of its fluorescence intensity (Figure 1c), and a linear decrease of its fluorescence lifetime at PS concentration higher than 0.5 µM (Figure 1d). We recorded the luminescence exponential decay of AGuIX@Tb and AGuIX@Tb-P1 (Figure S2). AGuIX@Tb fluorescence lifetime was estimated at 1 ms whereas AGuIX@Tb-P1 fluorescence lifetime was 1 µs. This decrease of fluorescence lifetime of Tb in presence of P1 supported the concept of energy transfer between Tb and P1. Moreover, we measured the luminescence of both nanoparticles after excitation at 351 nm with a 50 µs delay between excitation and photon detection at 545 nm. As shown in Figure S3, P1 luminescence was obtained between 630 and 690 nm, corresponding to the energy transfer between Tb and P1, concomitantly to the decrease of Tb emission. Collectively, the results obtained allowed us to conclude that Tb energy transfer to P1 is a FRET/non radiative transfer type characterised by a quenching constant, $Kq = 0.045 \times 10^9$ M^{-1} s^{-1}.

2.2. Nanoscintillator Response to X-ray Excitation

AGuIX@Tb emission spectrum after X-ray excitation presented a similar profile compared with UV/visible light excitation (Figure 1b). To validate the pipeline acquisition setup under X-ray sessions, X-ray spectroscopy experiments were performed at different tube currents and voltages. Spectra intensities were positively related to tube current from 0.5 to 12.5 mA at 320 kVp (Figure 2a) and to tube voltage from 25 to 320 kVp at 12.5 mA (Figure 2b). For each condition, the four characteristic emission peaks of Tb cations (Tb^{3+}) were detected at 485, 545, 590, and 620 nm, respectively. For these different Tb cation emission peaks, the correlation coefficients between tube currents and peak intensities were 0.97, 0.99, 0.99, and 0.84, respectively (Figure 2c). The maximum peak values were associated linearly with the tube voltage values (Figure 2d). Correlation coefficient values were found to be 0.96, 0.99, 0.99, and 0.96 for the four Tb peaks 485, 545, 590, and 620 nm, respectively.

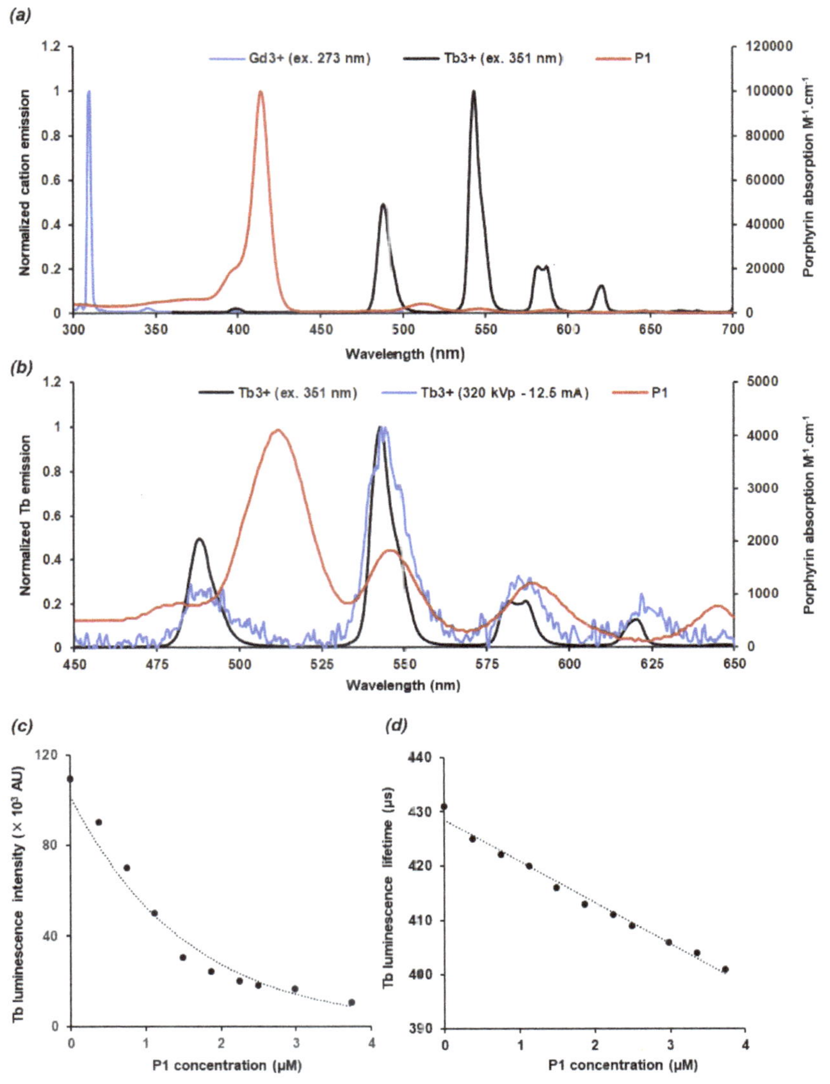

Figure 1. Photo-physical properties of AGuIX@Tb-P1 nanoparticle. (**a**) P1 absorption spectrum, AGuIX@Gd-P1 and AGuIX@Tb-P1 emission spectra (after 273 and 351 nm excitation, respectively, in water. Data presented in (**b**) is a zoomed view of the P1 Q bands, between 450 and 650 nm. Tb^{3+} emission overlap after UV (351 nm) or Tb^{3+} X-ray exposition; the tube parameters were set at 320 kVp, 12.5 mA, and 3 Gy·min^{-1}. Tb^{3+} luminescence intensity (**c**) and lifetime (**d**) were estimated as a function of increasing concentration of P1. An exponential decay of luminescence intensity and a linear decreased of the fluorescence lifetime was respectively obtained.

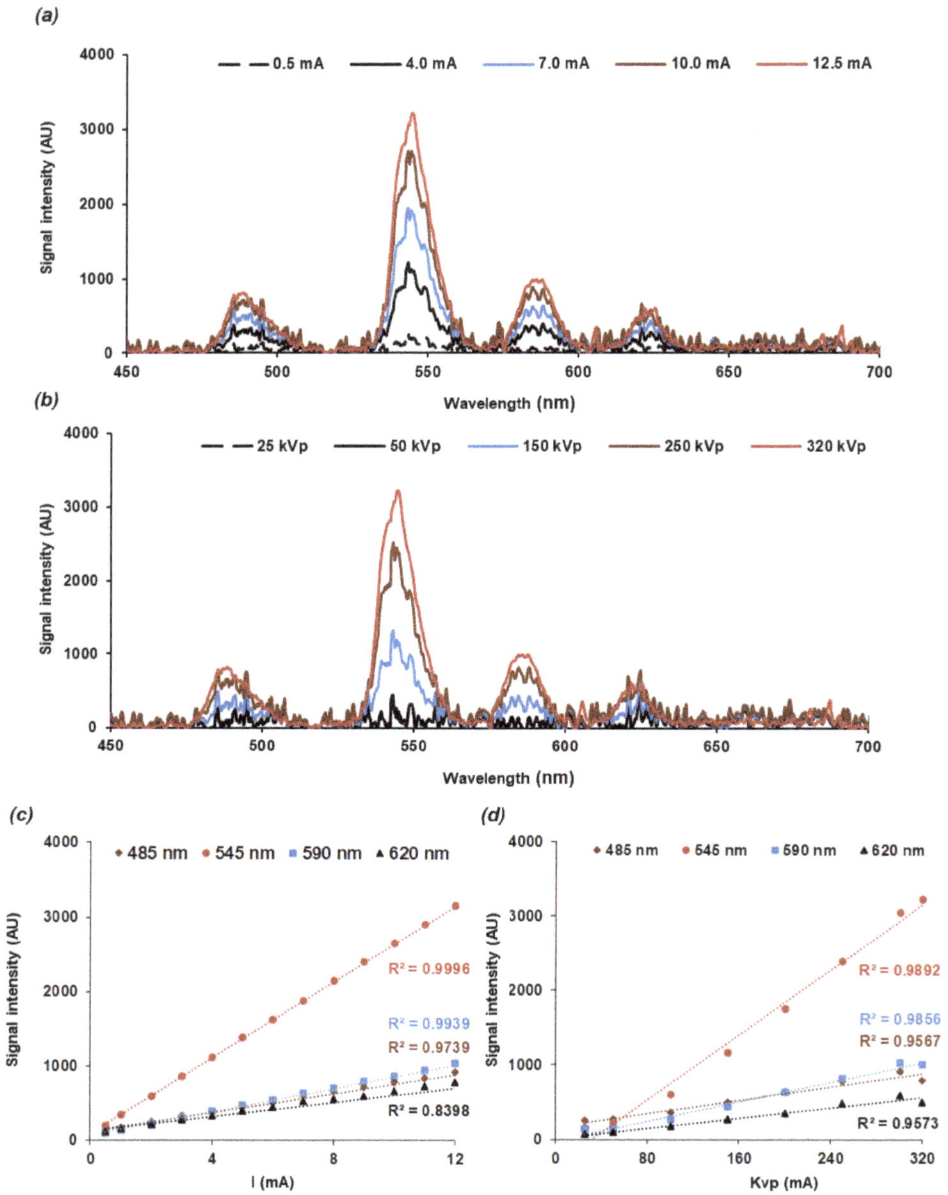

Figure 2. Tb scintillation response as a function of X ray parameters. (**a**) Tb^{3+} cation solution was irradiated by increasing tube intensity (mA) with a constant voltage set at 320 kVp. Tb^{3+} emission spectra were monitored in the range of 450 to 700 nm. The signal obtained (in arbitrary unit, AU) were plotted for each P1 Q band as a function of tube intensity (**c**). Similarly, the cation solution was irradiated by increasing tube voltages (kVp) with a tube current set at 12.5 mA (photons energy). (**b**) Luminescence signals were plotted for each P1 Q band as a function of tube voltage (**d**). In both cases, Tb luminescence was enhanced for each P1 Q band whatever tube intensity or voltage applied; the main increase variations were obtained for 545 nm P1 Q band.

2.3. Energy Transfer and Singlet Oxygen Production

We highlighted reactive species and singlet oxygen production using fluorescent probes, respectively, SOSG (Figure 3a) and APF (Figure 3b), during X-ray exposure. X-ray parameters were set to 320 kVp and 12.5 mA as it provided the highest Tb scintillation. Both APF and SOSG signals increased continuously during X-ray exposure. Addition of sodium azide (NaN_3), a singlet oxygen quencher, confirmed that the type II-PDT reaction (singlet oxygen generation) was mainly involved in X-PDT as SOSG and APF fluorescence signals were mostly inhibited.

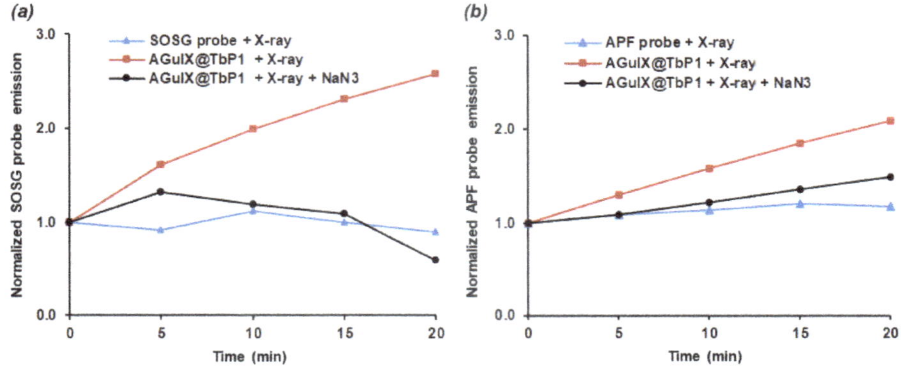

Figure 3. Kinetics of reactive species and singlet oxygen production under X-ray irradiation in water. Experiments were performed with 400 µM P1 equivalent AGuIX@Tb-P1 solution and probe fluorescence signals were revealed every 5-min during continuous X-ray irradiation. (**a**) 10 µM SOSG probe was used to react specifically to singlet oxygen. (**b**) 5 µM APF probe was used to assess reactive species production. Each fluorescence probe was irradiated alone. The production of singlet oxygen was inhibited by adding sodium azide (NaN_3) in the reaction mixture. Irradiation was performed with tube parameters set at 320 kVp, 12.5 mA, and 3 Gy/min.

2.4. Cytotoxicity of AGuIX@Tb-P1 on Glioblastoma Cell Growth

We assessed whether treatment of P1 alone or AGuIX@Tb led to U-251 MG cytotoxicity, using the MTT procedure. Since P1 is hydrophobic, we chose ZnPy3P1 which is soluble in culture medium. IC_{50} was estimated at 34.8 ± 9.9 µM after 24 h and 10.4 ± 3.4 µM after 72 h-treatment duration. In addition, IC_{50} was similar when GBM cells were exposed to AGuIX@Tb with IC_{50} estimated at 1.73 ± 0.3 and 1.56 ± 0.1 mM, after 24 and 72 h exposition, respectively. We finally tested the effect t of increasing concentration (1.0 to 20.0 µM; concentrations are expressed as P1 equivalent throughout the paper) of AGuIX@Tb-P1 and AGuIX@Gd-P1 (Figure S4). No cytotoxicity was observed whatever the dose tested and treatment duration.

2.5. NPs Cell Uptake

Cell uptake kinetics were established after cell exposure to 2.5 and 5 µM AGuIX@Tb-P1 over 48 h (Figure 4a). Cell uptake was quantified based on the fluorescence emission of P1. NPs accumulated within the cells, reaching a maximum at 48 h. Since the AGuIX doped with Gd or Tb are synthesized with a similar scheme and behave similar hydrodynamic diameter, we compared NP uptake after cell exposure to 1.0 or 2.5 µM AGuIX@Tb-P1 or AGuIX@Gd-P1 for 24 h. There was no significant difference suggesting that the replacement of Gd by Tb did not affect NP absorption within the cells (Figure 4b). Moreover, when cells were treated with 1 µM AGuIX@-complexes for 24 h (Figure 4b), uptake was low as compared to the results of cell absorption performed with 2.5 or 5.0 µM nanoparticles for the same culture delay (Figure 4a). Since NP absorption could lead to the increase of oxidative stress within the cells, we assessed whether nanoparticle uptake was related to reactive species generation, using DCF2-DA probe. A slight but significant increase was observed after cell exposure to 2.5 μM nanoparticle. However, cell exposure to 1 μM AGuIX@Gd-P1 (Figure 4c) was enough to modify redox statue within the cells. We evaluated whether AGuIX@Tb-P1 uptake is associated with stress-mediated cell death by quantifying propidium iodide positive cells, after cell exposure to 2.5 or 5 μM over 48 h. No change was observed, supporting that the accumulation of AGuIX@Tb-P1 was not related to cell death (Figure 4d).

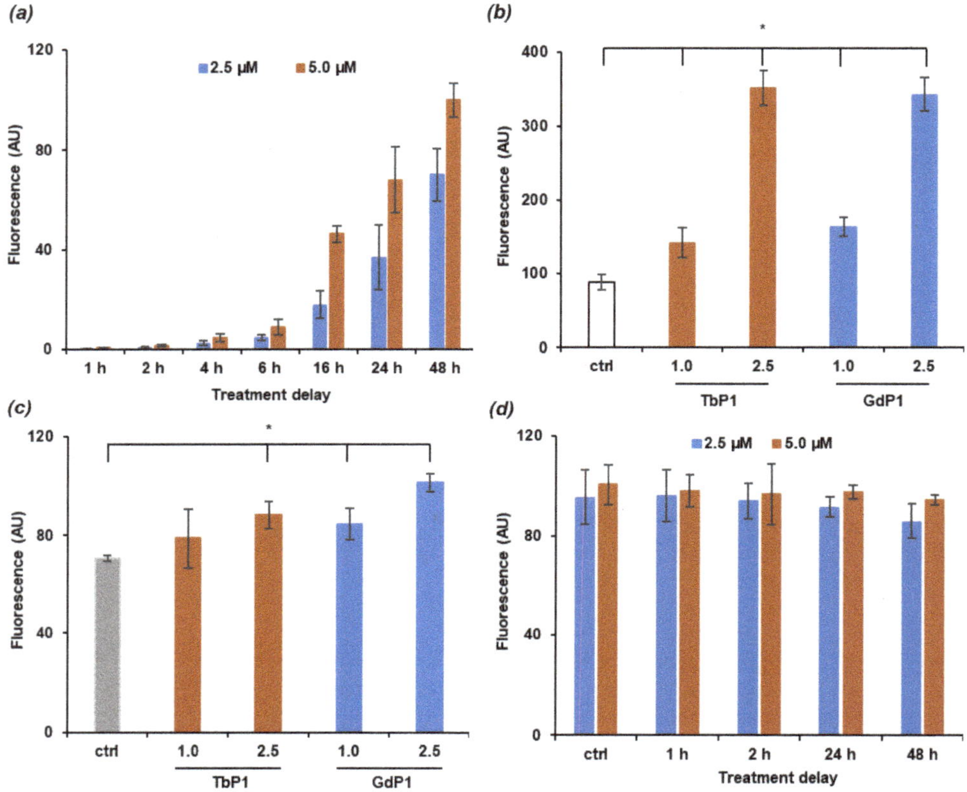

Figure 4. AGuIX@ complexes uptake by U-251 MG cells. (**a**) U-251 MG cell uptake kinetics with AGuIX@Tb-P1 were performed over 48 h. (**b**) Cell uptake between AGuIX@Tb-P1 (TbP1) and AGuIX@Gd-P1 (GdP1) was compared for 24 h, for P1 equivalent concentration of 1.0 and 2.5 μM. (**c**) Reactive species content was quantified using DCF2-DA fluorescent probe after cell exposure to 1 or 2.5 μM nanoparticles for 24 h. (**d**) AGuIX@Tb-P1 uptake-mediated cell death was estimated with propidium iodide dye after cell exposure to 2.5 and 5.0 μM AGuIX@Tb-P1 for 24 h. Results are means ± S.D. of triplicate determinations from three independent experiments. *, $p < 0.05$, relative to control cells (Kruskal-Wallis test and post-hoc by the Dunn's test). Abbreviation: ctrl, untreated cells; AU, Arbitrary unit.

2.6. Photodynamic Effect on U-251 MG Cell Survival

We assessed whether cell exposure to photodynamic treatment led to cell growth inhibition after 24 h exposure duration and could limit cell clone formation in anchorage-dependent clonogenic assays. U-251 MG cells were pre-treated with 1.0 and 2.5 μM AGuIX@Tb-P1 and exposed to a red light (630 nm, 0.7 W, irradiance at 4.54 mW·cm^{-2}) to a fluence range of 2.5 to 10.0 J cm^{-2}, corresponding to an exposition duration of 2 min to 38 min). At the dose of 1 μM, cell growth inhibition was obtained with a fluence applied at 10.0 J cm^{-2} (Figure 5a). We observed that clone formation was inhibited when the cells were pre-treated with 1 μM AGuIX@Tb-P1 and exposed to a fluence at 2.5 (50%-inhibition) or 5.0 J cm^{-2} (90%-inhibition) (Figure 5b). Similarly, cell growth inhibition was observed in pre-treated U-251 MG cells with 2.5 μM nanoparticle and a fluence at 2.5 J cm^{-2} (20%-inhibition). Cell growth arrest increased with fluence up to 10.0 J cm^{-2} (80% inhibition) (Figure 5a). Since PDT is associated with the generation of oxidative stress, we quantified reactive species content immediately after cell exposure to red light (Figure 5c), when cells were pre-treated with 2.5 μM AGuIX@Tb-P1 and exposed to a fluence at 2.5 J cm^{-2}, which corresponds to an immediate cell growth inhibition estimated at 20%. We used DCF2-DA probes which reacts with several reactive oxygen-derived species and gives a valuable estimation of the oxidative stress generated within the cells. We found that reactive species content was enhanced 2 times over 30 min post light exposition, supporting the concept that cell survival depends mainly on oxidative stress-mediated by light treatment. Finally, at nanoparticle concentrations higher than 2.5 μM (data not shown), and a fluence higher than 2.5 J cm^{-2} (Figure 5b), the applied treatment killed all the cells.

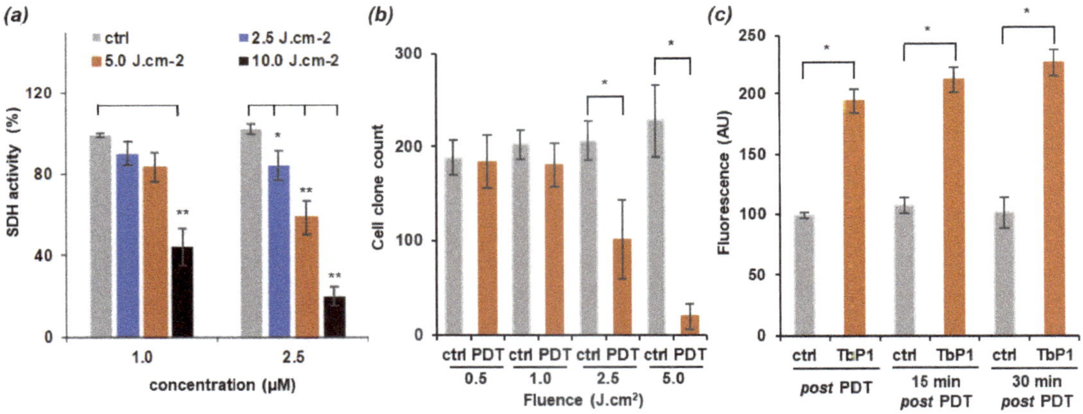

Figure 5. U-251 MG cell photodynamic effect. Cells were treated with AGuIX@Tb-P1 for 24 h before red-light exposition (2.5 to 10.0 J cm^{-2}). (**a**) cell viability was measured 24 h post treatment using the MTT assay. (**b**) Clonogenic assays after PDT treatment with increasing fluences. Clonogenic capabilities are expressed relative to control cells. Results are means ± S.D. (n = 12 wells/condition); * $p < 0.05$; (Kruskal-Wallis test and post-hoc by the Mann-Whitney test). (**c**) Cells were pre-treated with 2.5 μM AGuIX@Tb-P1 (TbP1), followed by photodynamic treatment at 2.5 J cm^{-2}. Reactive species were quantified by DCF2-DA probe using flow cytometry. In (**a**,**c**), results are means ± S.D. of three determinations from three independent experiments. * $p < 0.05$; ** $p < 0.001$ (Kruskal-Wallis test and post-hoc by the Dunn's test). Abbreviations: ctrl, cells treated with AGuIX@ conjugates without light exposure; SDH, succinate dehydrogenase; UA, arbitrary unit.

2.7. X-ray-Induced Photodynamic Effect on U-251 MG Cell Survival

Similarly, we studied whether X-ray irradiation promoted U-251 MG cell growth arrest by photodynamic-mediated effect. Reactive species generation was quantified immediately after cell exposure to X-ray irradiation (Figure 6a). X-ray ionization in itself generated reactive species and Tb scintillation could be involved in the redox change within the cells [5]. We assessed whether oxidative stress was generated in AGuIX@Tb and AGuIX@Tb-P1 pre-treated cells and irradiated at 2.0 Gy, with an energy set at 320 kVp, over 1 h post irradiation (Figure 6a), using DCF2-DA probe. Reactive species content was increased up to 1.5 time when cells were exposed to AGuIXTb-P1, whatever the delay. Conversely, at 30 min post exposition, reactive species content was enhanced in AGuIX@Tb pre-treated cells, but the increase was not significant when the results were compared to those obtained with untreated cells. Moreover, the obtained levels in pre-treated and untreated cells were close, after 1 h post irradiation.

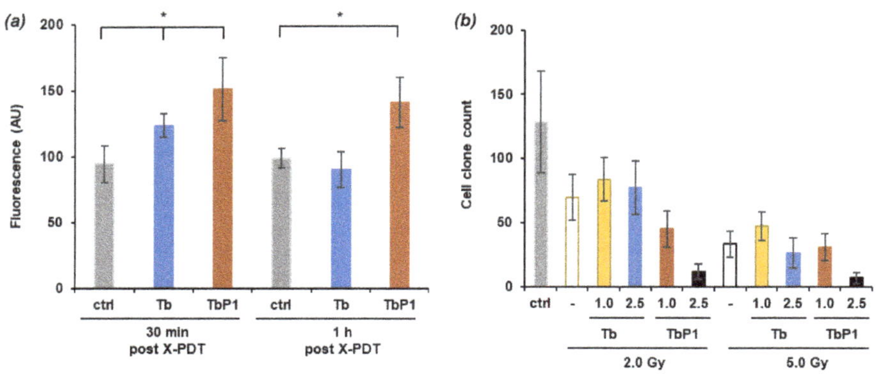

Figure 6. Effect of X-PDT on U-251 MG cell survival and reactive species content. (**a**) U-251 MG cells pre-treated with nanoparticles were exposed to X-ray irradiation at 2.0 Gy (320 kVp), then incubated with 50 μM DCF2-DA. The cells were harvested, and reactive species were quantified by flow cytometry. Results are means ± S.D of triplicate determination from three independent experiments. * $p < 0.05$ (Kruskal-Wallis test and post-hoc by the Dunn's test), comparison between nanoparticles-treated and -untreated cells (ctrl, control). (**b**), U-251 MG cells were treated with 1.0 or 2.5 μM P1 equivalent concentration of AGuIX@Tb-P1 (TbP1) and corresponding AGuIX@Tb (Tb) concentration, respectively, 16.6 or 45.1 μM (Tb), for 24 h, before X-ray irradiation at 2.0 or 5.0 Gy (320 kVp). Results are means ± S.D. of clones counting from at least 12 wells. Abbreviation: ctrl, irradiated cells without NPs pre-treatment. Abbreviation: AU, arbitrary unit.

In order to validate the concept of a photodynamic effect in X-PDT strategy, we performed anchorage-dependent clonogenic assays with AGuIX@NPs. Because Gd and Tb are neighbors in terms of atomic number, their behavior regarding X-ray interaction is considered as similar. Cells were pre-treated with 1 μM AGuIX@NPs and irradiated at X-ray doses ranging from 0.5 up to 5.0 Gy, either at 160 or 320 kVp, corresponding to the dose rate of 1.5 or 3.0 Gy·min^{-1}, respectively. These conditions were chosen since cell clone formation was mainly inhibited when cells were pre-treated with 2.5 μM AGuIX@Tb-P1 and irradiated for X-ray doses higher than 2.0 Gy at 320 kVp (Figure 6b). Experimental results were plotted in the quadratic semi log model (Figure 7a) and compared (Figure 7b). Curve parameters were computed for survival factor at 2.0 Gy (Table 1). Because we used X-ray dose exposures up to 5.0 Gy, which were close to those classically used in clinical practice (1.5 to 2.5 Gy), we considered the β-parameter equal to 0. In fact, β -parameter governs the slope for high dose exposure (sub-lethal damage), while the α-parameter reflects the enhanced benefit of X-ray-induced PDT on cell survival id est the direct lethal cell damage [39]. As shown in Table 1 and Figure 7b, significant survival difference ($p = 0.0018$) was found at 2.0 Gy between scintillating (AGuIX@Tb-P1) and non-scintillating (AGuIX@Gd-P1) nanoparticles, when X-ray energies were set to 320 kVp corresponding to a mean 116 keV X-ray energy; DMF was estimated of 0.650 versus 0.876. The DER at 2.0 Gy was estimated at 1.54. Finally, for the same Gy dose, the enhanced factor was 35% when cells were treated with AGuIX@Tb-P1. Moreover, we did not find any significant change between results from cells treated with AguIX@Gd-P1 and those obtained of the cells treated with AguIX@Gd for the same experimental conditions (Figure 7b). At energies set at 160 kVp, we did not find any difference between both nanoparticles either doped with Tb or Gd with or without P1. Accordingly, our results support the concept of AGuIX@Tb-P1 photodynamic-mediated effect, whereas nanoparticle chelated with Gd did not.

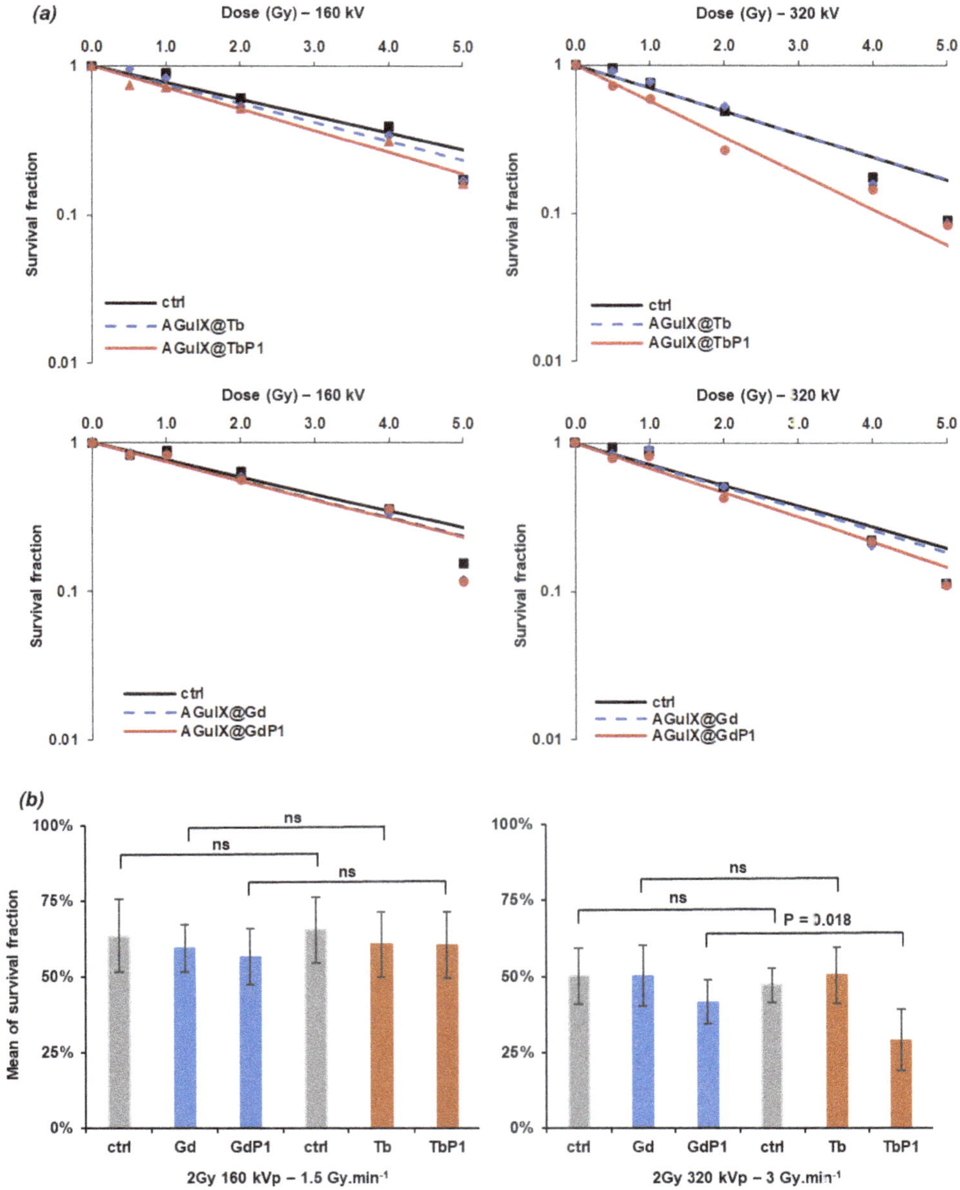

Figure 7. Survival curve of pre-treated U-251 MG cells with AGuIX@Tb-P1 or AGuIX@Gd-P1 exposed to X-ray irradiation. (**a**) LQ model established after numeration of cell clones obtained for AGuIX@-complexes pre-treated cells, then irradiated at 160 and 320 KVp for a range of 0.5 to 5.0 Gy. (**b**) Means of survival fraction at 2.0 Gy were compared using the Kruskal-Wallis test (with $\alpha = 0.05$), and post-hoc by the Mann-Whitney test ($\alpha = 0.05$) for unpaired groups ($n = 12$ wells per condition from at least 4 independent experiments). Abbreviation: ctrl, irradiated cells without nanoparticles pre-treatment.

Table 1. LQ model parameters for the range of 0–5.0 Gy irradiations.

Condition Tested	α Parameter	Dose Modifying Factor (DMF)	Survival Fraction ($SF_{2.0\ Gy}$)	Enhanced Factor (%)	DER
		160 kVp/1.5 Gy·min^{-1}			
Control	0.263	1.000	0.592		
AguIX@Gd	0.292	0.966	0.587		
AguIX@GdP1	0.296	0.862	0.558		
AguIX@Tb	0.291	0.885	0.559		
AguIX@TbP1	0.333	0.769	0.515		
		320 kVp/3.0 Gy·min^{-1}			
Control	0.341	1.000	0.504		
AguIX@Gd	0.340	1.000	0.507		
AguIX@GdP1	0.385	0.876	0.463	8.1	1.08
AguIX@Tb	0.354	1.000	0.493	2.2	1.02
AguIX@TbP1	0.559	0.650	0.327	35.1	1.54

Two dose rates were assessed at 1.5 and 3.0 Gy·min^{-1}. AGuIX@Gd-P1 or AGuIX@Tb-P1 concentrations were set to 1 μM P1 equivalent. Control, cells exposed to X-ray alone; DER, dose enhanced ratio.

3. Discussion

3.1. Gadolinium Substitution by Terbium in the AguIX@ Platform

X-ray-induced PDT represents an alternative to PDT leading to the possibility to access to tumors localized in deep brain tissue. As shown herein, we proposed to replace Gd in the original AGuIX@Gd by Tb and grafted P1 to demonstrate the interest of such nanoparticle in X-PDT. This strategy is based on several lines of evidence: as demonstrated previously, AGuIX@Gd-P1 accumulate within human cell-grafted tumor by EPR in orthotopic site [38,40], even though the nanoparticle delivery depends mainly on the development of angiogenesis processes. Moreover, both AGuIX@Gd and AGuIX@Gd-P1 are eliminated from the animal body by the renal route [34,38]. However, in the latter, nanoparticle clearance was associated to the liver/feces body elimination. Moreover, we showed that the replacement of Gd by Tb, in addition to the grafting of P1, did not change the hydrodynamic diameter of the nano-objects.

Tb was selected as a scintillating agent and we demonstrated the spectral overlap between Tb and P1, which is necessary for X-PDT (Figures 1 and 2). Thus, Tb can transfer the energy received after nanoparticle X-ray irradiation to P1 by FRET non radiative transfer. These results are consistent with previous findings, which highlighted the relation between the 545 nm Tb emission peak, P1 Q3 band and X-ray photons ranging from 20 to 130 kVp in activating photo-agents, through induced visible luminescence from rare-earth particles [31]. In PDT, a large part of visible light is absorbed by the photosensitizer which produces singlet oxygen, known as the PDT type II reaction [3]. In contrast, when X-PDT is used, only a small fraction of the X-ray emitted photons will be converted into scintillations [21,22]. Indeed, the interaction between material and high energy photons depends on the Z atomic number id est electronic density and incident energy [4,21]. Usually, in medical imaging, scintillation crystals are designed thick enough to increase the probability to completely stop incident photons via the photoelectric effect [41]. Conversely, in aqueous media, the effective density will be lower than expected, leading only to a small fraction of X-ray energy converted into visible light. Moreover, the photodynamic efficiency can depend on various factors such as the distribution of the scintillating nanoparticles within the tumor tissue, the composition of the tumor stromal microenvironment, the level of molecular oxygen in this microenvironment, and the illumination density resulting from the scintillating agents. Bulin et al. and Abliz et al. demonstrated the production of singlet oxygen via porphyrin X-ray-induced activation of Gd oxysulfide doped with Tb or Tb oxide, respectively [27,30]. However, low X-ray energies from 10 to 130 kVp (id est mean energies ranging from 5 to 70 keV at 20 mA) were tested, while energies used in radiotherapy are commonly from 160 to few hundred keV for preclinical in vivo experiments. In addition, clinical radiotherapy involves much higher energies, around 6 MV. Scintillation yield is dramatically lower at such levels of keV energies since the probability of photoelectric interaction becomes minimal [41]. Therefore, X-PDT under conventional linear accelerators should be less effective than with radiograph tubes. On the other hand, the bioluminescence of Tb as a scintillating agent was demonstrated, using high concentrations of the nanoscintillator and irradiation deposits without compatibility with the energies in pre-clinical studies or in the clinic [42–44].

3.2. *Irradiation of AGuIX@Tb-P1 Produces Singlet Oxygen*

As shown in Figure 3, AGuIX@Tb-P1 produced singlet oxygen under X-ray irradiation, as demonstrated with APF and SOSG probes. APF probe reacts with the hydroxyl radical and singlet oxygen; SOSG is specific to singlet oxygen [45–48]. Recently, it has been reported that the SOSG fluorescent signal occurs independently either of the presence of singlet oxygen or in the absence of photosensitizer during X-ray irradiation [49]. It has been also shown that the probe under UV excitation, generated an endoperoxide derivative which acts as a photosensitizer producing singlet oxygen [50]. No dramatic change was observed when the SOSG probe was tested without nanoparticles (Figure 3a), at a dose rate of 3.0 $Gy \cdot min^{-1}$ and with an energy set up at 320 kVp. When sodium azide, as a quencher of singlet oxygen, was added in the reaction mixture, the changes observed were mainly due to the production of singlet oxygen generation from P1. Alternatively, the APF probe was used instead of SOSG in the same experimental conditions (Figure 3b). In the experimental conditions used, we could not exclude that during X-ray irradiation of AGuIX@Tb-P1, the APF fluorescence signal increase was associated to the generation of reactive species (especially hydroxyl radical), as observed by the slight but continuous increase of the slope curve, even though in the presence of sodium azide (Figure 3a). Such an increase has been related to X-ray-mediated water radiolysis alone in the presence of the APF probe [51]. However, Bulin et al. [27] hypothesized that P1 acts as a radiosensitizer and therefore the molecule could involve the generation of reactive species under X-ray exposition.

3.3. *Irradiation of AGuIX@Tb-P1 Induces a Photodynamic Effect on U-251 MG Cell Growth*

We assessed whether AGuIX@Tb-P1 altered U-251 MG cell survival by PDT and X-PDT. We used U-251 MG cells whose behavior is close to that of GBM in situ [52]. Collectively, we demonstrated that P1 grafted into AGuIX@Tb limits the cytotoxicity of the molecule alone (Figure S4), in agreement with previous findings [53,54]. AGuIX@Tb-P1 pre-treated cell exposure to photodynamic or X-ray irradiation triggered cell death and limited cell clone formation (Figures 6b and 7). Interestingly, we found that the treatment efficacy was enhanced with X-PDT relative to radiation treatment when cells were exposed to AguIX@Tb-P1 instead of AguIX@Gd-P1 at 2.0 Gy (320 kVp), with a DMF estimated at 0.65 corresponding to an enhanced factor calculated at 35%. The benefit of X-PDT relative to radiation treatment was obtained with diverse nanoparticle designs, for examples, with P1 and Ce-doped with titanium oxide on A549 human lung cancer cell; with P1-grafted silicium oxide nanospheres on Hela cervical cells; with rat L9 glioma cell exposed to LaF_3:Tb particles with adsorbed meso-tetra(4-carboxyphenyl)porphyrin with low irradiation source; or with $SrAl_2O_4:Eu^{2+}$ nanoparticles; collectively, allowing to the conclusion that tumors cells undergo cell death by cumulative/synergistic effects of irradiation treatment and X-ray induced photo-treatment [55–58]. Finally, cancer cell exposure to AGuIX doped with Gd involved cell radiosensitization. However, we did not find any significant change when cells were treated with AGuIX@Gd, as compared to the results of untreated cells exposed to X-rays alone (Figure 7 and Table 1). This discrepancy could be explained since the radiosensitizing effects on cancer cells from diverse origin has been reported for Gd concentration from 100 μM up to 1 mM [34], whereas herein the concentration of lanthamide was 16 and 25 μM for Tb and Gd, respectively; in both cases, the P1 equivalent concentration tested was 1 μM.

In conclusion, we showed that replacement of Gd by Tb in the initial AGuIX design leads to a promising nanoparticle for X-PDT in in vivo experimental. Moreover, we demonstrated that chelated Tb nanoparticles react linearly to X-ray energy (at least up-to 320 kVp) and flow, and they were able to activate P1 to produce singlet oxygen. The constructed nanoparticle was not toxic, while remaining unexposed to light. In vitro experiments confirmed the interest of these AGuIX design, notably at a 3.0 Gy·min^{-1} dose rate. However, it has to be noted that such a strategy should not be considered with external radiotherapy but with low-energy devices (corresponding to a few hundred of keV), such as radiograph tubes, to ensure high energy conversion and singlet oxygen production while lowering exposure dose. Such investigations are currently being conducted in our group.

4. Materials and Methods

4.1. Reagents

Fluorescent probes, 3′-(p-aminophenyl) fluorescein (APF), 2′,5′(di-acetate) dichlorofluorescein (DCF2-DA), Singlet Oxygen Sensor Green™ (SOSG) and propidium iodide were from Molecular Probe (Merck-Sigma, St Quentin Fallavier, France). 3-(4,5-dimethylthiazol-2-yl)-2,5-diphenyl tetrazolium (MTT) was purchased from Acros Organics (Thermo Fisher Scientific, France). 5-(4-carboxyphenyl succinimide ester)-10,15,20-triphenylporphyrin (P1) and zinc(II) 5-(4-carboxyphenyl)-10,15,20-(tri-N-methyl-4-pyridyl) porphyrin trichloride (ZnPy3P1) were purchased from Porphychem (Porphychem SAS, Dijon, France). Other reagents were of analytical grade.

4.2. Synthesis and Preparation of the AGuIX@-Complexes

We used four different nanoparticles, based on the same AGuIX® polysiloxane core surrounded by 1,4,7,10-tetraazacyclododecane-1,4,7,10-tetraacetic acid (DOTA)/metal cation (3+) complexes, covalently grafted to the inorganic matrix. The cations used were Terbium (Tb^{3+}, Z = 65, A = 159 g mol^{-1}) and Gadolinium (Gd^{3+}, Z = 64, A = 157 g mol^{-1}). P1 was covalently grafted as a photosensitizer (AGuIX@Tb-P1 or AGuIX@Gd-P1).

Ultra-small siloxane particles were obtained in a two-step synthesis procedure as described previously [33]. In brief, 3-aminopropyl triethoxysilane (185 mmol) and 1,4,7,10-tetra-azacyclododecane-1-glutaric anhydride-4,7,10-triacetic acid (137 mmol, DOTAGA) were mixed to react in 630 mL of diethylene glycol (DEG) at room temperature for 20 h in order to create the DOTAGA silane. Tetraethyl orthosilicate (249 mmol) was added and the mixture was left for 1 h with stirring before addition of 6.3 L of ultrapure water for the condensation Sol-Gel reaction and hydrolysis. The mixture was heated successively at 80 °C for 6 h, then at 50 °C for 18 h. After incubation, the pH was adjusted to 2.0 with HCl (12N). The ultra-small siloxane particles were purified by tangential filtration on Vivaflow® membranes with a cut-off at 5 kDa (Sartorius Stedim Biotech, Aubagne, France). The final volume was 400 mL with the purification factor of 1000. The pH of the solution was adjusted to 7.4 with 1 M NaOH solution. The final amount of free DOTA was measured by europium titration, as described previously [31]. Free DOTA groups were estimated at 100 mM DOTA. Tb chelation was performed by addition of 1.5 mmole trichloroterbium hexahydrate in 15 mL of the particle solution. The pH was adjusted at 6.0 with 1 M NaOH, and the mixture heated until temperature raised 80 °C. Each day, the pH was measured and adjusted to 6.0, until there was no pH change. The ultra-small particles were filtered onto VivaSpin® membranes (cut-off at 5 kDa, Sartorius Stedim Biotech). The volume obtained was 15 mL and the purification factor was 100. The Tb chelation yield was determined by Inductively Coupled Plasma–Mass Spectroscopy and was estimated at 60%. Finally, the pH was adjusted at 7.4, before freeze-drying.

P1 was grafted following the protocol described previously [33,39]. In summary, 500 mM (1.5 g) of DOTA-free particles, obtained from the first step of particle synthesis, were dispersed in 3 mL of pure water and 1.5 mmol of trichloroterbium hexahydrate. The mixture's pH was adjusted to 6.0 with 1 M NaOH solution and heated to 80 °C for Tb chelation as described above. 12 mL of DEG pre-heated at 40 °C was added to the solution. 150 μmol P1 diluted in DMSO was added drop-by-drop under stirring. The final solvent ratio H_2O/DEG/DMSO was 13/53/34. The mixture was left stirring at 40 °C for 12 h in the dark and was filtrated through VivaSpin® membranes (cut-off at 5 kDa, Sartorius Stedim Biotech). The purification factor was 100 and the final volume was 15 mL. The Tb chelating yield, determined by Inductively Coupled Plasma–Mass Spectroscopy, was 47%. The P1 coupling yield was estimated at 29% by recording the absorbance at 520, 550 nm, 590 and 650 nm Q bands. The final Tb:P1 ratio was 16 moles Tb for 1 mole P1. All concentrations of nanoparticle containing P1 will be referred to thereafter as the concentration of P1. A stock AGuIX@Tb-P1 suspension was prepared in water and was 3 mM P1 equivalent; the solution of AGuIX@Tb was 37.5 mM (Tb equivalent) in water.

4.3. *Dynamic Light Scattering Size*

Direct measurements of the size distribution of the nanoparticles suspended in any medium were performed via Zetasizer NanoS DLS (Dynamic light scattering, laser He-Ne 633 nm, Malvern Instrument, Orsay, France). Prior to the experiment, the nanoparticles were diluted in 0.01 M NaCl (pH 7.4).

4.4. *Potential of the AGuIX Conjugates*

Potential ζ measurements were carried out with a Zetasizer Nano-Z (Malvern Instrument) equipped with a He-Ne laser at 633 nm. Before measurement, the nanoparticles were dispersed in 0.01 M NaCl and buffer solutions (Honeywell Fluka™ Buffer Solution, ThermoFischer Scientific, Ilkirch, France).

4.5. *Synthesis of AGuIX@Gd and AGuIX@Gd-P1*

The AGuIX@ complexes were synthesized and characterized as described previously [33], with a ratio of 1 mole of P1 for 25 moles of Gadolinium (Gd), as estimated by Inductively Coupled Plasma–Mass Spectroscopy analysis. Stock solution of AGuX@Gd or AguX@Gd-P1 was respectively 50 mM Tb equivalent, and 4 mM P1 equivalent in water.

4.6. *Photophysical Properties of AGuIX@Tb-P1*

Absorption spectra were recorded on a UV-3600 UV-visible double beam spectrophotometer (Shimadzu, Marne-La-Vallée, France). Fluorescence spectra were recorded on a Fluorolog FL3-222 spectrofluorometer (Horiba Jobin Yvon, Longjumeau, France) equipped with a 450 W Xenon lamp and thermostated cell compartment (25 °C), a UV-visible photomultiplier R928 (Hamamatsu Photonics, Hamamatsu, Japan), and an InGaAs infrared detector (DSS-16A020L Electro-Optical System Inc., Phoenixville, PA, USA). The excitation beam was diffracted by a double ruled grating SPEX monochromator (1200 grooves/mm blazed at 330 nm). The emission beam was diffracted by a double ruled grating SPEX monochromator (1200 grooves/mm blazed at 500 nm). Singlet oxygen emission was detected through a double ruled grating SPEX monochromator (600 grooves/mm blazed at 1 μm) and a long-wave pass (780 nm). All spectra were measured in four-face quartz vials. All the emission spectra (fluorescence and singlet oxygen luminescence) have been displayed with the same absorbance (less than 0.2) with the lamp and photomultiplier correction.

Spectral overlap, as well as Förster radius, was computed to characterize the energy transfer from the Tb cation (Tb^{3+}) to P1. Moreover, Tb luminescence decay profile was recorded using a Fluorolog spectrofluorometer; the excitation wavelength was set at 351 nm and the emission peaks were scanned in the 400–690 nm region. The luminescence lifetime of Tb alone or in mixture with P1 was recorded using lifetime Fluorolog. We assessed the 545 nm peak decay as it is the highest Tb fluorescence peak. If relevant, we computed the quenching constant (expressed as $L{\cdot}mol^{-1}\ s^{-1}$) as $Kq = K/\tau_0$, where K is the Stern-Volmer constant which was graphically determined; τ_0 is the Tb fluorescence lifetime without photosensitizer.

4.7. Tb Scintillation Assessment under X-ray

Samples were irradiated using a biological X-ray Irradiator X-RAD 320 (Precision X-ray INC., North Branford, CT, USA) with a tungsten anode. Photons were produced by X-ray tubes and produced continuous energy distribution. The tube parameters were set from 25 up to 320 kVp (id est mean photon energies of 8 and 116 keV) with the current set from 0.5 up to 12.5 mA [59]. A 2 mm Al filter was used to remove low energy photons. For Tb scintillation assessment, irradiation time was set at 90 s for each parameter.

An optical fiber was inserted inside the irradiation chamber, in front of the vial containing AGuIX@Tb solutions to gather emission fluorescence photons. Emission spectra were recorded with an USB2000 spectrometer (Ocean Optics Inc, Dunedin, FL, USA). This versatile high-resolution spectrometer (FWHM = 3.5 nm) is an optical instrument based on a diffraction grating and a one-dimensional CCD detector array. Integration time was set to 5 s, the spectrum bandwidth ranged from 340 to 820 nm and the optical fiber was placed across from a transparent vial (UVette® 220–1600 nm; cat.no. 952010051, Eppendorf, Hamburg, Germany). Emission spectra were recorded at different times to assess photonic density configurations on the Tb scintillation performance. Each measurement was repeated 7 times and all spectra were subtracted with the same solution spectrum obtained without irradiation. When varying X ray energy, we set the tube current to maximum and we set the voltage to 320 kVp when we assessed the tube current on the AGuIX@Tb response. A linear correlation coefficient was computed to characterize the relation between AGuIX@Tb peaks intensities, exciting photons energy (X ray kVp) and X ray flow (X ray mA), respectively. The energy transfer from the nanoparticles to a photo-agent was assessed with setting irradiation parameters to 320 kVp and 12.5 mA (a 3.0 $Gy{\cdot}min^{-1}$ dose rate in our experimental conditions).

4.8. Singlet Oxygen Production during Red Light Exposition or X-ray Irradiation

The reaction mixture was prepared in 30 mM Tris/HCl (pH 7.4) containing 400 μM AGuIX@Tb-P1 or 45 mM AGuIX@Tb and 5 μM APF or 10 μM SOSG probe. Singlet oxygen quenching was achieved by addition of sodium azide (NaN_3; stock solution, 1 M) prepared in the same buffer, to a final concentration at 10 mM. Irradiation was set to 320 kVp, 12.5 mA, and source-surface distance adjusted to yield a 3.0 Gy min^{-1} dose rate. Fluorescence emission was detected spectroscopically at 515 and 525 nm for APF and SOSG, respectively. Home-made software allowed long acquisition times and synchronization between laser illumination and signal recording. Integration time was set to 100 ms and time points were acquired every 5 min during 20 min. Moreover, P1 at 100 μM was irradiated without a nanoscintillator with the same parameters to validate the absence of the direct excitation by X-rays.

4.9. Biological Experiments

4.9.1. Cell Culture

Human U-251 MG (ECACC 09063001, Salisbury, UK) glioblastoma-derived cells were cultivated in Roswell Park Memorial Institute medium (RPMI) without phenol red, containing 10% (*v*/*v*) heat-inactivated (30 min at 56 °C) fetal calf serum (Invitrogen, Paisley, UK), 1% (*v*/*v*) non-essential amino acid (Invitrogen), 0.5% (*v*/*v*) essential amino acid (Invitrogen), 1 mM sodium pyruvate (Invitrogen), 1% (*v*/*v*) vitamin (Invitrogen) 0.1 $mg{\cdot}mL^{-1}$ of L-serine, 0.02 $mg{\cdot}mL^{-1}$ L-asparagine (Merck-Sigma), and 1% (*v*/*v*) antibiotics (10,000 $U{\cdot}mL^{-1}$ penicillin, 10 $mg{\cdot}mL^{-1}$ streptomycin) (Merck-Sigma). The cells were seeded routinely at 10^5 cells mL^{-1} and cultivated at 37 °C in a 5% CO_2 humidified atmosphere (Incubator Binder, Tübingen, Germany).

4.9.2. Cell Growth Assessment

Impact of AGuIX@Tb or ZnPy3P1 on U-251 MG cell survival was assessed by the MTT procedure, based on the measurement of mitochondrial succinate dehydrogenase activity (EC 1.3.5.1) [60]. Cells were seeded in 96 well-plates at 10^4 cells/well and left growing for 24 h. Cells were treated with increasing concentrations of ZnPy3P1 (up to 400 μM) and AGuIX@Tb (0.1 to 2 mM) over three days. In addition, glioblastoma cells were exposed to increasing concentrations of AGuIX@Tb-P1 or AGuIX@Gd-P1 (1 to 10 μM) for 72 h. At the time, the medium was discarded and replaced by 100 μL of complete medium containing 0.5 mg mL^{-1} MTT. The plates were incubated for 2 h at 37 °C and formazan crystals obtained were dissolved by adding 100 μL of DMSO. The plates were read at 540 nm (Multiskan Ascent spectrophotometer, Thermo Fisher Scientific, Illkirch, France). Results are expressed relative to those obtained from untreated cells (control), taken as 100. They represented quadruplicate determinations from two independent experiments ($n = 8$).

4.9.3. Nanoparticles Cell Uptake

U-251 MG cells were seeded in 6 well-plates at 10^5 cells/well and left to grow over 48 h. Cells were treated with increasing concentrations of nanoparticles (0.5 to 5.0 μM AGuIX@Tb-P1 or AGuIX@Gd-P1) over 48 h. After incubation, cell layers were washed with 2 mL of Dulbecco's Phosphate buffer saline (DPBS, Merck-Sigma) and cells were suspended with 0.5 mL of 0.05% (*w*/*v*) Trypsin/0.02% (*w*/*v*) EDTA solution (Invitrogen) per well for 5 min at 37 °C. Complete medium (0.5 mL RPMI containing 10% (*v*/*v*) fetal calf serum) was then added. The cell suspension obtained was centrifuged at 1000× *g* for 5 min at 4 °C. Cell pellets were washed with 1 mL of DPBS and centrifuged again. Cells were suspended finally in 0.5 mL of DPBS and left on ice. Fluorescence of P1 was measured in 5000 cells/sample by flow cytometry (Gallios Analyzer, Beckman Coulter France, Roissy, France), with excitation/emission settings at 638 nm and 660/30 nm. Results obtained from nanoparticle uptake were expressed relative to those obtained reaching the maximum of nanoparticle absorption taken as 100. Results are expressed as mean ± SD from triplicate determinations from 3 independent experiments.

4.9.4. PDT and X-PDT Conditions

Cell exposure to a red light was performed at 630 nm with a Laser diode (Biolitec, Biomedical Technology, Iena, Germany) at 0.7 W, corresponding to an irradiance at 4.54 mW·cm^{-2}. Cell layers were exposed to 0.5 to 10.0 J cm^{-2}, corresponding to an exposition duration of 2 min to 38 min. X-ray irradiation was performed using the X-ray Irradiator X-RAD 320 (Precision X-ray Inc., North Branford, CT, USA), using tube parameters set at 160 and 320 kV (mean photons energies of 58 and 116 keV) with current set at 10 mA. A 2 mm Al filter was used to remove low energy photons. The dose rates were 1.5 and 3.0 Gy min^{-1} for 160 and 320 kVp respectively. The doses delivered were 0.5, 2.0, 4.0, and 5.0 Gy for both dose rates.

4.9.5. Reactive Species Quantification during PDT or X-PDT

U-251 MG cells were seeded in 6 well-plates at 2×10^5 cells/well. Cells were left to grow over 48 h, then, treated in the presence of 1 μM AGuIX@-complexes for 24 h. Cells were washed with 2 mL of DPBS. Each well was filled again with complete medium before light exposition or X-ray irradiation. Reactive species generation measurements were achieved post treatment over 1 h. At each time point, cell medium was changed by 2 mL of pre-warmed medium containing 50 μM DCF2-DA for 30 min at 37 °C. Cells were then successively harvested by trypsination, washed with DPBS, and suspended in 0.5 mL DPBS before flow cytometry analysis. Reactive species generation was measured in 5000 cells/sample by flow cytometry with excitation/emission settings at 488 nm and 520/30 nm. Cell death was quantified also using propidium iodide in kinetic uptake experiments. Cells were treated with 20 μM propidium iodide (diluted in DMSO) added to the medium. Propidium iodide-positive cells were numbered by flow cytometric analysis with excitation/emission settings at 488 nm and 620/630 nm (FL3), respectively. Results represent the median of fluorescent peak. Results are expressed relative to those obtained from untreated cells, taken as 100. Results are expressed as mean ± SD of triplicate determinations from 3 independent experiments.

4.9.6. Anchorage-Dependent Clonogenic Assay

The clonogenic assay was performed in 6 well-plates with U-251 MG cells seeded at 500 cells/well. Cells were then treated in the presence of 16.6 μM of AGuIX@Tb, 25 μM AGuIX@Gd, 1 μM AGuIX@Tb-P1 or 1 μM AGuIX@Gd-P1 (P1 equivalent concentration) for 24 h at 37 °C. After incubation, cells were washed with 2 mL of DPBS. 2 mL of complete medium were added in each well before X-ray radiation or red-light exposition. Cells were left to grow at 37 °C for 7 days. At time, cell clones were successively washed with 2 mL DPBS, fixed with 1 mL of 4% (*v*/*v*) formol (pH 7.4) at room temperature for 15 min, washed with 1 mL of DPBS, and stained for 30 min with 0.05% (*w*/*v*) crystal Violet solution prepared in DPBS and containing 25% (*v*/*v*) methanol. Finally, cells were washed three times with 2 mL of bi-distilled water. Cell clones obtained were analysed after picture capture (GelCount™, Oxford Optronix, Abingdon, UK) and ImageJ quantification (N.I.H., Bethesda, MA, USA). Image analysis was performed with well area taken as 862 mm^2. Cell clone counting was improved by background subtraction. Data from untreated and treated cell conditions were compared and expressed as the mean ± SD ($n = 12$).

Survival fraction (SF) was calculated using the linear quadratic (LQ) model, based on the equation: $SF_D = \exp(-(\alpha D + \beta D2))$, where survival fraction is defined as SF_D = (plating efficiency at the dose D)/(platting efficiency at 0 Gy); D corresponds to the Gy dose; α and β, are determined from the established semi log curve, as SF_D = f (Gy dose). The effects of radiation alone or X-PDT, related to untreated cells or exposed to each AGuIX@NP, were compared by calculating the dose modifying factor (DMF) and the survival fraction at 2.0 Gy (SF_2). We also determined the Dose Enhanced Ratio (DER) at 2.0 Gy. DER is defined as the ratio of $SF_{2.0\,Gy}$ calculated for untreated cells to that calculated of treated cells after irradiation (DER = $SF_{2.0\,Gy}$ (control cells)/$SF_{2.0\,Gy}$ (treated cells)). We finally calculated the enhanced factor, expressed in percentage, as EF (%) = 100 × ($SF_{2.0\,Gy}$ (control cells) − $SF_{2.0\,Gy}$ (treated cells))/$SF_{2.0\,Gy}$ (control cells)).

4.9.7. Statistical Analysis

Results obtained were analyzed using the Kruskal-Wallis test ($\alpha = 0.05$) and post-hoc Dunn's test ($\alpha = 0.05$) for paired groups. Any difference was considered significant at $p < 0.05$. Results obtained from clonogenic assays were analyzed using the Kruskal-Wallis test (with $\alpha = 0.05$), and post-hoc by the Mann-Whitney test ($\alpha = 0.05$) for unpaired groups.

Supplementary Materials: The following are available online at https://www.mdpi.com/article/10.3390/ph14050396/s1. Figure S1: Determination of potential of AGuIX@Tb-P1 and AGuIX@Tb, Figure S2: Luminescence decay of AGuIX@Tb and AGuIX@Tb-P1 in water, Figure S3: Luminescence of AGuIX@Tb-P1, Figure S4: SDH activities after U-251 MG cell exposure to AGuIX@ nanoparticles.

Author Contributions: Conceptualization, J.D., B.H., F.L., O.T., M.B.-H. and H.S.; Data curation, P.A.; Formal analysis, J.D., M.I., B.D., D.B., P.A., P.R. and H.S.; Funding acquisition, J.D., P.A., F.L., O.T. and M.B.-H.; Investigation, J.D., M.I., B.D., D.B., P.R. and H.S.; Methodology, A.D.; Project administration, O.T. and M.B.-H.; Software, A.D.; Supervision, F.L., C.F., O.T. and M.B.-H.; Validation, J.D., P.A., C.F. and H.S.; Visualization, J.D., M.I., B.D. and H.S.; Writing—original draft, J.D. and H.S.; Writing—review & editing, J.D., F.L., C.F., O.T., M.B.-H. and H.S. All authors have read and agreed to the published version of the manuscript.

Funding: The work was supported by funds of the CRAN, the *French Ligue Nationale Contre le Cancer* and European EURONANOMED II programme (447/15/GM). The authors would like to thank CPER IT2MP and the *Laura: les couleurs de la vie* association for their support in material funding.

Institutional Review Board Statement: Not applicable.

Informed Consent Statement: Not applicable.

Data Availability Statement: The datasets used and/or analysed during the current study are available from the corresponding author on reasonable request.

Acknowledgments: We thank Huguette Louis from the cytometry core facility of UMS2008 IBSLor (Université de Lorraine-CNRS-INSERM).

Conflicts of Interest: The AGuIX design is protected by patent WO2011135101 (F.L. and O.T.). Elsewhere, the authors report no conflicts of interest in this work.

Abbreviations

AGuIX@Tb or Gd, Ultra small AGuIX-designed nanoparticle chelated with Terbium or Gadolinium; AGuIX@Tb-P1 or Gd-P1 Ultra small AGuIX-Tb or Gd grafted with Porphyrin; APF, 3′-(p-aminophenyl) fluorescein; DCF-2DA, 2′,5′(di-acetate) dichlorofluorescein; DER, dose enhanced ratio; DMF, dose modified factor; DPBS, Dulbecco's phosphate buffer saline, Gd, Gadolinium; GBM, glioblastoma multiforme; MTT, 3-(4,5-dimethylthiazol-2-yl)-2,5-diphenyl tetrazolium; P1, 5-(4-carboxyphenyl succinimide ester)-10,15,20-triphenyl porphyrin; PDT, Photo Dynamic Therapy; RPMI, Roswell Park Memorial Institute medium; SOSG, Singlet Oxygen Sensor Green™; Tb, Terbium; X-PDT, X-ray-induced Photo Dynamic Therapy; Zn3PyP1, zinc(II) 5-(4-carboxyphenyl)-10,15,20-(tri-N-methyl-4-pyridyl) porphyrin trichloride.

References

1. Davis, M.E. Glioblastoma: Overview of disease and treatment. *Clin. J. Oncol. Nurs.* **2016**, *20*, S2–S8. [CrossRef]
2. Stupp, R.; Hegi, M.E.; Mason, W.P.; Bent, M.J.V.D.; Taphoorn, M.J.B.; Janzer, R.C.; Ludwin, S.K.; Allgeier, A.; Fisher, B.; Belanger, K.; et al. Effects of radiotherapy with concomitant and adjuvant temozolomide versus radiotherapy alone on survival in glioblastoma in a randomised phase III study: 5-year analysis of the EORTC-NCIC trial. *Lancet Oncol.* **2009**, *10*, 459–466. [CrossRef]
3. Calixto, G.M.F.; Bernegossi, J.; De Freitas, L.M.; Fontana, C.R.; Chorilli, M. Nanotechnology-based drug delivery systems for photodynamic therapy of cancer: A review. *Molecules* **2016**, *21*, 342. [CrossRef]
4. Bechet, D.; Mordon, S.R.; Guillemin, F.; Barberi-Heyob, M. Photodynamic therapy of malignant brain tumours: A complementary approach to conventional therapies. *Cancer Treat. Rev.* **2014**, *40*, 229–241. [CrossRef] [PubMed]
5. Larue, L.; Ben Mihoub, A.; Youssef, Z.; Colombeau, L.; Acherar, S.; André, J.-C.; Arnoux, P.; Baros, F.; Vermandel, M.; Frochot, C. Using X-rays in photodynamic therapy: An overview. *Photochem. Photobiol. Sci.* **2018**, *17*, 1612–1650. [CrossRef] [PubMed]
6. Pinel, S.; Thomas, N.; Boura, C.; Barberi-Heyob, M. Approaches to physical stimulation of metallic nanoparticles for glioblastoma treatment. *Adv. Drug Deliv. Rev.* **2019**, *138*, 344–357. [CrossRef] [PubMed]
7. Maeda, H.; Sano, Y.; Takeshita, J.; Iwai, Z.; Kosaka, H.; Marubayashi, T.; Matsukado, Y. A pharmacokinetic simulation model for chemotherapy of brain tumor with an antitumor protein antibiotic, neocarzinostatin: Theoretical considerations behind a two-compartment model for continuous infusion via an internal carotid artery. *Cancer Chemother. Pharmacol.* **1981**, *5*, 243–249. [CrossRef]
8. Maeda, H. Toward a full understanding of the EPR effect in primary and metastatic tumors as well as issues related to its heterogeneity. *Adv. Drug Deliv. Rev.* **2015**, *91*, 3–6. [CrossRef] [PubMed]
9. Beck, T.J.; Kreth, F.W.; Beyer, W.; Mehrkens, J.H.; Obermeier, A.; Stepp, H.; Stummer, W.; Baumgartner, R. Interstitial photodynamic therapy of nonresectable malignant glioma recurrences using 5-aminolevulinic acid induced protoporphyrin IX. *Lasers Surg. Med.* **2007**, *39*, 386–393. [CrossRef]
10. Krishnamurthy, S.; Powers, S.K.; Witmer, P.; Brown, T. Optimal light dose for interstitial photodynamic therapy in treatment for malignant brain tumors. *Lasers Surg. Med.* **2000**, *27*, 224–234. [CrossRef]
11. Stummer, W.; Beck, T.; Beyer, W.; Mehrkens, J.H.; Obermeier, A.; Etminan, N.; Stepp, H.; Tonn, J.-C.; Baumgartner, R.; Herms, J.; et al. Long-sustaining response in a patient with non-resectable, distant recurrence of glioblastoma multiforme treated by interstitial photodynamic therapy using 5-ALA: Case report. *J. Neuro-Oncol.* **2007**, *87*, 103–109. [CrossRef]
12. Kostron, H.; Fiegele, T.; Akatuna, E. Combination of FOSCAN® mediated fluorescence guided resection and photodynamic treatment as new therapeutic concept for malignant brain tumors. *Med. Laser Appl.* **2006**, *21*, 285–290. [CrossRef]
13. Stepp, H.; Beck, T.; Pongratz, T.; Meinel, T.; Kreth, F.-W.; Tonn, J.C.; Stummer, W. ALA and malignant glioma: Fluorescence-guided resection and photodynamic treatment. *J. Environ. Pathol. Toxicol. Oncol.* **2007**, *26*, 157–164. [CrossRef]
14. Eljamel, M.S.; Goodman, C.; Moseley, H. ALA and Photofrin® Fluorescence-guided resection and repetitive PDT in glioblastoma multiforme: A single centre Phase III randomised controlled trial. *Lasers Med. Sci.* **2007**, *23*, 361–367. [CrossRef] [PubMed]
15. Zilidis, G.; Aziz, F.; Telara, S.; Eljamel, M.S. Fluorescence image-guided surgery and repetitive Photodynamic Therapy in brain metastatic malignant melanoma. *Photodiagn. Photodyn. Ther.* **2008**, *5*, 264–266. [CrossRef] [PubMed]
16. Johansson, A.; Palte, G.; Schnell, O.; Tonn, J.-C.; Herms, J.; Stepp, H. 5-aminolevulinic acid-induced protoporphyrin IX levels in tissue of human malignant brain tumors. *Photochem. Photobiol.* **2010**, *86*, 1373–1378. [CrossRef]

17. Aziz, F.; Telara, S.; Moseley, H.; Goodman, C.; Manthri, P.; Eljamel, M.S. Photodynamic therapy adjuvant to surgery in metastatic carcinoma in brain. *Photodiagn. Photodyn. Ther.* **2009**, *6*, 227–230. [CrossRef]
18. Jiang, F.; Robin, A.M.; Katakowski, M.; Tong, L.; Espiritu, M.; Singh, G.; Chopp, M. Photodynamic therapy with photofrin in combination with Buthionine Sulfoximine (BSO) of human glioma in the nude rat. *Lasers Med. Sci.* **2003**, *18*, 128–133. [CrossRef] [PubMed]
19. Zimmermann, A.; Ritsch-Marte, M.; Kostron, H. mTHPC-mediated photodynamic diagnosis of malignant brain tumors. *Photochem. Photobiol.* **2001**, *74*, 611–616. [CrossRef]
20. Stummer, W.; Pichlmeier, U.; Meinel, T.; Wiestler, O.D.; Zanella, F.; Reulen, H.-J. Fluorescence-guided surgery with 5-aminolevulinic acid for resection of malignant glioma: A randomised controlled multicentre phase III trial. *Lancet Oncol.* **2006**, *7*, 392–401. [CrossRef]
21. Van Eijk, C.W. Inorganic scintillators in medical imaging. *Phys. Med. Biol.* **2002**, *47*, R85–R106. [CrossRef] [PubMed]
22. Ren, X.-D.; Hao, X.-Y.; Li, H.-C.; Ke, M.-R.; Zheng, B.-Y.; Huang, J.-D. Progress in the development of nanosensitizers for X-ray-induced photodynamic therapy. *Drug Discov. Today* **2018**, *23*, 1791–1800. [CrossRef]
23. Chen, W.; Zhang, J. Using nanoparticles to enable simultaneous radiation and photodynamic therapies for cancer treatment. *J. Nanosci. Nanotechnol.* **2006**, *6*, 1159–1166. [CrossRef]
24. Kamkaew, A.; Chen, F.; Zhan, Y.; Majewski, R.L.; Cai, W. Scintillating nanoparticles as energy mediators for enhanced photodynamic therapy. *ACS Nano* **2016**, *10*, 3918–3935. [CrossRef]
25. Retif, P.; Pinel, S.; Toussaint, M.; Frochot, C.; Chouikrat, R.; Bastogne, T.; Barberi-Heyob, M. Nanoparticles for radiation therapy enhancement: The key parameters. *Theranostics* **2015**, *5*, 1030–1044. [CrossRef] [PubMed]
26. Tang, Y.; Hu, J.; Elmenoufy, A.H.; Yang, X. Highly efficient FRET system capable of deep photodynamic therapy established on X-ray excited mesoporous LAF3:TB scintillating nanoparticles. *ACS Appl. Mater. Interfaces* **2015**, *7*, 12261–12269. [CrossRef] [PubMed]
27. Bulin, A.-L.; Vasil'Ev, A.; Belsky, A.; Amans, D.; LeDoux, G.; Dujardin, C. Modelling energy deposition in nanoscintillators to predict the efficiency of the X-ray-induced photodynamic effect. *Nanoscale* **2015**, *7*, 5744–5751. [CrossRef] [PubMed]
28. Yang, C.-C.; Wang, W.-Y.; Lin, F.-H.; Hou, C.-H. Rare-earth-doped calcium carbonate exposed to X-ray irradiation to induce reactive oxygen species for tumor treatment. *Int. J. Mol. Sci.* **2019**, *20*, 1148. [CrossRef] [PubMed]
29. Li, X.; Xue, Z.; Jiang, M.; Li, Y.; Zeng, S.; Liu, H. Soft X-ray activated NaYF4:Gd/Tb scintillating nanorods for in vivo dual-modal X-ray/X-ray-induced optical bioimaging. *Nanoscale* **2017**, *10*, 342–350. [CrossRef]
30. Abliz, E.; Collins, J.E.; Bell, H.; Tata, D.B. Novel applications of diagnostic X-rays in activating a clinical photodynamic drug: Photofrin II through X-ray induced visible luminescence from "rare-earth" formulated particles. *J. X-ray Sci. Technol.* **2011**, *19*, 521–530. [CrossRef] [PubMed]
31. Elmenoufy, A.H.; Tang, Y.; Hu, J.; Xu, H.; Yang, X. A novel deep photodynamic therapy modality combined with CT imaging established via X-ray stimulated silica-modified lanthanide scintillating nanoparticles. *Chem. Commun.* **2015**, *51*, 12247–12250. [CrossRef] [PubMed]
32. Mignot, A.; Truillet, C.; Lux, F.; Sancey, L.; Louis, C.; Denat, F.; Boschetti, F.; Bocher, L.; Gloter, A.; Stephan, O.; et al. A top-down synthesis route to ultrasmall multifunctional Gd-based silica nanoparticles for theranostic applications. *Chemistry* **2013**, *19*, 6122–6136. [CrossRef]
33. Le Duc, G.; Roux, S.; Paruta-Tuarez, A.; Dufort, S.; Brauer, E.; Marais, A.; Truillet, C.; Sancey, L.; Perriat, P.; Lux, F.; et al. Advantages of gadolinium based ultra-small nanoparticles vs molecular gadolinium chelates for radiotherapy guided by MRI for glioma treatment. *Cancer Nanotechnol.* **2014**, *5*, 4. [CrossRef] [PubMed]
34. Sancey, L.; Lux, F.; Kotb, S.; Roux, S.; Dufort, S.; Bianchi, A.; Cremillieux, Y.; Fries, P.; Coll, J.-L.; Rodriguez-Lafrasse, C.; et al. The use of theranostic gadolinium-based nanoprobes to improve radiotherapy efficacy. *Br. J. Radiol.* **2014**, *87*, 20140134. [CrossRef]
35. McMahon, S.J.; Hyland, W.B.; Muir, M.F.; Coulter, J.A.; Jain, S.; Butterworth, K.T.; Schettino, G.; Dickson, G.R.; Hounsell, A.R.; O'Sullivan, J.M.; et al. Biological consequences of nanoscale energy deposition near irradiated heavy atom nanoparticles. *Sci. Rep.* **2011**, *1*, 18. [CrossRef]
36. Sicard-Roselli, C.; Brun, E.; Gilles, M.; Baldacchino, G.; Kelsey, C.; McQuaid, H.; Polin, C.; Wardlow, N.; Currell, F. A new mechanism for hydroxyl radical production in irradiated nanoparticle solutions. *Small* **2014**, *10*, 3338–3346. [CrossRef]
37. Du, Y.; Sun, H.; LeDuc, G.; Doussineau, T.; Ji, K.; Wang, Q.; Lin, Z.; Wang, Y.; Liu, Q.; Tillement, O.; et al. Radiosensitization effect of AGuIX, a Gadolinium-based nanoparticle, in nonsmall cell lung cancer. *ACS Appl. Mater. Interfaces* **2020**, *12*, 56874–56885. [CrossRef] [PubMed]
38. Gries, M.; Thomas, N.; Acherar, S.; Frochot, C.; Lux, F.; Tillement, O.; Barberi-Heyob, M.; Daouk, J.; Rocchi, P.; Choulier, L.; et al. Multiscale selectivity and in vivo biodistribution of NRP-1-targeted theranostic aguix nanoparticles for PDT of glioblastoma. *Int. J. Nanomed.* **2020**, *15*, 8739–8758. [CrossRef] [PubMed]
39. Franken, N.A.; Oei, A.L.; Kok, H.P.; Rodermond, H.M.; Sminia, P.; Crezee, J.; Stalpers, L.J.; Barendsen, G.W. Cell survival and radiosensitisation: Modulation of the linear and quadratic parameters of the LQ model (Review). *Int. J. Oncol.* **2013**, *42*, 1501–1515. [CrossRef]
40. Toussaint, M.; Pinel, S.; Bastogne, T.; Frochot, C.; Tillement, O.; Lux, F.; Barberi-Heyob, M.; Auger, F.; Durieux, N.; Thomassin, M.; et al. Proton MR spectroscopy and diffusion mr imaging monitoring to predict tumor response to interstitial photodynamic therapy for glioblastoma. *Theranostics* **2017**, *7*, 436–451. [CrossRef] [PubMed]

41. Clement, S.; Deng, W.; Camilleri, E.; Wilson, B.C.; Goldys, E.M. X-ray induced singlet oxygen generation by nanoparticle-photosensitizer conjugates for photodynamic therapy: Determination of singlet oxygen quantum yield. *Sci. Rep.* **2016**, *6*, 19954. [CrossRef]
42. Kulmala, S.; Hakanen, A.; Laine, E.; Haapakka, K. X-ray irradiation-induced optical luminescence of terbium(III) chelates in aqueous solutions. *J. Alloy. Compd.* **1995**, *225*, 279–283. [CrossRef]
43. Kulmala, S.; Laine, E.; Hakanen, A.; Raerinne, P.; Haapakka, K. Mechanism of terbium(III)-enhanced lyoluminescence of X-ray irradiated sodium chloride. *Anal. Chim. Acta* **1994**, *294*, 1–11. [CrossRef]
44. Jürgensen, A. XEOL spectroscopy of lanthanides in aqueous solution. *Can. J. Chem.* **2017**, *95*, 1198–1204. [CrossRef]
45. Setsukinai, K.-I.; Urano, Y.; Kakinuma, K.; Majima, H.J.; Nagano, T. Development of novel fluorescence probes that can reliably detect reactive oxygen species and distinguish specific species. *J. Biol. Chem.* **2003**, *278*, 3170–3175. [CrossRef] [PubMed]
46. Price, M.; Reiners, J.J.; Santiago, A.M.; Kessel, D. Monitoring singlet oxygen and hydroxyl radical formation with fluorescent probes during photodynamic therapy. *Photochem. Photobiol.* **2009**, *85*, 1177–1181. [CrossRef] [PubMed]
47. Takahashi, J.; Misawa, M. Characterization of reactive oxygen species generated by protoporphyrin IX under X-ray irradiation. *Radiat. Phys. Chem.* **2009**, *78*, 889–898. [CrossRef]
48. Misawa, M.; Takahashi, J. Generation of reactive oxygen species induced by gold nanoparticles under X-ray and UV Irradiations. *Nanomed. Nanotechnol. Biol. Med.* **2011**, *7*, 604–614. [CrossRef]
49. Liu, H.; Carter, P.J.H.; Laan, A.C.; Eelkema, R.; Denkova, A.G. Singlet oxygen sensor green is not a suitable probe for $^1O_2$ in the presence of ionizing radiation. *Sci. Rep.* **2019**, *9*, 8393. [CrossRef] [PubMed]
50. Gollmer, A.; Arnbjerg, J.; Blaikie, F.H.; Pedersen, B.W.; Breitenbach, T.; Daasbjerg, K.; Glasius, M.; Ogilby, P.R. Singlet oxygen sensor green: Photochemical behavior in solution and in a mammalian cell. *Photochem. Photobiol.* **2011**, *87*, 671–679. [CrossRef] [PubMed]
51. Popovich, K.; Tomanová, K.; Čuba, V.; Procházková, L.; Pelikánová, I.T.; Jakubec, I.; Mihóková, E.; Nikl, M. LuAg:Pr^{3+}-porphyrin based nanohybrid system for singlet oxygen production: Toward the next generation of PDTX drugs. *J. Photochem. Photobiol.* **2018**, *179*, 149–155. [CrossRef]
52. Jacobs, V.L.; Valdes, P.A.; Hickey, W.F.; De Leo, J.A. Current review of in vivo GBM rodent models: Emphasis on the CNS-1 tumour model. *ASN Neuro* **2011**, *3*, AN20110014. [CrossRef]
53. Luo, W.; Liu, R.-S.; Zhu, J.-G.; Li, Y.-C.; Liu, H.-C. Subcellular location and photodynamic therapeutic effect of chlorin e6 in the human tongue squamous cell cancer Tca8113 cell line. *Oncol. Lett.* **2015**, *9*, 551–556. [CrossRef] [PubMed]
54. Pavani, C.; Uchoa, A.F.; Oliveira, C.S.; Iamamoto, Y.; Baptista, M.S. Effect of zinc insertion and hydrophobicity on the membrane interactions and PDT activity of porphyrin photosensitizers. *Photochem. Photobiol. Sci.* **2009**, *8*, 233–240. [CrossRef] [PubMed]
55. Yang, C.-C.; Sun, Y.-J.; Chung, P.-H.; Chen, W.-Y.; Swieszkowski, W.; Tian, W.; Lin, F.-H. Development of Ce-doped TiO_2 activated by X-ray irradiation for alternative cancer treatment. *Ceram. Int.* **2017**, *43*, 12675–12683. [CrossRef]
56. Kuo, W.-J.; Wang, Y.-C.; Chen, M.-H.; Tung, F.-I.; Liu, T.-Y. Study of a novel vehicle developed for enhancing the efficacy of radiation therapy. *Ceram. Int.* **2017**, *43*, S789–S796. [CrossRef]
57. Chen, M.-H.; Jenh, Y.-J.; Wu, S.-K.; Chen, Y.-S.; Hanagata, N.; Lin, F.-H. Non-invasive photodynamic therapy in brain cancer by use of Tb^{3+}-doped LAF_3 nanoparticles in combination with photosensitizer through X-ray irradiation: A proof-of-concept study. *Nanoscale Res. Lett.* **2017**, *12*, 1–6. [CrossRef] [PubMed]
58. Wang, G.D.; Nguyen, H.T.; Chen, H.; Cox, P.B.; Wang, L.; Nagata, K.; Hao, Z.; Wang, A.; Li, Z.; Xie, J. X-ray induced photodynamic therapy: A combination of radiotherapy and photodynamic therapy. *Theranostics* **2016**, *6*, 2295–2305. [CrossRef] [PubMed]
59. Huda, W.; Scalzetti, E.M.; Levin, G. Technique factors and image quality as functions of patient weight at abdominal CT. *Radiology* **2000**, *217*, 430–435. [CrossRef]
60. Mosmann, T. Rapid colorimetric assay for cellular growth and survival: Application to proliferation and cytotoxicity assays. *J. Immunol. Methods* **1983**, *65*, 55–63. [CrossRef]

pharmaceuticals

MDPI

Article

Synergistic Photoantimicrobial Chemotherapy of Methylene Blue-Encapsulated Chitosan on Biofilm-Contaminated Titanium

Chiu-Nan Lin [1], Shinn-Jyh Ding [1,2,*] and Chun-Cheng Chen [2,3,*]

1 Institute of Oral Science, Chung Shan Medical University, Taichung City 402, Taiwan; charliennnn@gmail.com
2 Department of Stomatology, Chung Shan Medical University Hospital, Taichung City 402, Taiwan
3 School of Dentistry, Chung Shan Medical University, Taichung City 402, Taiwan
* Correspondence: sjding@csmu.edu.tw (S.-J.D.); fw3256@gmail.com (C.-C C.); Tel.: +886-4-24718668 (S.-J.D.)

Abstract: Intensive efforts have been made to eliminate or substantial reduce bacterial adhesion and biofilm formation on titanium implants. However, in the management of peri-implantitis, the methylene blue (MB) photosensitizer commonly used in photoantimicrobial chemotherapy (PACT) is limited to a low retention on the implant surface. The purpose of this study was to assess enhancive effect of water-soluble quaternary ammonium chitosan (QTS) on MB retention on biofilm-infected SLA (sandblasted, large grid, and acid-etched) Ti alloy surfaces in vitro. The effectiveness of QTS + MB with different concentrations in eliminating Gram-negative *A. actinomycetemcomitans* or Gram-positive *S. mutans* bacteria was compared before and after PACT. Bacterial counting and lipopolysaccharide (LPS) detection were examined, and then the growth of human osteoblast-like MG63 cells was evaluated. The results indicated that the synergistic QTS + MB with retention ability significantly decreased the biofilm accumulation on the Ti alloy surface, which was better than the same concentration of 1 wt% methyl cellulose (MC). More importantly, the osteogenic activity of MG63 cells on the disinfected sample treated by QTS + MB-PACT modality was comparable to that of sterile Ti control, significantly higher than that by MC + MB-PACT modality. It is concluded that, in terms of improved retention efficacy, effective bacteria eradication, and enhanced cell growth, synergistically, PACT using the 100 μg/mL MB-encapsulated 1% QTS was a promising modality for the treatment of peri-implantitis.

Keywords: dental implant; peri-implantitis; photoantimicrobial chemotherapy; methylene blue; chitosan

Citation: Lin, C.-N.; Ding, S.-J.; Chen, C.-C. Synergistic Photoantimicrobial Chemotherapy of Methylene Blue-Encapsulated Chitosan on Biofilm-Contaminated Titanium. *Pharmaceuticals* **2021**, *14*, 346. https://doi.org/10.3390/ph14040346

Academic Editor: Serge Mordon

Received: 15 March 2021
Accepted: 8 April 2021
Published: 9 April 2021

Publisher's Note: MDPI stays neutral with regard to jurisdictional claims in published maps and institutional affiliations.

1. Introduction

When the pristine surface of a dental titanium implant is exposed to the oral cavity after installation, bacterial adhesion occurs in a similar way to teeth [1,2]. After adhesion, bacteria will form microcolonies, which are then embedded within an extracellular polysaccharide matrix to construct complex three-dimensional biofilm structures that are difficult to remove [2,3]. More importantly, bacterial contamination endangers the osseointegration of titanium implants [4,5] and ultimately leads to bone loss around the implant. According to the literature [6], after 10 years of use without systematic supportive treatment, peri-implant disease has become a common clinical problem. Van Velzen et al. evaluated the 10-year survival rate of SLA (sandblasting, large-grit and acid-etching) titanium dental implants and the incidence of peri-implant diseases in patients with complete and partial edentulousness, and pointed out the occurrence of peri-implantitis in 7% of dental implants [7]. As the prevalence of peri-implantitis has increased significantly, as many as 33% of implants have experienced progressive bone loss [8]. Thus, in order to successfully treat peri-implantitis, exhaustive eradication of pathogenic microorganisms from the surface of the implant is indispensable [9–12]. The current clinical approach is to use mechanical debridement to destroy biofilms and reduce bacterial adhesion [13], supplemented by adjunct antibacterial therapies, such as the use of antiseptics and antibiotics [11,14,15]. However, the efficacy of these strategies is limited by the complex topography of the implant surface, the emergence of drug-resistant microorganisms, and the survival of a few viable bacteria. Therefore, the strategies used cannot completely cause the osseointegration of the implant to a large extent [16]. In fact, there is currently no gold standard method for the treatment of peri-implantitis [15]. Undoubtedly, it necessitates developing newly effective treatments targeting bacterial eradication for clinical use.

For this reason, extensive investigations have been explored to develop photodynamic therapy (PDT), which was introduced as a medical treatment in 1904 [17]. PDT uses a specific wavelength of light to activate a photosensitive drug (photosensitizer, PS) that preferentially binds to cells or microorganisms in the presence of oxygen in the air. This may lead to the formation of several reactive oxygen species, resulting in cell death [15,18]. Among various applications, such as antimicrobial elimination, anticancer therapy, and wound healing [19–21], the former is the so-called antimicrobial photodynamic therapy (aPDT) or photoantimicrobial chemotherapy (PACT) [22]. The PACT modality has been used alone or in combination with other treatment options, such as mechanical debridement.

As for photosensitizers, the commonly used methylene blue (MB) is a hydrophilic, water-soluble phenothiazine derivative and the first synthetic compound used as an antiseptic in clinical practice [23]. Moreover, it has been approved by the Food and Drug Administration (FDA) for intravenous administration of methemoglobinemia [24] and has been used to target oral bacteria [25]. In the management of peri-implantitis, MB-PACT treatment usually aims to eliminate bacterial adhesion and biofilm on the implant surface, thereby achieving re-osseointegration. However, satisfactory therapeutic strategies or scientifically based treatment recommendations are still not available [17,26]. For example, in terms of clinical practice, the retention capacity of the flowable photosensitizer should be considered, and the loss of photosensitizer during use should be avoided to improve the therapeutic effect. To overcome this problem, viscous polymers can be adopted to achieve the adhesion of photosensitive drug to mucosa (such as oral buccal mucosa, nasal mucosa) [27,28]. Bioadhesion strategies can increase the retention time and availability of drugs [29]. In order to improve the MB retention during PACT, López-Jiménez et al. used hydroxymethyl cellulose as a mucoadhesive in the Periowave system [30].

Among the viscous polymers, naturally occurring chitosan polysaccharides are structurally similar to glycosaminoglycans [31]. Chitosan polysaccharide has non-antigenicity, high hydrophilicity and good film-forming properties, and is a natural choice for biomedical applications (such as scaffolds and mucosal carriers) [32–35]. It has been reported that due to the interaction between the cationic chain of chitosan and the negatively charged residues on the bacterial surface, the broad antibacterial spectrum of chitosan is effective against Gram-positive and Gram-negative bacteria and fungi [36,37]. On the other hand, positively charged chitosan may be a good choice as a PS carrier and bioadhesive, which can increase the retention at the application site [38]. This is because on the basis of charge repulsion, cationic drugs will be rapidly released from the cationic hydrogel [39]. Moreover, the hydrophilic chitosan facilitates the close contact between the PS surface and the aqueous environment of microorganisms [40], which has broad prospects for the delivery of PS. Shrestha et al. pointed out that the use of chitosan-conjugated photosensitizers had a synergistic advantage over the use of photosensitizers alone, which can improve the anti-biofilm efficacy of Gram-positive bacteria and Gram-negative bacteria [40]. Choi et al. found that pretreatment with chitosan before MB-medicated PDT can effectively improve the photodynamic effect of MB at a low concentration of 40 μg/mL on the eradication of *H. pyloriat* [41]. Therefore, it is speculated that the MB encapsulation in viscous chitosan can stay on the surface of the infected implant for a long time and achieve a higher photo-inactivation efficacy against bacteria. However, the commonly used chitosan polysaccharide has the disadvantage of being only dissolved in an acidic solution, which results in a relatively low pH solution of less than 3. More importantly, from a clinical point of view, the pH of the solution is preferably 7 because of oral environment. In this regard, some water-soluble chitosan derivatives, such as quaternary ammonium chitosan

(QTS), can replace acid-soluble chitosan polysaccharides [35]. Herein, we chose QTS as the adhesive (or carrier) and analyzed the antimicrobial effect of QTS + MB-PDT on Gram-positive and Gram-negative bacteria. It is hypothesized that the synergistic combination of MB-conjugated QTS (QTS + MB) could have the dual functions of enhancing the photo-inactivation effect and improving the adhesion stability of the photosensitizer, which in turn maintains the osseointegration of the implant. The overall experimental process and purpose were schematically illustrated in Figure S1.

2. Results

2.1. Cytotoxicity

Figure 1 shows the effects of different formulations on L929 cytotoxicity. Obviously, the viability value of the positive control (DMSO) was less than 10% at the two culture time points, and showed significant lower ($p < 0.05$) cell viability than other test samples. When cells were seeded with MB and/or QTS, a concentration-dependent decrease in cellular activity was noted. At 1 min of culture, as the concentration increased, the viability of the cells inoculated with MB alone decreased from 81% to 62%. Similar to the downward trend, the cell viability of QTS was between 63% and 49%. However, the value of L929 cells cultured with MC1 agent was higher than 90%, indicating no signs of cytotoxicity, but the co-presence of MB100 caused the viability value down to 71%. In the case of the MB100 agent with different QTS concentrations, the value was significantly ($p < 0.05$) lower than the corresponding one without MB100. After 10 min of culture, it is worth noting that, except for the positive control, the viability values of all samples increased slightly.

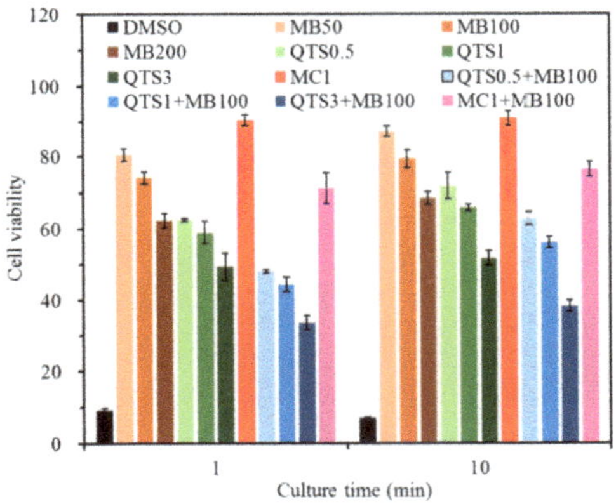

Figure 1. Cytotoxicity of L929 cells cultured with different test agents for 1 and 10 min. The cell viability was normalized to the negative control (culture medium) in terms of absorbance ($n = 3$).

2.2. Viscosity

The viscosity of various solutions varies with the shear rate, as shown in Figure 2. Phosphate buffer solution (PBS) and MB100 did not change remarkably with the increase of shear rate. When the shear rate was great than 5 1/s, their viscosity was similar. However, the viscosity of the two solutions was obviously lower than that of the three QTS and MC1 solutions. The viscosity of the three QTS solutions decreased as the shear rate increased to 5 1/s. A higher concentration of QTS caused a larger viscosity. For example, the apparent viscosities of QTS0.5 and QTS1 were about 0.20 and 0.31 Pa·s at the shear rate of 5 1/s, respectively. The constant viscosity of MC1 was about 0.23 Pa·s.

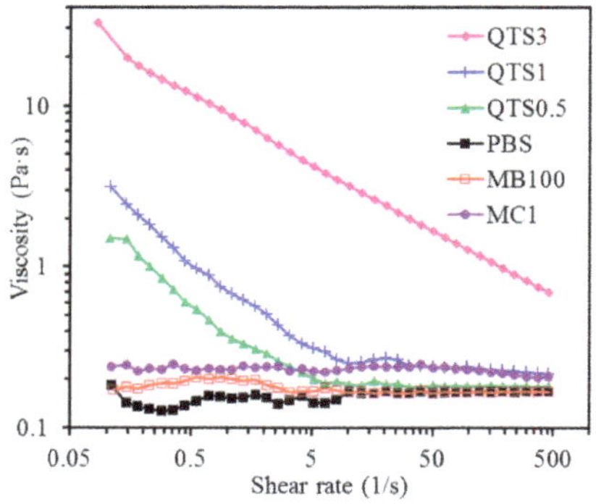

Figure 2. The viscosity changes of various solution samples under different shear rates.

2.3. Retention Efficacy

It can be clearly seen that MB100 alone flows directly from the top to the bottom of the Ti alloy plate, as shown in Figure 3a. On the contrary, as the concentration of QTS increased, the QTS + MB solution can be retained on the inclined surface (Figure 3b–d), especially the QTS3 + MB100 droplet was completely remained on the vertical plate. A part of MC1 + MB100 solution was stuck on the plate (Figure 3e). From the perspective of the retention of the solution, the order of the viscous agent' ability to retain MB100 was QTS3 > QTS1 > MC1 > QTS0.5.

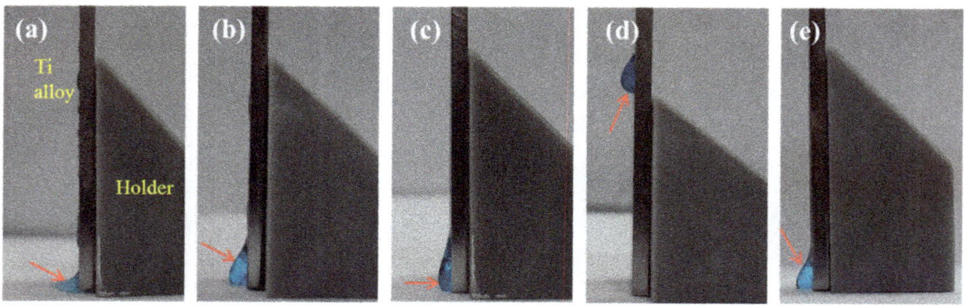

Figure 3. Images photographed for the retention efficacy of (**a**) methylene blue (MB)100, (**b**) quaternary ammonium chitosan (QTS)0.5 + MB100, (**c**) QTS1 + MB100, (**d**) QTS3 + MB100, and (**e**) MC1 + MB100 photosensitizers after the test solution was placed on the top of Ti alloy plate with 90-degree angle. Arrow indicates the flowing test solution.

2.4. Bacteriostatic Ratio

In the absence of laser irradiation, as the concentration of MB increased, *A. actinomycetemcomitans* (Figure 4a) and *S. mutans* (Figure 4b) attached to Ti alloy surfaces were more inactivated or killed. For example, the killing amount of *A. actinomycetemcomitans* in the MB100 and MB200 groups was 49% and 60%, respectively. Similar to these findings, regardless of the type of bacteria, the QTS group showed a trend in a significantly different manner. More importantly, the combinational MB and QTS modality exhibited a synergistic killing effect, which was significantly higher ($p < 0.05$) than that of the corresponding MB alone or QTS alone. However, MC1+MB100 had the same effect as MB100. By laser irradiation, when the MB group was compared with the corresponding MB-PDT group, in addition to the concentration-dependent bacteriostatic ratio, enhanced antibacterial efficacy was also found. When using MB100-PACT, approximately 89% of *A. actinomycetemcomitans* and 91% of *S. mutans* were eliminated. It is worth noting that the bacteriostatic ratio of the three QTS+MB100-PACT groups was higher than that of the MB100-PACT and MC1 +

MB-PACT groups. In addition, QTS1 + MB100-PACT and QTS3 + MB100-PACT had similar results on *A. actinomycetemcomitans* and *S. mutans*, with a bacterial death rate of over 98%.

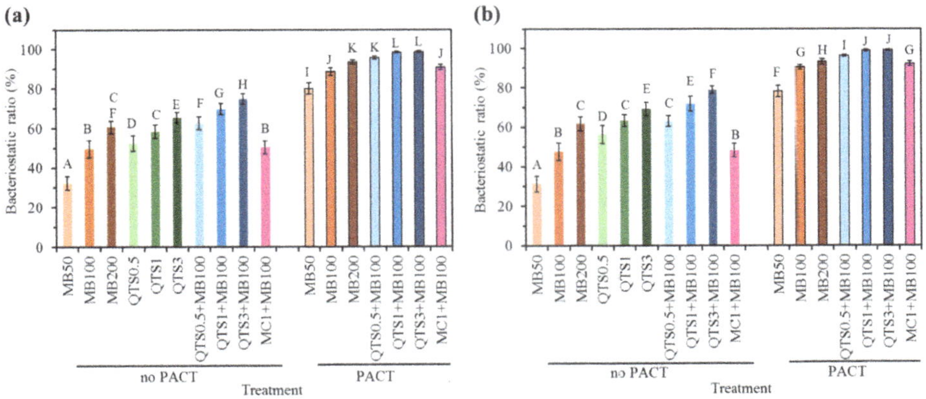

Figure 4. Bacteriostatic ratios of (**a**) *A. actinomycetemcomitans*-contaminated and (**b**) *S. mutans*-contaminated samples after various treatments with and without photoantimicrobial chemotherapy (PACT). Different capital letters showed statistically significant differences at $p < 0.05$ ($n = 10$).

2.5. Bacterial Colonies

The observation of bacterial colonies on the sample surface further depicted the elimination efficacy of different treatments, including MB100, QTS1, and MC1. Before treatment (Figure 5a), a large number of oval-shaped *A. actinomycetemcomitans* colonies uniformly adhered to the contaminated surface. Upon cleaning with QTS1 alone (Figure 5b) or MB100 alone (Figure 5c), the number of bacteria was reduced to some degree. Furthermore, the number of *A. actinomycetemcomitans* colonies on the decontaminated surface treated by MB100-PACT (Figure 5d) and MC1 + MB100 (Figure 5f) surfaces was remarkably lower than the number on the contaminated surface. Not surprisingly, the QTS1 + MB100-PACT treatment resulted in a sparse distribution of colonies (Figure 5e).

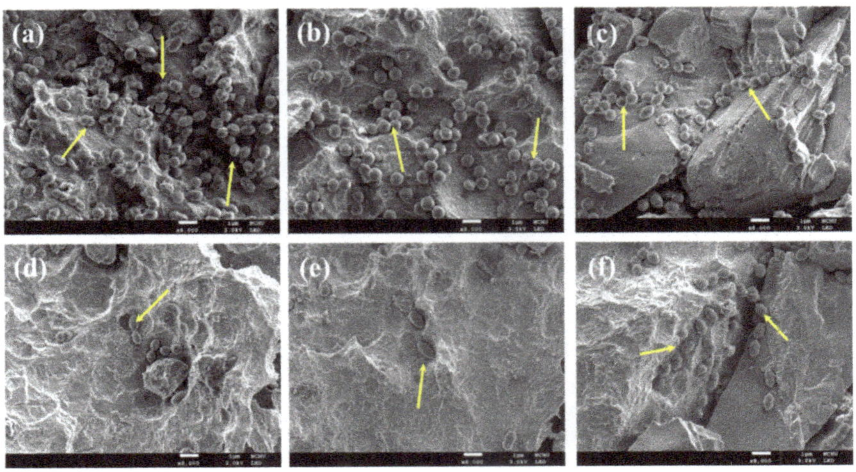

Figure 5. SEM images of the *A. actinomycetcmcomitans*-contaminated SLA Ti alloy surfaces (**a**) before and after treatment with (**b**) QTS1 alone, (**c**) MB100 alone, (**d**) MB100-PACT, (**e**) QTS1 + MB100-PACT, and (**f**) MC1 + MB100-PACT. The arrows indicate the presence of bacteria.

Regarding the colonies of *S. aureus*, densely packed sphere-like microcolonies were shown on the surface of the biofilm-contaminated samples (Figure 6a). Similar to the results of A. actinomycetemcomitans-contaminated surface, the treatment by QTS1 alone (Figure 6b) and MB100 alone (Figure 6c) reduced the number of colonies on the sample surfaces. In contrast, through various PDT modalities (Figure 6d–f), the number of bacterial colony-forming units (CFU) was greatly reduced when compared with the control and solution groups alone.

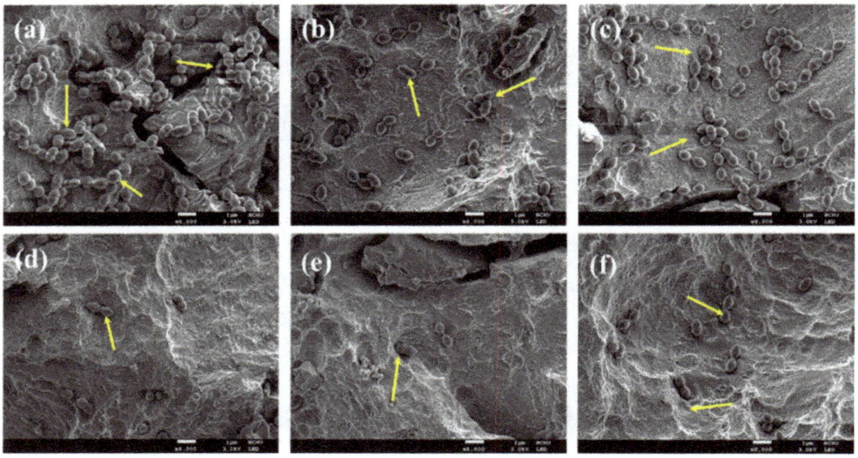

Figure 6. SEM images of the *S. mutans*-contaminated SLA Ti alloy surfaces (**a**) before and after treatment with (**b**) QTS1 alone, (**c**) MB100 alone, (**d**) MB100-PACT, (**e**) QTS1 + MB100-PACT, and (**f**) MC1 + MB100-PACT. The arrows indicated the presence of bacteria. The arrows indicate the presence of bacteria.

2.6. Residual LPS Amount

The effect of PACT treatment on the amount of LPS derived from *A. actinomycetcmcomitans* remaining on the sample surfaces is shown in Figure 7. There was no doubt that the LPS amount in the contaminated Ti control was higher. Conversely, the LPS amount remaining on various MB-PACT-treated surfaces was significantly ($p < 0.05$) reduced. The higher the MB concentration in the PACT treatment, the lower the residual amount of LPS exhibited. When using QTS to encapsulate MB, it was obvious that the synergetic QTS + MB100-PACT modality diminished the amount of LPS, more than the corresponding MB100-PACT group, especially QTS1+MB100-PACT. Compared with the MB100-PACT group, incorporation of 0.5%, 1%, and 3% QTS significantly ($p < 0.05$) reduced the residual amount of LPS by 24%, 57%, and 33%, respectively. However, there was no significant ($p > 0.05$) difference between MB100-PACT and MC1+MB-PACT.

2.7. Cell Morphology

The initial cell morphology attached to the disinfected surfaces of samples treated with MB100-PACT, QTS1 + MB100-PACT, and MC1 + MB100-PACT is shown in Figure 8, in addition to the sterile Ti alloy (Figure 8a,e). The cells on the surface of the uncontaminated Ti alloy were well adhered and fully spread with a well-defined morphology after 6 h of culture. When MG63 cells were cultured on surfaces treated with MB100-PACT (Figure 8b,f) and MC1 + MB-PACT (Figure 8d,h), it seemed that cell attachment was adversely affected, showing less spreading. In the case of QTS1+MB100-PACT (Figure 8c,g), MG63 cells adhered well to the decontaminated surface, indicating the extending filopodia. Nevertheless, there were still a few bacteria on all decontaminated surfaces.

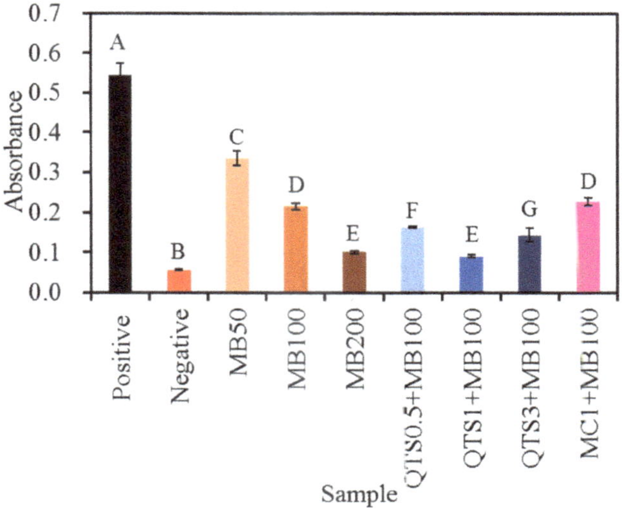

Figure 7. Residual lipopolysaccharide (LPS) level derived from *A. actinomycetcmcomitans* on sample surfaces after various photoantimicrobial chemotherapy (PACT) treatments. The contaminated Ti alloy without PACT and the sterile Ti alloy were used as a positive control and negative control, respectively. Different capital letters showed statistically significant differences at $p < 0.05$ ($n = 5$).

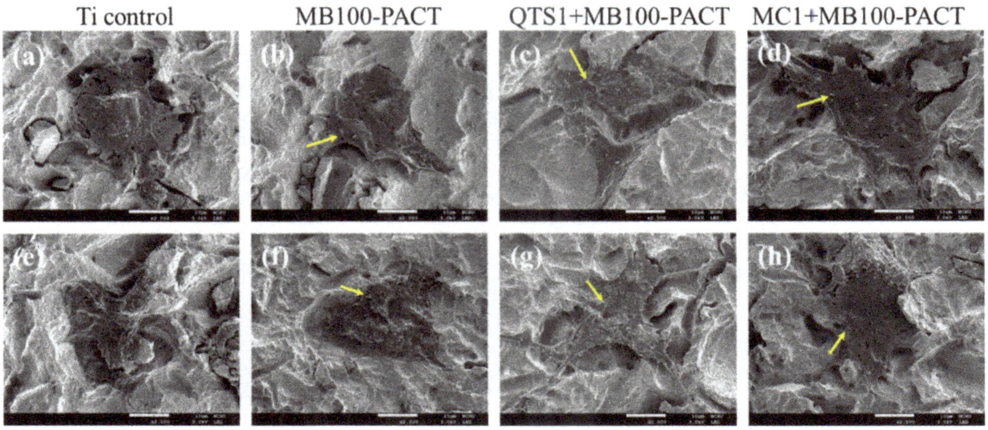

Figure 8. SEM images of MG63 cells after 6 h of culture on the (**a**,**e**) sterile Ti surface, (**b**–**d**) *A. actinomycetcmcomitans*-contaminated, or (**f**–**h**) *S. mutants*-contaminated surfaces treated by (**b**,**f**) MB100-PACT, (**c**,**g**) QTS1 + MB100-PACT, and (**d**,**h**) MC1 + MB100-PACT.

2.8. Cell Proliferation

After culturing MG63 cells on PACT-decontaminated sample surfaces and the sterile Ti alloy surfaces, the number of viable cells increased during the incubation interval due to the increase in absorbance (Figure 9). Regarding the samples infected with *A. actinomycetcmcomitans*, the cell proliferation of all PDT-treated groups was significantly ($p < 0.05$) lower than that of the sterile Ti control (Figure 9a). On the 7th day, the cell proliferation of the QTS1 + MB100 group was significantly ($p < 0.05$) higher than that of the MB100 and MC1 + MB100 groups. For samples contaminated with *S. mutants*, the proliferation of MG63 cells in the sterile Ti control was also higher than that of all PACT groups (Figure 9b). In addition, the QTS1+MB100 group was superior to the MC1+MB100 group except for one-day culture.

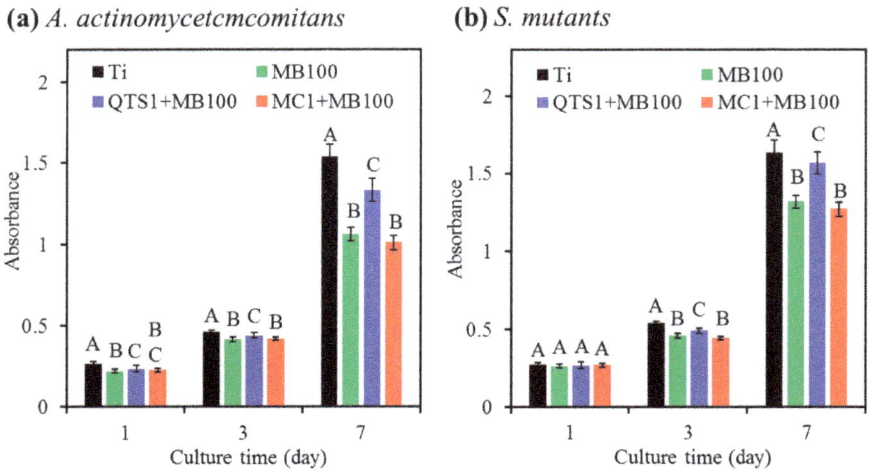

Figure 9. The cell viability was normalized to the negative control (culture medium) in terms of absorbance ($n = 3$). Proliferation analysis of MG63 cells cultured on the surfaces after PACT (MB100, QTS1 + MB100 and MC1 + MB100) treated samples contaminated with (**a**) *A. actinomycetcmcomitans* and (**b**) *S. mutants*. The sterile Ti alloy was used as a control. Different capital letters showed statistically significant differences at $p < 0.05$ ($n = 5$).

2.9. Alkaline Phosphatase (ALP) Activity

After 7 days and 14 days of culture, the ALP activity of MG63 cells on the PACT-treated samples is shown in Figure 10. The sterile Ti alloy control had the highest ALP expression than all PACT treatments against the two bacteria at all culture intervals, except for the QTS1 + MB100 group against *S. mutants* on day 14. Compared with MC1+MB100 and MB100, PACT treatment with QTS1 + MB1 produced a significantly ($p < 0.05$) higher ALP amount. On the *A. actinomycetcmcomitans*-disinfected surface for 14 days of cell culture, the ALP amount in the QTS+MB100-PACT modality was 9% higher than that of the MC1 + MB100-PACT modality (Figure 10a), which was 20% on the *S. mutants*-disinfected surface (Figure 10b).

2.10. Mineralization

The results of calcium deposits secreted by MG63 cells on the biofilm-disinfected surfaces treated by different PACT modalities are shown in Figure 11, indicating the increased Ca amount with the increasing culture time. After 7 days of culture, there was no significant difference ($p > 0.05$) between all groups including the sterile Ti alloy control. On the contrary, on the 14th day, it is worth noting that the MG63 cells grown on the sterile control surface and biofilm-disinfected surface treated with QTS1 + MB100-PACT secreted a similar content of calcium deposits, and there was no significant difference ($p > 0.05$). On the other hand, at the same MB100 concentration of MB100, QTS1 agent could significantly ($p < 0.05$) cause higher Ca deposits than MC1 agent. The cells on the *A. actinomycetcmcomitans*-disinfected surface by QTS + MB100-PACT increased by about 13% compared with the MC1+MB100-PACT treatment (Figure 11a), while the cells on *S. mutants*-disinfected surface increased by 28% (Figure 11b).

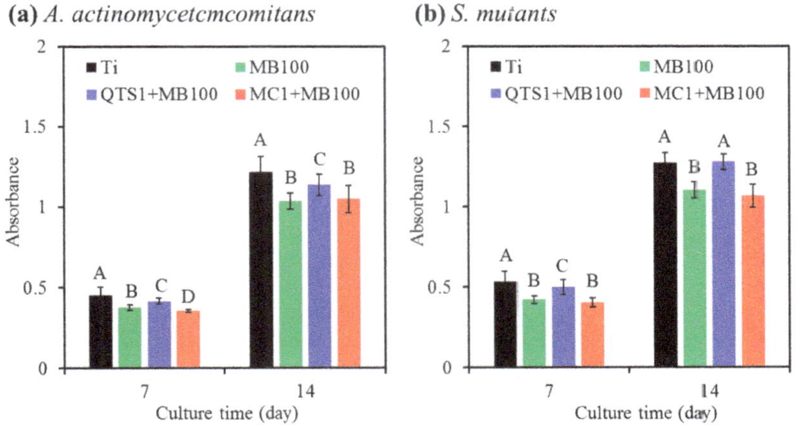

Figure 10. ALP activity of MG63 cells cultured on the surfaces after PACT (MB100, QTS1 + MB100 and MC1 + MB100) treated samples contaminated with (**a**) *A. actinomycetcmcomitans* and (**b**) *S. mutants*. The sterile Ti alloy was used as a control. Different capital letters showed statistically significant differences at $p < 0.05$ ($n = 5$).

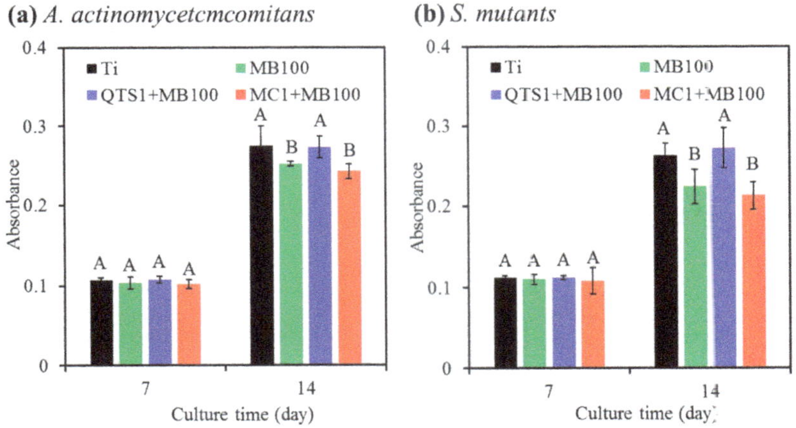

Figure 11. Mineralization analysis of MG63 cells cultured on the surfaces after PACT (MB100, QTS1 + MB100 and MC1 + MB100) treated samples contaminated with (**a**) *A. actinomycetcmcomitans* and (**b**) *S. mutants*. The sterile Ti alloy was used as a control. Different capital letters showed statistically significant differences at $p < 0.05$ ($n = 5$).

3. Discussion

Although dental implant advancements have aimed to facilitate bone healing, the prevention of bacterial infections is also a critical factor in the success of a surgical treatment [41]. Implant-associated infections produce serious post-surgical problems that adversely affect osseointegration and result in implant device failures. Therefore, the prevention and elimination of bacterial adhesion and colonization plays an important role in the successful implantation. PACT has been considered as a reliable adjunct to conventional therapeutic modalities, such as mechanical (curettage and root planing) and surgical (transplant) methods, or can be used alone in the peri-implantitis treatment [5,42]. PACT neither causes any resistance to microorganisms, nor is affected by the resistance of existing drugs [43]. However, effective removal of biofilm and bacterial toxins from the surface of infected implants is still an unresolved clinical issue. Many studies intend to develop effective modality to eradicate bacterial colonies and biofilms on the surface of implants for restoring osseointegration.

The antimicrobial efficacy of PACT is based on the large accumulation of PS in or on the cytoplasmic membrane [44,45]. Based on clinical need, PS should have sufficient retention on the surface of infected implant for intracellular uptake, thereby progressing photo-inactivation. However, due to its fluid characteristics can lead to ineffective therapy, the retention of PS remained to be resolved. Modifying photosensitizers with viscous biopolymers will be an attractive strategy to solve this problem. Indeed, as evidenced in this study, by incorporation of a viscosity-enhancing agent, such as MC, MB retention can be given to the peri-implantitis area. MC is a water-soluble derivative of polysaccharide cellulose, and it is also a non-toxic, biocompatible FDA-approved material [46]. The introduction of viscous QTS molecules as adhesive promoters in this study can also effectively overcome the shortcomings of MB retention. This antibacterial carrier may be easily applicable to large or complex implant surfaces, and may adhere to the material substrate (or teeth). One of the current purposes was to synergistically combine MB with antibacterial and viscous QTS to construct a new type of photosensitizing platform, whereby the QTS+MB-PACT modality can effectively treat peri-implantitis.

In order to understand the concentration effect, water-dissolvable QTS at concentrations of 0.5, 1, and 3 wt% were used. The viscosity of the three QTS and MC solutions was significantly higher than that of MB. Not surprisingly, the viscosity of QTS3 was higher than QTS1 and QTS0.5. As expected, the addition of QTS1 and QTS3 effectively improve the retention of MB on the implant surface due to the viscosity effect, binding the MB together like glue. On the other hand, the viscous MC1 also kept MB part on the surface of the Ti alloy sample. However, Ludmila et al. pointed out that, in addition to preferential accumulation in bacterial targets, ideal PS should also be eliminated quickly after administration [47]. Therefore, high-viscosity QTS, such as 3% may not be suitable for use because it was difficult to remove. The long-term retention of QTS + MB on the implant surface may affect periodontal tissues and cells, resulting in the cytotoxicity [21], as described below.

Low toxicity is one of the essential characteristics of ideal photosensitizers, especially towards mammalian cells [15,48]. Therefore, when the photosensitizer was conjugated with polymers, the dose-dependent cytotoxicity of the modified PS should be examined to verify its bio-safety. From a clinical point of view, a short interval (a few minutes) in the PACT treatment of peri-implantitis was commonly employed, so this study used 1 min and 10 min of culture time. The cell viability of 50 and 100 μg/mL MB and MC1 with and without MB were all higher than 70%, while other test agents were all lower than 70%. According to ISO 10993-5 standard definition, more than 70% viability is considered non-cytotoxic [49]. Tate et al. pointed out that MC concentration as high as 8% will not cause the death of primary rat cortical astrocytes [50]. Rolim et al. reported that MB concentrations exceeding 163.5 μM (equivalent to 52.3 μg/mL) can induce cytotoxicity [51]. It is reasonable to consider that the higher the concentration of MB and QTS, the lower the cell viability, indicating that its potential cytotoxicity was higher [52,53]. Regardless of QTS or MC mucoadhesive, the combination with MB100 significantly reduced cell viability. Soukos and Goodson suggested that due to incomplete penetration of MB in oral biofilms in a clinical setting, PACT may take up to 15 min to proceed [15]. Therefore, to safely apply PS in the treatment of peri-implantitis without damaging adjacent normal tissues, low concentrations of MB were preferential used to prevent possible toxic effects of residual photosensitizers remaining dyes [21].

It is well recognized that biofilm formation is initiated by bacterial adhesion, which in turn constitutes a three-dimensional structure and develops a mature entity [54]. Thus, it is indeed necessary to evaluate the number of bacteria and its remaining LPS on the Ti alloy surface after various treatments. When the Ti alloy sample was exposed to the bacterial suspension for 24 h, marked bacterial colonization occurred. On the contrary, the use of cationic MB alone can reduce the growth of bacteria, but from the results of CFU assay, there was no sign of effective eradication effect, as confirmed by an earlier report [55]. MB can attach the bacterial membranes of Gram-positive and Gram-negative bacteria by interaction with the anionic region on the bacterial cell wall [5]. In addition, MB may be toxic to some extent. It is reasonable to think that strong antibacterial activity often resulted in high cytotoxicity, which was in line with the results reported by Miyata [56]. MB alone has a concentration-dependent bactericidal activity [5,55], but it fails to completely eradicate bacteria. Therefore, MB-PACT was used instead of MB to eliminate bacteria and biofilms more effectively. Not surprisingly, the percentage reduction of the two viable bacteria in MB-PACT was significantly improved compared with MB alone. Kim et al. pointed out that when MB alone caused 53% of cell death, the use of 100 μg/mL MB with the Periowave diode laser can reduce 77% of *A. actinomycetemcomitans* attached on the SLA titanium surface [57]. The higher MB concentration in MB-PACT lead to higher bactericidal activity, which was consistent with the previous studies [18,58]. Another interesting study was to test the potential role of QTS as an antimicrobial adjuvant, which was believed to help prevent bacterial infection and adhesion on the surface of the implant. According to a previous study [35], the concentration-dependent bactericidal activity of QTS against the Gram-positive bacteria and Gram-negative bacteria examined is due to the production of reactive oxygen species (ROS). Moreover, cationic QTS may destroy the integrity of the biofilm and kill the microbial cells by piercing the surface of negatively charged microbial cells [37]. Results of SEM images corroborated the current findings of the bacteria count.

Since MB alone and QTS alone had a potent antimicrobial efficacy, it can be speculated that the combination of the two agents had an additive effect on eliminating bacteria. In fact, the cationic QTS + MB solution provided advantages by allowing the construction of a viscous photosensitizing platform as a multifunctional role including the MB-PDT efficacy and the inherent antibacterial ability of QTS. Camacho-Alonso et al. pointed out that the chitosan + MB-PACT group resulted in the lower CFU/mL count of Enterococcus faecalis compared with MB-PACT alone in experimentally infected root canals of extracted human teeth, but there was no significance between these groups [59]. Darabpour et al. reported that chitosan nanoparticles destroyed the biofilm structure of *Staphylococcus aureus* and *Pseudomonas aeruginosa*, which in turn made MB deeper and higher permeability, thereby enhancing the eradication efficacy of MB-PACT [60]. Carpio-Perochena et al. also found that the combination of carboxymethyl chitosan/rose bengal and PACT can reduce more viable bacteria than carboxymethyl chitosan alone [61]. Regarding the effect of cellulose, MC1 did not reveal an additional antibacterial efficacy from the comparison between the MB100-PACT and MC1+MB100-PACT groups. In short, compared with the same concentration of MC, QTS helped to improve the antibacterial efficacy.

In addition to inactivating pathogenic bacteria, it is also important to clean the LPS (the major cell wall component of Gram-negative bacteria) from the surface of the implant because the LPS remaining on the material surfaces may not only facilitate subsequent microbial adhesion [62] but also jeopardize the proliferation of connective tissue cells [63]. Biofilm formation on the surface of the implant can protect bacteria and promote the persistence of infection, which is difficult to completely avoid on SLA-pretreated implants [64]. In addition, biofilm makes antibiotic treatment insufficient to eradicating the infection. Thus, residual LPS analysis should be conducted to unveil the importance of QTS + MB-PACT. Contrary to the high amount of LPS on the contaminated surface, PACT treatment resulted in a significant reduction in LPS caused by *A. actinomycetemcomitans*. There was a positive correlation between the increase in MB concentration and the decrease in LPS, which was in full agreement with the previous study [5]. On the other hand, there was no significant difference in residual LPS and bacteriostatic ratio between MB100-PACT and MC1+MB-PACT. As mentioned earlier, it was due to the insufficient antibacterial ability of MC.

Regarding the QTS + MB-PACT modality, the more positive charge of QTS + MB would have a higher affinity for LPS [65], which may be the reason why QTS + MB100 was higher than MB100 in the eradication of LPS after photoactivation. Therefore, benefit can be expected to use this novel therapeutic modality to enhance the treatment efficiency of peri-implantitis. Interestingly, in the QTS+MB-PACT modality, QTS1 eliminated LPS more effective than QTS3. Although QTS adhesive had viscous ability, it can firmly bond MB onto the robust Ti alloy surface, but too high QTS concentration (such as 3 wt%) may prevent the laser penetration through MB due to the optical shielding or inhibit the MB penetration to the bacterial membrane. Generally, PS molecules or PS-carrying particles can be transferred into cells through transmembrane and along-membrane diffusion, by non-specific endocytosis or even pinocytosis, or can also be internalized through phagocytosis [44]. Therefore, under the appropriate QTS concentration, the effective retention of MB near the cell membrane will have better photobactericidal activity.

In order to achieve long-term success in clinical practice, Ti implants must support the growth of cells and tissues. The presence of bacteria could complicate the process of osseointegration. In this sense, it was indispensable for verifying whether the contaminated Ti alloy treated with various PACTs used in this study can maintain the osteoblast function as uncontaminated Ti alloy control. A concentration of 100 μg/mL MB is used in the therapy of peri-implantitis, such as commercial Periowave modality [57,66]. Almeida et al. found in a rat model of periodontal disease that 100 μg/mL MB-PDT can effectively reduce bone loss in the furcation region of the first molar [66]. Based on the current results of cytotoxicity and antibacterial efficacy, this concentration was used to evaluate the in vitro osteogenic activity using MC63 cells.

Cell attachment and proliferation are the initial phase of cell–material interaction [67]. The results of the current study clearly indicated that compared with MC1 + MB100-PACT and MB100-PACT methods, QTS1 + MB100-PACT enhanced the cell attachment and proliferation of MG63 osteoblasts. This may be because that antibacterial QTS synergistically reduced bacterial colonies and LPS amount on the disinfected surface, making it less harmful to cell growth [63,68]. Eick et al. found that the cell attachment of human alveolar osteoblasts on disinfected SLA Ti implant treated with effective PDT was equivalent to the Ti surface without bacterial biofilm exposure [68]. After the cells grow, osteoblasts will continue to differentiate and secrete ALP (an earlier differentiation biomarker), and then deposit calcium [69]. In fact, with increasing culture time, the ALP and calcium content of MG63 cells increased significantly. Similar to the trend of cell proliferation, the ALP expression response to the decontaminated samples showed the order of QTS1 + MB100-PACT > MC1 + MB100-PACT ≅ MB100-PACT. The biocompatibility of MC can be used to explain why there was difference in antibacterial efficacy and osteoblast function between MC1 + MB100-PACT and MB100-PACT.

For re-osseointegration of decontaminated implant, the ability of cells to produce mineralized matrix is important. As a result, on the surface of infected samples treated with QTS1 + MB100-PDT, the calcium deposits of MG63 osteoblast were significantly higher than those of MC + MB100-PACT and MB100-PACT. More importantly, although the disinfection was not fully conducted, its osseointegration was comparable to the uncontaminated Ti alloy control. This result can be explained as a large number of osteoblasts can coexist with a small number of bacteria and grow well [5,70]. As stated by Gristina [70], the fate of biomaterials is the competition between bacterial adhesion and biofilm growth and tissue integration, which depends on the number of existing bacteria. As mentioned earlier, it has been proven that QTS+MB-PACT can remarkably reduce bacterial colonies, which was beneficial to osteoblast function on the surface of the decontaminated implant. The osteogenesis-related increase with the increasing culture time can be used to interpret the point. In short, due to the synergistic effect of QTS and MB photoactivation, MB-conjugated QTS could perform the dual functions of enhancing bacterial elimination efficacy and improving osseointegration of Ti implant. This PACT modality based on the QTS and MB synergetic system may be a promising approach of killing bacteria.

4. Materials and Methods

4.1. Preparation of QTS

According to previous research [35], quaternary ammonium chitosan (QTS) was prepared using chitosan polysaccharide (Sigma-Aldrich, St. Louis, MO, USA). Briefly, the as-received chitosan polysaccharide was dissolved in acetic acid, and then acetic anhydride (Echo Chemicals, Miaoli, Taiwan) was added to the solution. The reaction was blocked with NaOH, and the mixture was dialyzed against water and freeze-dried to obtain water-soluble N-acetylated chitosan. Glycidyltrimethylammonium chloride (GTMAC; Tokyo Chemical, Tokyo, Japan) as a quaternizing agent was added to 2.5% N-acetylated chitosan in water and reacted for 8 h. The product was placed in cold acetone and kept in the refrigerator for 24 h. After discarding the acetone, the remaining sample was dissolved in methanol for 1 h, and then precipitated in a 4:1 acetone–ethanol solution, followed by filtration and drying at 60 °C to obtain QTS, which was stored in 4 °C refrigerator before use. The chemical structure of QTS has been determined in an earlier study [35].

4.2. Preparation of Solution

Three concentrations (50, 100, and 200 μg/mL) of methylene blue (Riedel-deHaen, Buffalo, NY, USA) were prepared in phosphate buffer solution (PBS) with a pH 7 [5] and stored in the dark at 4 °C for usage later. Three concentrations of 0.5, 1, and 3 w/v% QTS solutions were obtained by dissolving the powder in PBS. In addition, MB was also dissolved in different QTS concentrations in a beaker wrapped with aluminum foil to a final concentration of 100 μg/mL under thoroughly stirring for 1 h to homogenize the composition, which was called QTS + MB100 mixed solution. One percent methyl cellulose (MC; Showa, Tokyo, Japan) in PBS was also used as an adhesive and control, compared with QTS. For simplicity, the sample code "MB100" represented the use of 100 μg/mL MB solution, and "QTS1" was for 1% QTS.

4.3. L929 Cytotoxicity

The L929 cytotoxicity of various agent concentrations in Dulbecco's modified Eagle medium (DMEM; HyClone, Logan, UT, USA) was conducted according to ISO 10993-5 standard by a MTT 3-(4,5-dimethylthiazol-2-yl)-2,5-diphenyltetrazolium bromide; Sigma-Aldrich) assay. The L929 cell suspension (10^4 cells per well) was cultured in 24-well microplates under a 5% CO_2 humidified atmosphere at 37 °C for 1 day, and then exchanged with DMEM containing test agents for a culture time of 1 and 10 min. DMEM alone was used as a negative control, whereas DMEM containing 10% dimethyl sulfoxide (DMSO; Sigma-Aldrich) was used as a positive control. Ultimately, microplates were read at 563 nm using a BioTek Epoch spectrophotometer (Winooski, VT, USA). The absorbance results were recorded for three independent measurements. The cell viability was normalized to the negative control in terms of absorbance.

4.4. Viscosity

The Discovery HR-2 rheometer (TA Instruments, New Castle, DE, USA) with parallel plate geometry was used to evaluate the viscosity (η) of PBS, MB100, MC1, and three QTS solutions at room temperature. The viscosity of the test solution was a function of the shear rate between 0.1 and 500 1/s, with ten data points per decade.

4.5. Preparation of Titanium Alloy

A 10 × 10 × 3 mm^3 Ti6Al4V alloy disc (ASTM F136-84; Titanium Industries, Parsippany, NJ, USA) was used as the implant substrate and wet-ground with a 3M Wetordry 1200 grit SiC sandpaper (St. Paul, MN, USA). After sandblasting with 100 μm Al_2O_3 particles (Korox, Bego, Bremen, Germany) for 10 s, the disc was acid-etched in $HCl/H_2SO_4/H_2O$ (1:1:100) at 100 °C for 30 min to produce the SLA surface [64]. Through ultrasonically cleaning in acetone and ethanol for separate 20 min, the disc was rinsed in deionized water and then dried in an oven at 60 °C. Figure S1 schematically illustrated the experimental procedure including solution analyses, PDT steps, antibacterial evaluation, and osteoblast function examination.

4.6. Solution Flowability

To verify the retention ability of various agents on the implant surfaces, the MB100 photosensitizer solution with and without QTS or MC1 was dropped onto the surfaces of the Ti alloy plate with a 90-degree inclination angle. After 1 min, an image was taken by camera to observe the flowability of the test agent.

4.7. Bacteria Seeding

Gram-negative *Aggregatibacter actinomycetemcomitans* (*A. actinomycetemcomitans*; IDH 781) and Gram-positive *Streptococcus mutans* (*S. mutans*; ATCC 700610) bacteria were used as the bacterial species tested. The bacteria were cultured in the Bacto tryptic soy broth (Beckton Dickinson, Sparks, MD, USA) and grown to the optical density of about 1.0 at 600 nm detected by using a Beckman Coulter DU-640 spectrophotometer (Fullerton, CA, USA). They were diluted by broth to a density of 2×10^6 colony-forming units (CFU)/mL. The SLA Ti alloy discs were randomly allocated to 24-well culture plates and then sterilized by soaking in a 75% ethanol solution and exposure to UV light overnight. After that, 1 mL of the bacteria suspension was inoculated onto the disc surface and cultured in an incubator at 37 °C for 1 day to attach. The contaminated samples were gently washed with PBS (pH 7.4) twice to remove non-adherent bacteria. The contaminated Ti alloy disc without any treatment was used as a control, while the MB, QTS, QTS+MB100, and MC1+MB100 groups with and without PACT treatment were assigned as the experimental groups.

4.8. Photodynamic Treatment

The contaminated disc sample was coated with 100 μL of test solution for 1 min of reaction and then washed by PBS. Afterwards, some samples were irradiated with a low-level AlGaInP diode laser (Aculas-HB, Konftec, New Taipei City, Taiwan) with an output diameter and 8 mm. The laser beam with maximum output power of 80 mW (4.8 J/cm^2) was irradiated in a continuous mode at 660 nm for 1 min [58], and the irradiation distance was 10 mm at an incidence angle of 90°.

4.9. Antibacterial Efficacy

4.9.1. Bacterial Counting

For bacterial counting, the conventional spread plate method was used to count the number of bacterial colonies distributed on the surface of the sample. After bacterial adhesion, contaminated sample washing and test solution spraying, some samples were treated by PACT, as described above. After that, 1 mL of PBS was added to each sample, and then the adhered bacteria were ultrasonically detached in a 150 W ultrasonic bath (DC150H, Taiwan Delta New Instrument Co. Ltd., New Taipei City, Taiwan) for 5 min at a frequency of 40 kHz. A 100 μL aliquot of the bacterial suspension was collected from each well, and four serial ten-fold dilutions were performed using PBS. Then, 100 μL of the bacterial dilution was spread on a 15 mL of Trypticase soy agar (Conda, Madrid, Spain) plate (9 cm in diameter) for 1 day of incubation at 37 °C. The total numbers of CFU in each dish were counted and the bacteriostatic ratio (%) was calculated as follows:

$$\text{Bacteriostatic ratio (\%)} = (\text{Ncontrol} - \text{Nexperiment})/\text{Ncontrol} \times 100\%,$$

where Ncontrol and Nexperiment are the number of bacterial colonies on contaminated Ti alloy control (CFU/sample) and experimental groups (CFU/sample), respectively [71]. Ten measurements were taken for each group.

4.9.2. Bacterial Colony Observation

To further observe the changes of bacterial colonies after treatment, the contaminated Ti control and the disinfected groups of QTS1, MB100, MB100-PACT, QTS1 + MB100-PACT, and MC1 + MB100-PACT were viewed using a scanning electron microscopy (SEM; JEOL JSM-7800 F, Tokyo, Japan). The samples were washed two times with PBS and fixed in

4% paraformaldehyde (Sigma-Aldrich) for 20 min, then dehydrated for 20 min at each concentration using a graded ethanol series, mounted on a stub, and then coated with a gold layer.

4.9.3. LPS Detection

After PDT treatment and PBS washing twice, the ToxinSensor chromogenic Limulus Amebocyte Lysate endotoxin assay kit (GenScript, Piscataway, NJ, USA) was used to detect the amount of LPS remaining on the sample surface [58]. A sterile Ti alloy sample was used a negative control, which a Ti alloy sample contaminated with A. actinomycetemcomitans was used as a positive control. The absorbance at 545 nm was the mean of five independent measurements using a BioTek Epoch plate reader (Winooski, VT, USA).

4.10. MG63 Cell Culture

MG63 human osteoblast-like cells (BCRC 60279; Hsinchu, Taiwan) were used to examine the effects of various PACT modalities on cell function of decontaminated Ti samples. The cells were suspended in DMEM supplemented with 10% fetal bovine serum (FBS) (Gibco, Langley, OK, USA) and 1% penicillin (10,000 U/mL)/streptomycin (10,000 μg/mL) solution (Gibco) in 5% CO_2 at 37 °C. A cell suspension (10^4 cells/well) in a 24-well plate was seeded on each sample, and an uncontaminated Ti alloy sample was used as a control.

4.10.1. Cell Attachment

To observe cell morphology on the sample surface after initial 6 h incubation, the cells were washed three times with PBS and fixed in 2% glutaraldehyde (Wako, Tokyo, Japan) at 4 °C for 2 h. After dehydration in the graded ethanol series and drying using critical point dryer device (LADD 28000; Williston, VT, USA), the cell sample was coated with a gold layer and observed by SEM.

4.10.2. Cell Proliferation

After 1, 3, and 7 days of incubation, cell proliferation was assayed using the MTT assay, according to the previous study [5]. The results were obtained through five independent measurements and reported in terms of absorbance at 563 nm detected with a BioTek Epoch plate reader.

4.10.3. Alkaline Phosphatase Activity

To examine the early cell differentiation on the 7th and 14th days, an alkaline phosphatase (ALP) activity assay was conducted using the TRACP & ALP assay kit (Takara, Shiga, Japan) [72]. Five samples were used in each group, and the average was obtained at 405 nm absorbance using a BioTek Epoch plate reader.

4.10.4. Calcium Deposit Quantification

Alizarin Red S staining method was used to quantify calcium deposits secreted by MG63 cells. After 7 and 14 days of incubation, the cells were washed with PBS and fixed in 4% paraformaldehyde at 4 °C for 10 min, and then stained in 0.5% Alizarin Red S (Sigma–Aldrich) in PBS for 10 min [72]. After that, the calcium mineral precipitate was extracted with a 10% cetylpyridinium chloride solution for 30 min. The absorbance of extract was detected at 562 nm using a BioTek Epoch plate reader. Five samples were used to obtain the average.

4.11. Statistical Analysis

A one-way analysis of variance (ANOVA) statistical analysis was used to assess significant differences between means. Duncan's multiple comparisons were utilized to determine the significance of the standard deviation in the data between samples. Statistical data was analyzed using the SPSS 14.0 software for Windows (SPSS Inc., Chicago, IL, USA).

In all cases, when the p-value was less than 0.05, the result was considered statistically significant.

5. Conclusions

From a viewpoint of clinical practice, there is a great need to treat peri-implantitis and gain re-osseointegration. Within the limit of this in vitro model, viscous QTS played two roles in MB-mediated PACT: one was a photosensitizer carrier, and the other is an antibacterial agent. This QTS adhesive effectively improved the retention of MB on the implant surface. Compared with MB-mediated PACT, QTS + MB-PACT can more effectively remove bacteria attached to the SLA-pretreated titanium surface and produce osseointegration. It is worth noting that QTS + MB modality was better than MC+MB modality in terms of antibacterial efficacy and osteogenesis. All in all, considering the compromise between cytotoxicity, retention ability, in vitro antibacterial efficacy, and osteogenesis into account, synergistic photoantimicrobial chemotherapy with 1% QTS and 100 μg/mL MB was a potential modality for the treatment of peri-implantitis, which was better than 1% MC+MB100. Further studies are required, including examination of multispecies biofilm model on screw-shaped implants and osteointegration in vivo, before QTS+MB-PACT can be used for clinical treatment.

Supplementary Materials: The following are available online at https://www.mdpi.com/article/10.3390/ph14040346/s1. Figure S1: Schematic flowchart of the experimental process and purpose.

Author Contributions: Conceptualization, C.-N.L., S.-J.D., and C.-C.C.; methodology, C.-N.L., S.-J.D., and C.-C.C.; validation, S.-J.D. and C.-C.C.; formal analysis, C.-N.L., S.-J.D., and C.-C.C.; investigation, C.-N.L.; data curation, S.-J.D. and C.-C.C.; writing—original draft preparation, C.-N.L.; writing—review and editing, S.-J.D.; supervision, C.-C.C. All authors have read and agreed to the published version of the manuscript.

Funding: This research received no external funding.

Institutional Review Board Statement: Not applicable.

Informed Consent Statement: Not applicable.

Data Availability Statement: The data presented in this study are available in this article and supplementary materials.

Conflicts of Interest: The authors declare no conflict of interest.

References

1. Charalampakis, G.; Ramberg, P.; Dahlén, G.; Berglundh, T.; Abrahamsson, I. Effect of Cleansing of Biofilm Formed on Titanium Discs. *Clin. Oral Impl. Res.* **2015**, *26*, 931–936. [CrossRef] [PubMed]
2. Arciola, C.R.; Campoccia, D.; Montanaro, L. Implant Infections: Adhesion, Biofilm Formation and Immune Evasion. *Nat. Rev. Microbiol.* **2018**, *16*, 397–409. [CrossRef] [PubMed]
3. Aveyard, J.; Bradley, J.W.; McKay, K.; McBride, F.; Donaghy, D.; Raval, R.; D'Sa, R.A. Linker-Free Covalent Immobilization of Nisin Using Atmospheric Pressure Plasma Induced Grafting. *J. Mater. Chem. B* **2017**, *5*, 2500–2510. [CrossRef] [PubMed]
4. Yuan, K.; Chan, Y.J.; Kung, K.C.; Lee, T.M. Comparison of Osseointegration on Various Implant Surfaces after Bacterial Contamination and Cleaning: A Rabbit Study. *Int. J. Oral Maxillofac. Impl.* **2014**, *29*, 32–40. [CrossRef]
5. Huang, T.C.; Chen, C.J.; Chen, C.C.; Ding, S.J. Enhancing Osteoblast Functions on Biofilm-Contaminated Titanium Alloy by Concentration-Dependent Use of Methylene Blue-Mediated Antimicrobial Photodynamic Therapy. *Photodiagn. Photodyn. Ther.* **2019**, *27*, 7–18. [CrossRef]
6. Roos-Jansåker, A.M.; Lindahl, C.; Renvert, H.; Renvert, S. Nine-to Fourteen-Year Follow-Up of Implant Treatment. Part II: Presence of Peri-Implant Lesions. *J. Clin. Periodontol.* **2006**, *33*, 290–295. [CrossRef] [PubMed]
7. Van Velzen, F.J.J.; Ofec, R.; Schulten, E.A.J.M.; ten Bruggenkate, C.M. 10-Year Survival Rate and the Incidence of Peri-Implant Disease of 374 Titanium Dental Implants with a SLA Surface: A Prospective Cohort Study in 177 fully and Partially Edentulous Patients. *Clin. Oral Impl. Res.* **2015**, *26*, 1121–1128. [CrossRef]
8. Fransson, C.; Lekholm, U.; Jemt, T.; Berglundh, T. Prevalence of Subjects with Progressive Bone Loss at Implants. *Clin. Oral Impl. Res.* **2005**, *16*, 440–446. [CrossRef] [PubMed]
9. Mombelli, A.; Lang, N.P. The Diagnosis and Treatment of Peri-Implantitis. *Periodontology 2000* **1998**, *17*, 63–76. [CrossRef]

10. Pier-Francesco, A.; Adams, R.J.; Waters, M.G.J.; Williams, D.W. Titanium Surface Modification and its Effect on the Adherence of *Porphyromonas Gingivalis*: An in Vitro Study. *Clin. Oral Impl. Res.* **2006**, *17*, 633–637. [CrossRef]
11. Chen, C.J.; Ding, S.J.; Chen, C.C. Effects of Surface Conditions of Titanium Dental Implants on Bacterial Adhesion. *Photomed. Laser Surg.* **2016**, *34*, 379–388. [CrossRef]
12. Kao, H.; Chen, C.C.; Huang, Y.R.; Chu, Y.H.; Csík, A.; Ding, S.J. Metal Ion-Dependent Tailored Antibacterial Activity and Biological Properties of Polydopamine-Coated Titanium Implants. *Surf. Coat. Technol.* **2019**, *378*, 124998. [CrossRef]
13. Slots, J.; Ting, M. *Actinobacillus Actinomycetemcomitans* and *Porphyromonas Gingivalis* in Human Periodontal Disease: Occurrence and Treatment. *Periodontology 2000* **1999**, *20*, 82–121. [CrossRef]
14. Bonito, A.J.; Lux, L.; Lohs, K.N. Impact of Local Adjuncts to Scaling and Root Planing the Periodontal Disease Therapy: A Systematic Review. *J. Periodontol.* **2005**, *76*, 1227–1236. [CrossRef]
15. Soukos, N.S.; Goodson, J.M. Photodynamic Therapy in the Control of Oral Biofilms. *Periodontology 2000* **2011**, *55*, 143–166. [CrossRef] [PubMed]
16. Konopka, K.; Goslinski, T. Photodynamic Therapy in Dentistry. *J. Dent. Res.* **2007**, *86*, 694–707. [CrossRef] [PubMed]
17. Sculean, A.; Aoki, A.; Romanos, G.; Schwarz, F.; Miron, R.J.; Cosgarea, R. Is Photodynamic Therapy an Effective Treatment for Periodontal and Peri-Implant Infections? *Dent. Clin. N. Am.* **2015**, *59*, 831–858. [CrossRef] [PubMed]
18. Briggs, T.; Blunn, G.; Hislop, S.; Ramalhete, R.; Bagley, C.; McKenna, D.; Coathup, M. Antimicrobial Photodynamic Therapy—a Promising Treatment for Prosthetic Joint Infections. *Lasers Med. Sci.* **2018**, *33*, 523–532. [CrossRef]
19. Silva, J.C.E.; Lacava, Z.G.M.; Kuckelhaus, S.; Silva, L.P.; Neto, L.F.M.; Sauro, E.E.; Tedesco, A.C. Evaluation of the Use of Low Level Laser and Photosensitizer Drugs in Healing. *Lasers Surg. Med.* **2004**, *34*, 451–457. [CrossRef] [PubMed]
20. Baham, P.; Herron, C.; Street, C.; Darveau, R. Antimicrobial Photodynamic Therapy May Promote Periodontal Healing Through Multiple Mechanisms. *J. Periodontol.* **2009**, *80*, 1790–1798. [CrossRef] [PubMed]
21. Takasaki, A.A.; Aoki, A.; Mizutani, K.; Schwarz, F.; Sculean, A.; Wang, C.Y.; Koshy, G.; Romanos, G.; Ishikawa, I.; Izumi, Y. Application of Antimicrobial Photodynamic Therapy in Periodontal and Peri-Implant Diseases. *Periodontol. 2000* **2009**, *51*, 109–140. [CrossRef]
22. Wainwright, M. Photoantimicrobials and PACT: What's in an Abbreviation? *Photochem Photobiol Sci.* **2019**, *18*, 12–14. [CrossRef] [PubMed]
23. Oz, M.; Lorke, D.E.; Hasan, M.; Petroianu, G.A. Cellular and Molecular Actions of Methylene Blue in the Nervous System. *Med. Res. Rev.* **2009**, *31*, 93–117. [CrossRef] [PubMed]
24. Khdair, A.; Gerard, B.; Handa, H.; Mao, G.; Shekhar, M.P.V.; Panyam, J. Surfactant-Polymer Nanoparticles Enhance the Effectiveness of Anticancer Photodynamic Therapy. *Mol. Pharm.* **2008**, *5*, 795–807. [CrossRef]
25. Wilson, M. Lethal Photosensitisation of Oral Bacteria and its Potential Application in the Photodynamic Therapy of Oral Infections. *Photochem. Photobiol. Sci.* **2004**, *3*, 412–418. [CrossRef] [PubMed]
26. Bürgers, R.; Witecy, C.; Hahnel, S.; Gosau, M. The Effect of Various Topical Peri-Implantitis Antiseptics on *Staphylococcus Epidermidis, Candida Albicans*, and *Streptococcus Sanguinis*. *Arch. Oral Biol.* **2012**, *57*, 940–947. [CrossRef]
27. Longer, M.A.; Robinson, J.R. Fundamental Aspects of Bioadhesion. *Pharm. Int.* **1986**, *7*, 114–117.
28. Shaikh, R.; Singh, T.R.R.; Garland, M.J.; Donnelly, R.F. Mucoadhesive Drug Delivery Systems. *J. Pharma. Bioallied. Sci.* **2011**, *3*, 89–100.
29. Ahuja, A.; Ali, J.; Khar, K.R. Mucoadhesive Drug Delivery Systems. *Drug Dev. Ind. Pharm.* **1997**, *23*, 489–515. [CrossRef]
30. López-Jiménez, L.; Fusté, E.; Martínez-Garriga, B.; Arnabat-Domínguez, J.; Vinuesa, T.; Viñas, M. Effects of Photodynamic Therapy on *Enterococcus Faecalis* Biofilms. *Lasers Med. Sci.* **2015**, *30*, 1519–1526. [CrossRef]
31. Sechriest, V.F.; Miao, Y.J.; Niyibizi, C.; Westerhausen-Larson, A.; Matthew, H.W.; Evans, C.H.; Fu, F.H.; Suh, J.K. GAG-Augmented Polysaccharide Hydrogel: A Novel Biocompatible and Biodegradable Material to Support Chondrogenesis. *J. Biomed. Mater. Res.* **2000**, *49*, 534–541. [CrossRef]
32. Ding, S.J. Biodegradation Behavior of Chitosan/Calcium Phosphate Composites. *J. Non-Crystal. Solids* **2007**, *353*, 2367–2373. [CrossRef]
33. Wang, P.; Shen, Y.; Zhao, L. Chitosan Nanoparticles Loaded with Aspirin and 5-Fluororacil Enable Synergistic Antitumour Activity Through the Modulation of NF-κB/COX-2 Signaling Pathway. *IET Nanobiotechnol.* **2020**, *14*, 479–484. [CrossRef]
34. Uzair, B.; Akhtar, N.; Sajjad, S.; Bano, A.; Fasim, F.; Zafar, N.; Leghari, S.A.K. Targeting Microbial Biofilms: By Arctium Lappa l. Synthesised Biocompatible CeO_2-NPs Encapsulated in Nano-Chitosan. *IET Nanobiotechnol.* **2020**, *14*, 217–223. [CrossRef]
35. Lin, M.C.; Chen, C.C.; Wu, I.T.; Ding, S.J. Enhanced Antibacterial Activity of Calcium Silicate-Based Hybrid Cements for Bone Repair. *Mater. Sci. Eng. C* **2020**, *110*, 110727. [CrossRef]
36. Raafat, D.; Sahl, H.G. Chitosan and its Antimicrobial Potential—A Critical Literature Survey. *Microb. Biotechnol.* **2009**, *2*, 186–201. [CrossRef]
37. Lin, C.H.; Chien, H.F.; Lin, M.H.; Chen, C.P.; Shen, M.; Chen, C.T. Chitosan Inhibits the Rehabilitation of Damaged Microbes Induced by Photodynamic Inactivation. *Int. J. Mol. Sci.* **2018**, *19*, 2598.
38. Bhattarai, N.; Gunn, J.; Zhang, M. Chitosan-Based Hydrogels for Controlled, Localized Drug Delivery. *Adv. Drug Deliv. Rev.* **2010**, *62*, 83–99. [CrossRef] [PubMed]
39. Lee, W.F.; Chiu, R.J. Investigation of Charge Effects on Drug Release Behavior for Ionic Thermosensitive Hydrogels. *Mater. Sci. Eng. C* **2002**, *20*, 161–166. [CrossRef]

40. Shrestha, A.; Hamblin, M.R.; Kishen, A. Characterization of a Conjugate between Rose Bengal and Chitosan for Targeted Antibiofilm and Tissue Stabilization Effects as a Potential Treatment of Infected Dentin. *Antimicrob. Agents Chemother.* **2012**, *56*, 4876–4884. [CrossRef] [PubMed]
41. Choi, S.S.; Lee, H.K.; Chae, H.S. Synergistic in Vitro Photodynamic Antimicrobial Activity of Methylene Blue and Chitosan Against *Helicobacter Pylori 26695. Photodiagn. Photodyn. Ther.* **2014**, *11*, 526–532. [CrossRef] [PubMed]
42. Romeo, U.; Nardi, G.M.; Libotte, F.; Sabatini, S.; Palaia, G.; Grassi, F.R. The Antimicrobial Photodynamic Therapy in the Treatment of Peri-Implantitis. *Int. J. Dent.* **2016**, 7692387. [CrossRef] [PubMed]
43. Hamblin, M.R. Antimicrobial Photodynamic Inactivation: A Bright new Technique to Kill Resistant Microbes. *Curr. Opin. Microbiol.* **2016**, *33*, 67–73. [CrossRef] [PubMed]
44. Akilov, O.E.; O'Riordan, K.; Kosaka, S.; Hasan, T. Photodynamic Therapy against Intracellular Pathogens: Problems and Potentials. *Med. Lasers Appl.* **2006**, *21*, 251–260. [CrossRef]
45. Maisch, T.; Baier, J.; Franz, B.; Maier, M.; Landthaler, M.; Szeimies, R.M.; Bäumler, W. The Role of Singlet Oxygen and Oxygen Concentration in Photodynamic Inactivation of Bacteria. *Proc. Natl. Acad. Sci. USA* **2007**, *104*, 7223–7228. [CrossRef] [PubMed]
46. Stalling, S.S.; Akintoye, S.O.; Nicoll, S.B. Development of Photocrosslinked Methylcellulose Hydrogels for Soft Tissue Reconstruction. *Acta Biomater.* **2009**, *5*, 1911–1918. [CrossRef]
47. Baltazar, L.M.; Ray, A.; Santos, D.A.; Cisalpino, P.S.; Friedman, A.J.; Nosanchuk, J.D. Antimicrobial Photodynamic Therapy: An Effective Alternative Approach to Control Fungal Infections. *Front. Microbiol.* **2015**, *6*, 202. [CrossRef]
48. Sharma, S.K.; Dai, T.; Kharkwal, G.B.; Huang, Y.Y.; Huang, L.; Bil de Arce, V.J.; Tegos, G.P.; Hamblin, M.R. Drug Discovery of Antimicrobial Photosensitizers Using Animal Models. *Curr. Pharm. Des.* **2011**, *17*, 1303–1319. [CrossRef] [PubMed]
49. Wei, C.K.; Ding, S.J. Dual-Functional Bone Implants with Antibacterial Ability and Osteogenic Activity. *J. Mater. Chem. B* **2017**, *5*, 1943–1953. [CrossRef]
50. Tate, M.C.; Shear, D.A.; Hoffman, S.W.; Stein, D.G.; LaPlaca, M.C. Biocompatibility of Methylcellulose-Based Constructs Designed for Intracerebral Gelation Following Experimental Traumatic Brain Injury. *Biomaterials* **2001**, *22*, 1113–1123. [CrossRef]
51. Rolim, J.P.; de-Melo, M.A.; Guedes, S.F.; Albuquerque-Filho, F.B.; de Souza, J.R.; Nogueira, N.A.; Zanin, I.C.; Rodrigues, L.K. The Antimicrobial Activity of Photodynamic Therapy Against Streptococcus Mutans Using Different Photosensitizers. *J. Photochem. Photobiol. B* **2012**, *106*, 40–46. [CrossRef]
52. Lee, Y.S.; Wurster, R.D. Methylene Blue Induces Cytotoxicity in Human Brain Tumor Cells. *Cancer Lett.* **1995**, *88*, 141–145. [CrossRef]
53. Jung, J.; Wen, J.; Sun, Y. Amphiphilic Quaternary Ammonium Chitosans Self-Assemble Onto Bacterial and Fungal Biofilms and Kill Adherent Microorganisms. *Colloids Surf. B Biointerfaces* **2019**, *174*, 1–8. [CrossRef] [PubMed]
54. Klausen, M.; Heydorn, A.; Ragas, P.; Lambertsen, L.; Aaes-Jørgensen, A.; Molin, S.; Tolker-Nielsen, T. Biofilm Formation by *Pseudomonas Aeruginosa* wild Type, Flagella and Type IV Pili Mutants. *Mol. Microbiol.* **2003**, *48*, 1511–1524. [CrossRef] [PubMed]
55. Fontana, C.R.; Abernethy, A.D.; Som, S.; Ruggiero, K.; Doucette, S.; Marcantonio, R.C.; Boussios, C.I.; Kent, R.; Goodson, J.M.; Tanner, A.C.; et al. The Antibacterial Effect of Photodynamic Therapy in Dental Plaque-Derived Biofilms. *J. Periodontal. Res.* **2009**, *44*, 751–759. [CrossRef] [PubMed]
56. Miyata, S.; Miyaji, H.; Kawasaki, H.; Yamamoto, M.; Nishida, E.; Takita, H.; Akasaka, T.; Ushijima, N.; Iwanaga, T.; Sugaya, T. Antimicrobial Photodynamic Activity and Cytocompatibility of Au25(Capt)18 Clusters Photoexcited by Blue LE D Light Irradiation. *Int. J. Nanomed.* **2017**, *12*, 2703–2716. [CrossRef] [PubMed]
57. Kim, S.H.; Park, S.H.; Chang, B.S.; Lee, S.Y.; Lee, J.K.; Um, H.S. Antimicrobial Effect of Photodynamic Therapy Using Methylene Blue and Red Color Diode Laser on Biofilm Attached to Sandblasted and Acid-Etched Surface of Titanium. *Laser Dent. Sci.* **2017**, *1*, 83–90. [CrossRef]
58. Huang, T.C.; Chen, C.J.; Ding, S.J.; Chen, C.C. Antimicrobial Efficacy of Methylene Blue-Mediated Photodynamic Therapy on Titanium Alloy Surfaces in Vitro. *Photodiagn. Photodyn. Ther.* **2019**, *25*, 7–16. [CrossRef] [PubMed]
59. Camacho-Alonso, F.; Julián-Belmonte, E.; Chiva-García, F.; Martínez-Beneyto, Y. Bactericidal Efficacy of Photodynamic Therapy and Chitosan in Root Canals Experimentally Infected with *Enterococcus Faecalis*: An in Vitro Study. *Photomed. Laser Surg.* **2017**, *35*, 184–189. [CrossRef] [PubMed]
60. Darabpour, E.; Kashef, N.; Mashayekhan, S. Chitosan Nanoparticles Enhance the Efficiency of Methylene Blue-Mediated Antimicrobial Photodynamic Inactivation of Bacterial Biofilms: An in Vitro Study. *Photodiagn. Photodyn. Ther.* **2016**, *14*, 211–217. [CrossRef] [PubMed]
61. del Carpio-Perochena, A.; Kishen, A.; Shrestha, A.; Bramante, C.M. Antibacterial Properties Associated with Chitosan Nanoparticle Treatment on Root Dentin and 2 Types of Endodontic Sealers. *J. Endod.* **2015**, *41*, 1353–1358. [CrossRef] [PubMed]
62. Davey, M.E.; O'Toole, G.A. Microbial Biofilms: From Ecology to Molecular Genetics. *Microbiol. Mol. Biol. Rev.* **2000**, *64*, 847–867. [CrossRef]
63. Kreisler, M.; Kohnen, W.; Christoffers, A.B.; Götz, H.; Jansen, B.; Duschner, H.; d'Hoedt, B. In Vitro Evaluation of the Biocompatibility of Contaminated Implant Surfaces Treated with an Er:YAG Laser and an Air Powder System. *Clin. Oral Impl. Res.* **2005**, *16*, 36–43. [CrossRef] [PubMed]
64. Chen, C.J.; Chen, C.C.; Ding, S.J. Effectiveness of Hypochlorous Acid to Reduce the Biofilms on Titanium Alloy Surfaces in Vitro. *Int. J. Mol. Sci.* **2016**, *17*, 1161. [CrossRef] [PubMed]

65. Shrestha, A.; Cordova, M.; Kishen, A. Photoactivated Polycationic Bioactive Chitosan Nanoparticles Inactivate Bacterial Endotoxins. *J. Endod.* **2015**, *41*, 686–691. [CrossRef]
66. Almeida, J.M.; Theodoro, L.H.; Bosco, A.F.; Nagata, M.J.; Oshiiwa, M.; Garcia, V.G. In Vivo Effect of Photodynamic Therapy on Periodontal Bone Loss in Dental Furcations. *J. Periodontal.* **2008**, *79*, 1081–1088. [CrossRef]
67. Ding, S.J.; Chu, Y.H.; Wang, D.Y. Enhanced Properties of Novel Zirconia-Based Osteo-Implant Systems. *Appl. Mater. Today* **2017**, *9*, 622–632. [CrossRef]
68. Eick, S.; Meier, I.; Spoerlé, F.; Bender, P.; Aoki, A.; Izumi, Y.; Salvi, G.E.; Sculean, A. In Vitro-Activity of Er:YAG Laser in Comparison with other Treatment Modalities on Biofilm Ablation from Implant and Tooth Surfaces. *PLoS ONE* **2017**, *12*, e0171086. [CrossRef] [PubMed]
69. Wu, I.T.; Kao, P.F.; Huang, Y.R.; Ding, S.J. In Vitro and in Vivo Osteogenesis of Gelatin-Modified Calcium Silicate Cement with Washout Resistance. *Mater. Sci. Eng. C* **2020**, *117*, 111297. [CrossRef]
70. Gristina, A.G. Biomaterial-Centered Infection: Microbial Adhesion Versus Tissue Integration. *Science* **1987**, *237*, 1588–1595. [CrossRef]
71. Wu, B.C.; Wei, C.K.; Hsueh, N.S.; Ding, S.J. Comparative Cell Attachment, Cytotoxicity and Antibacterial Activity of Radiopaque Dicalcium Silicate Cement and White-Coloured Mineral Trioxide Aggregate. *Int. Endod. J.* **2015**, *48*, 268–276. [CrossRef] [PubMed]
72. Huang, Y.R.; Wu, I.T.; Chen, C.C.; Ding, S.J. In Vitro Comparisons of Microscale and Nanoscale Calcium Silicate Particles. *J. Mater. Chem. B* **2020**, *8*, 6034–6047. [CrossRef] [PubMed]

pharmaceuticals

Article

The Comparison of In Vitro Photosensitizing Efficacy of Curcumin-Loaded Liposomes Following Photodynamic Therapy on Melanoma MUG-Mel2, Squamous Cell Carcinoma SCC-25, and Normal Keratinocyte HaCaT Cells

Marta Woźniak [1,*], Martyna Nowak [1], Anastasiia Lazebna [2], Kamil Więcek [2], Izabella Jabłońska [1], Krzysztof Szpadel [1], Aleksandra Grzeszczak [3], Jerzy Gubernator [3,*] and Piotr Ziółkowski [1]

1 Department of Pathology, Wroclaw Medical University, 50-367 Wroclaw, Poland; martyna.nowak96@o2.pl (M.N.); jabizabella@gmail.com (I.J.); szpadelk@gmail.com (K.S.); piotr.ziolkowski@umed.wroc.pl (P.Z.)
2 Department of Biotechnology, University of Wroclaw, 50-383 Wroclaw, Poland; stoneqer@gmail.com (A.L.); kamil.wiecek24@gmail.com (K.W.)
3 Department of Lipids and Liposomes, Faculty of Biotechnology, University of Wroclaw, 50-383 Wroclaw, Poland; aleksandra.grzeszczak@uwr.edu.pl
* Correspondence: marta.wozniak@umed.wroc.pl (M.W.); jerzy.gubernator@uwr.edu.pl (J.G.)

Citation: Woźniak, M.; Nowak, M.; Lazebna, A.; Więcek, K.; Jabłońska, I.; Szpadel, K.; Grzeszczak, A.; Gubernator, J.; Ziółkowski, P. The Comparison of In Vitro Photosensitizing Efficacy of Curcumin-Loaded Liposomes Following Photodynamic Therapy on Melanoma MUG-Mel2, Squamous Cell Carcinoma SCC-25, and Normal Keratinocyte HaCaT Cells. *Pharmaceuticals* **2021**, *14*, 374. https://doi.org/10.3390/ph14040374

Academic Editor: Serge Mordon

Received: 30 March 2021
Accepted: 15 April 2021
Published: 17 April 2021

Abstract: The research focused on the investigation of curcumin encapsulated in hydrogenated soy phosphatidylcholine liposomes and its increased photoactive properties in photodynamic therapy (PDT). The goal of this study was two-fold: to emphasize the role of a natural photoactive plant-based derivative in the liposomal formulation as an easily bioavailable, alternative photosensitizer (PS) for the use in PDT of skin malignancies. Furthermore, the goal includes to prove the decreased cytotoxicity of phototoxic agents loaded in liposomes toward normal skin cells. Research was conducted on melanoma (MugMel2), squamous cell carcinoma (SCC-25), and normal human keratinocytes (HaCaT) cell lines. The assessment of viability with MTT (3-(4,5-dimethylthiazolyl-2)-2,5-diphenyltetrazolium bromide) evaluated cell death after exposure to blue light irradiation after 4 h of pre-incubation with free and encapsulated curcumin. Additionally, the wound healing assay, flow cytometry, and immunocytochemistry to detect apoptosis were performed. The malignant cells revealed increased phototoxicity after the therapy in comparison to normal cells. Moreover, liposome curcumin-based photodynamic therapy showed an increased ratio of apoptotic and necrotic cells. The study also demonstrated that nanocurcumin significantly decreased malignant cell motility following PDT treatment. Acquired results suggest that liposomal formulation of a poor soluble natural compound may improve photosensitizing properties of curcumin-mediated PDT treatment in skin cancers and reduce toxicity in normal keratinocytes.

Keywords: curcumin; natural photosensitizer; photodynamic therapy; skin cancer treatment; squamous cell carcinoma; melanoma; normal keratinocytes; liposomes

1. Introduction

Skin cancers are among the most widespread types of neoplasm, affecting people from less pigmented, Caucasian populations, usually more than 50 years of age [1]. Those malignancies are divided into two main subgroups composed of more lethal melanoma and more prevalent non-melanoma skin cancers (NMSC). Currently, non-melanoma skin cancers are generally represented by tumours from transformed keratinocytes: cutaneous squamous cell carcinoma (cSCC) and basal cell carcinoma (BCC). There are also other skin-related neoplasms, including Kaposi's sarcoma, Merkel cell and adnexal carcinoma, cutaneous lymphoma, or dermatofibrosarcoma protuberans. Nevertheless, those conditions are not as common as NMSCs [2,3]. Like most tumours, skin cancer treatment involves widely used methods, such as surgical intervention, radiotherapy, chemotherapy, and immunotherapy. However, despite the effectiveness of the former, it cannot be applied in all cases due to adverse malignancy's localization and potential patient's co-morbidities [2]. Moreover, radiotherapy and chemotherapy can not only be toxic and lead to side effects but also might be ineffective in cells with developed resistance [4], whereas immunotherapy currently seems too complex for general use [5]. Therefore, further development of more efficient therapeutic strategies is still required.

One of the novel approaches that could potentially overcome those obstacles is the usage of photodynamic therapy (PDT). The main principle of this method is to apply and accumulate the chosen substance with photosensitive properties (called further photosensitizer (PS)) within tumour tissue. Later, after local irradiation with a specific wavelength laser, excited PS can transform surrounding molecules into highly active reactive oxygen forms (ROS). Depending on the mechanisms, excited PS may transfer electrons on organic compounds (via type I reaction), creating radicals such as hydrogen peroxide, or transmits its energy on molecular oxygen by developing a singlet oxygen ($^1O_2$) (via type II reaction) [6]. Accumulated ROS can damage plenty of biomolecules, including lipids, proteins, DNA, and carbohydrates. However, due to the limited diffusing capabilities of radicals, their damaging properties rely on PS. Whether PS exhibits a higher affinity to concentrate closely to mitochondria's membrane or enzymes, radicals' activities may have a different impact on cells. ROS and singlet oxygen from an activated photosensitizer can induce cell death, mainly by damaging lipids of plasma and organelles membranes, triggering caspases cascade, and inactivating anti-apoptotic proteins. Depending on the efficiency of PS, photokilling can occur by rough conditions of necrosis or (preferably) in milder cases apoptosis or/and autophagy [7–9].

Photodynamic therapy is gaining growing interest due to its low invasiveness, high selectivity, and comparable lower costs to other treatments. Nonetheless, presently, PDT is not applicable in the treatment of metastatic cells [10]. The most common PS evoke a low therapeutic effect against highly pigmented melanoma cells [5], and the method itself can be burdened with pain, especially in combination with commonly used 5-aminolevulinic acid (5-ALA) [4]. For all those reasons, PDT enhancement is currently extensively investigated, especially with novel, high-efficient, and less toxic photosensitizers.

Curcumin is a natural polyphenol extracted from turmeric (*Curcuma longa*), with well-documented anti-tumour, anti-inflammatory, and photoactive properties [11–13]. This golden polyphenol has already been used for its anti-inflammatory effects, as a treatment in various dermatological conditions [14].

Due to its exceptional attributes, this plant-derived substance could potentially play a dualistic role in PDT functioning simultaneously as PS and a direct therapeutic molecule. Experiments conducted on animal models and in vitro suggest that curcumin can downregulate various molecular responses in boosting up inflammatory and pro-survival pathways, such as those related to transcription factors like Nf-κB or AP-1 [15,16]. Thus, curcumin could potentially not only increase the chances of apoptosis in defective cells but also stimulate the production of cell killing radicals, making it a promising compound to use in PDT therapies.

Among all the previously mentioned benefits, the extremely poor water solubility and low bioavailability of this natural plant derivative limit its clinical use in cancer treatment. Moreover, basic skin properties made it an excellent barrier decreasing percutaneous penetration of curcumin. For this reason, the development of stable formulations of drug carriers that improve skin penetration and therapeutic effectiveness with reduced side effects is an essential challenge for many researchers [17]. Nowadays, various nanocarriers, which could greatly enhance the bioavailability of drugs, are under intensive development and some of them had already been functionalized for active targeting of skin cancers, including those based on gels or liposomes that are modified with aptamers [18,19].

According to several studies, nano-formulations of photosensitizers improve the pharmacokinetic effects and therapeutic advantage of free compounds [20–22]. Besides that, lipid formulations are proposed as an alternative strategy to potentiate the effect of PDT against resistant melanoma cells [23]. Liposomes are a versatile drug delivery system. They are not toxic and (when pegylated) exhibit longer circulation time among all drug carriers. Liposomes can encapsulate hydrophilic, hydrophobic, and amphiphilic molecules. Liposomes have many advantages, such as controlled release properties, cell affinity, tissue compatibility, reducing drug toxicity, and improving drug stability. As most drug carriers, liposomes can accumulate in inflammatory tissues by using the enhanced permeability and retention (EPR) effect. This accumulation can be further increased by decreasing particle size as well as pretreatment by some drugs and substances [24–27].

In general, at least in the animal model, an essential increase of drug concentration is observed in the tumour tissue when liposomal drugs are applied. In the case of curcumin, which is a hydrophobic, the use of liposomes may diminish issues with its low solubility and bioavailability, enhancing pharmacokinetics and accumulation in cancer tissues [16,20,21].

In this study, a relatively novel curcumin formulation has been used, in which curcumin is encapsulated in liposomes composed from hydrogenated soy phosphatidylcholine (HSPC), which exhibited high stability, due to a relatively rigid liposomes' bilayer and, therefore, low curcumin diffusion properties. This formulation proved its superiority in comparison to a free substance on pancreatic cancer cell lines and can be regarded as an essential improvement of the traditional route of curcumin supply [28–31].

Herein, a comparison of the phototoxic and anti-cancerous effects of curcumin and its stable HSPC liposomal formulation on skin cancer cell lines was conducted, including SCC-25 representing cutaneous squamous cell carcinoma, MUG-Mel2 representing a melanoma cell line, and normal human keratinocytes HaCaT representing control normal skin cells (Figure 1). To evaluate the effects of encapsulated curcumin as a photosensitizer in PDT on different skin cell lines, MTT (3-(4,5-dimethylthiazolyl-2)-2,5-diphenyltetrazolium bromide) dark cytotoxicity, phototoxicity assay, immunocytochemical staining against markers of apoptosis, Bcl-2, and Bax, measuring apoptosis with flow cytometry and a wound-healing assay, were performed.

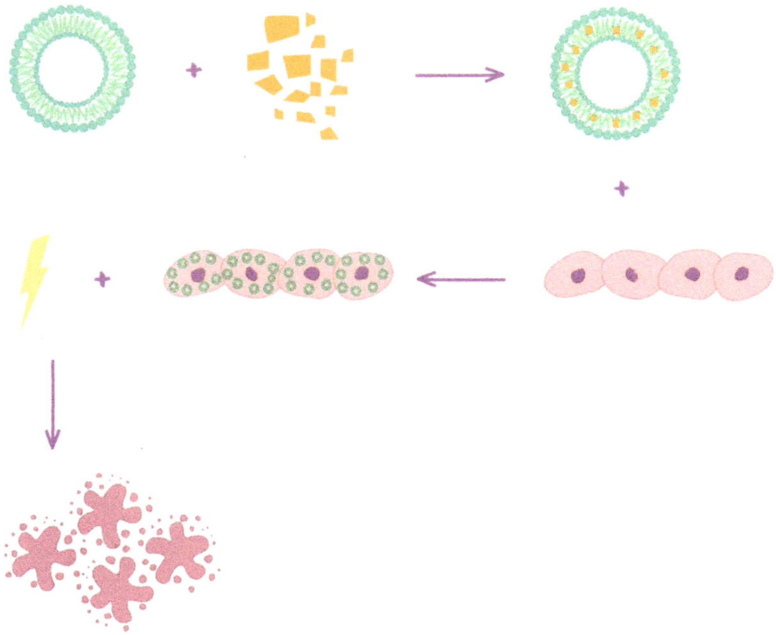

Figure 1. Scheme of encapsulated curcumin in liposomes and mechanism of photodynamic therapy.

2. Results

2.1. The Effect of Curcumin and Liposomal Curcumin Based PDT on MUG-Mel2, SCC-25, and HaCaT Cells Viability Measured by an MTT Assay

The effect of curcumin and liposome-curcumin-based PDT was performed on skin cancer cell MUG-Mel2 (melanoma cells), SCC-25 (squamous cell carcinoma), and normal keratinocyte cells HaCaT. Effectiveness of free and encapsulated curcumin was compared in doses of 5 and 10 μM after blue light low irradiation (2.5 J/cm^2). Results indicated that liposome curcumin-mediated PDT caused a significantly higher reduction of viability in both cancer cell lines than a free natural compound. Curcumin mediated-PDT in 10 μM concentration caused decreased viability in SCC-25 (34%) and MUG-Mel2 (27%). Liposomes with curcumin mediated-PDT inhibited cancer cells' growth more than a free compound after irradiation reaching IC50. Liposomal-curcumin-PDT exhibit cytotoxicity of 53% in MUG-Mel2 and 58% in SCC-25 at the same dose—10 μM and low irradiation dose (2.5 J/cm^2) while the viability of HaCaT was decreased only by 11%. Interestingly, HaCaT cells maintained viability of around 90% after different treatments. Liposomal curcumin in the concentration of 10 μM was chosen in all subsequent biological studies (Figure 2).

2.2. The Effect of Liposomal Curcumin Based PDT on MUG-Mel2, SCC-25, and HaCaT Cells in the Wound-Healing Process

To check whether liposomal curcumin-based PDT decreases HaCaT, SCC-25, and MUG-Mel2 cells' motility, the wound healing test was performed. The assay shows the migration of cells by evaluating a primaeval scratch's closure in a 24 h observation. The results show that PDT with liposomal curcumin caused the strongest effect of migration properties in MUG-Mel2 cancer cells. After 24 h from the treatment (liposomal curcumin and the light), there was no migration observed. In SCC-25 cells, the wound was minimally closed, whereas, in normal HaCaT cells, the wound closed almost entirely within 24 h of incubation after therapy. The results are presented in Figure 3.

2.3. The Effect of Liposomal Curcumin-Based PDT on MUG-Mel2, SCC-25, and HaCaT Cells on Bax and Bcl-2 Expression

Immunocytochemical staining allows examining whether the proposed therapy with liposomal curcumin and irradiation has a cytotoxic effect on cancer cells. To assess whether the treatment causes apoptosis in cancer cells, apoptosis-related proteins bax and bcl-2 were used for the immunocytochemical analysis and then evaluation of immunoreactivity was performed. An increase in the expression of bax and decreased expression of bcl-2 in cancer cells, MUG-Mel2 and SCC-25, was observed (Figure 4). In both cancer cell lines, pro-apoptotic bax protein showed strong expression after treatment of cells with liposomal curcumin and irradiation. The expression of bcl-2 was weak or moderate. Nonetheless, HaCat cells did not significantly change the expression of the previously described proteins after irradiation only, liposomal curcumin only, and PDT treatment.

2.4. The Impact of Liposomal Curcumin on Cells Lines' Apoptosis

Flow cytometry analysis was applied to evaluate cell death caused by liposomal curcumin in SCC-25, MUG-Mel2, and HaCaT cells (Figure 5). As shown in Figure 4A,B after 24 h of treatment, early and late apoptosis and necrosis in SCC-25 and MUG-Mel2 cells were observed. The combination of liposomal curcumin and PDT increased apoptosis to 40% and 30% in SCC-25 and MUG-Mel2 cells, respectively. Interestingly, after 24 h from irradiation, in SCC-25, cell death is mainly caused by early and late apoptosis, whereas, in MUG-Mel2, cell death is caused by late apoptosis and necrosis. In control cells, HaCaT, a slight increase in the apoptosis ratio in cells after treatment (10%) was observed.

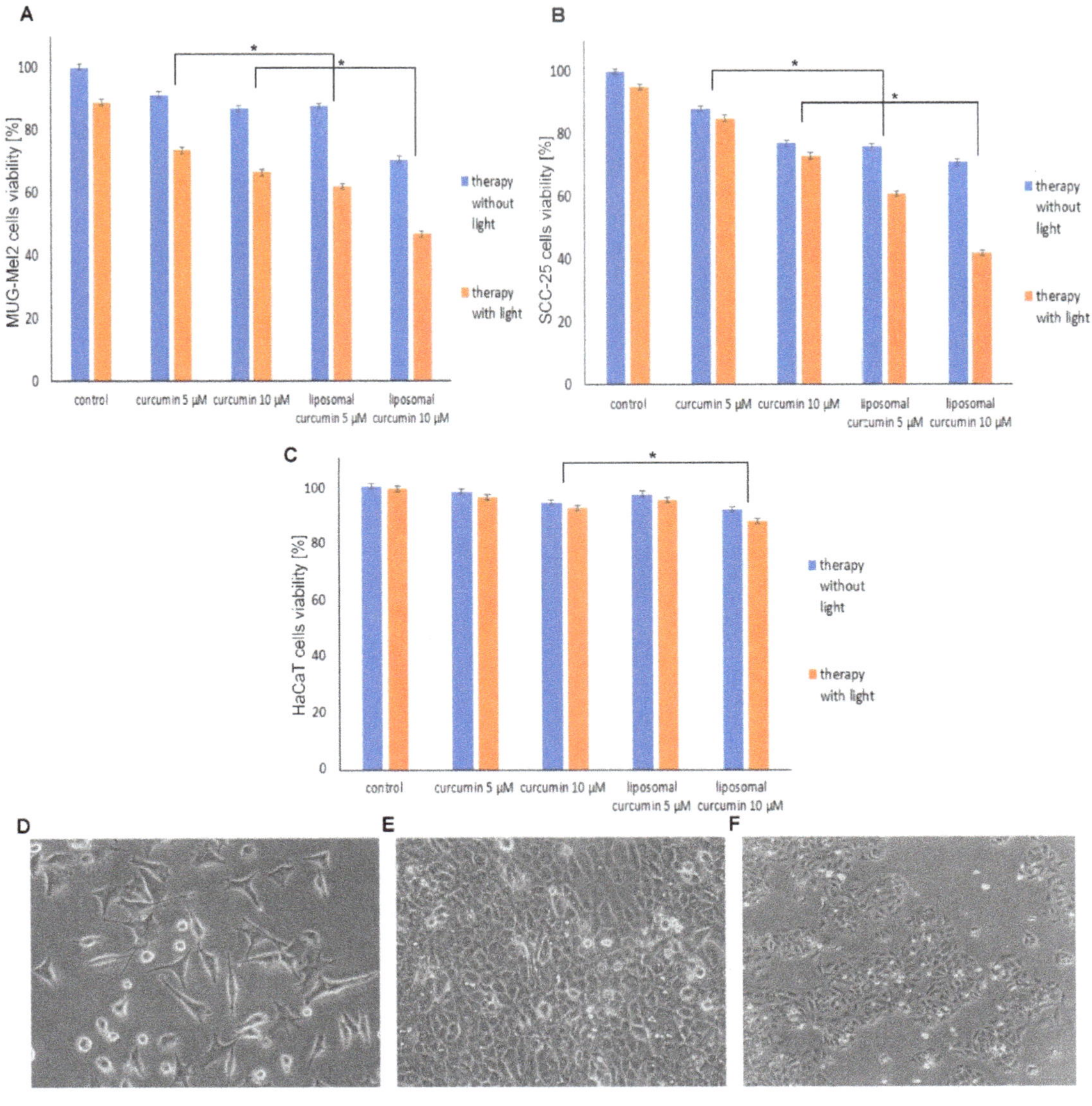

Figure 2. Results of cell viability after 4 h of incubation with 5, 10 µM curcumin and liposomal curcumin with and without the irradiation (2.5 J/cm^2) evaluated by the MTT (3-(4,5-dimethylthiazolyl-2)-2,5-diphenyltetrazolium bromide) assay. (**A**) Viability-dependent bars for MUG-Mel2 cells after incubation with curcumin and liposomal curcumin without/with light. (**B**) Viability-dependent bars for SCC-25 cells after incubation with curcumin and liposomal curcumin without/with the light. (**C**) Viability-dependent bars for HaCaT cells after incubation with curcumin and liposomal curcumin without/with light. Encapsulated curcumin is significantly more cytotoxic than free curcumin in cancer cells. (**D–F**) Representative images of MUG-Mel-2, SCC-25, HaCaT cells morphology detected by phase-contrast microscopy. Results represent the mean from three different experiments. * $p < 0.05$.

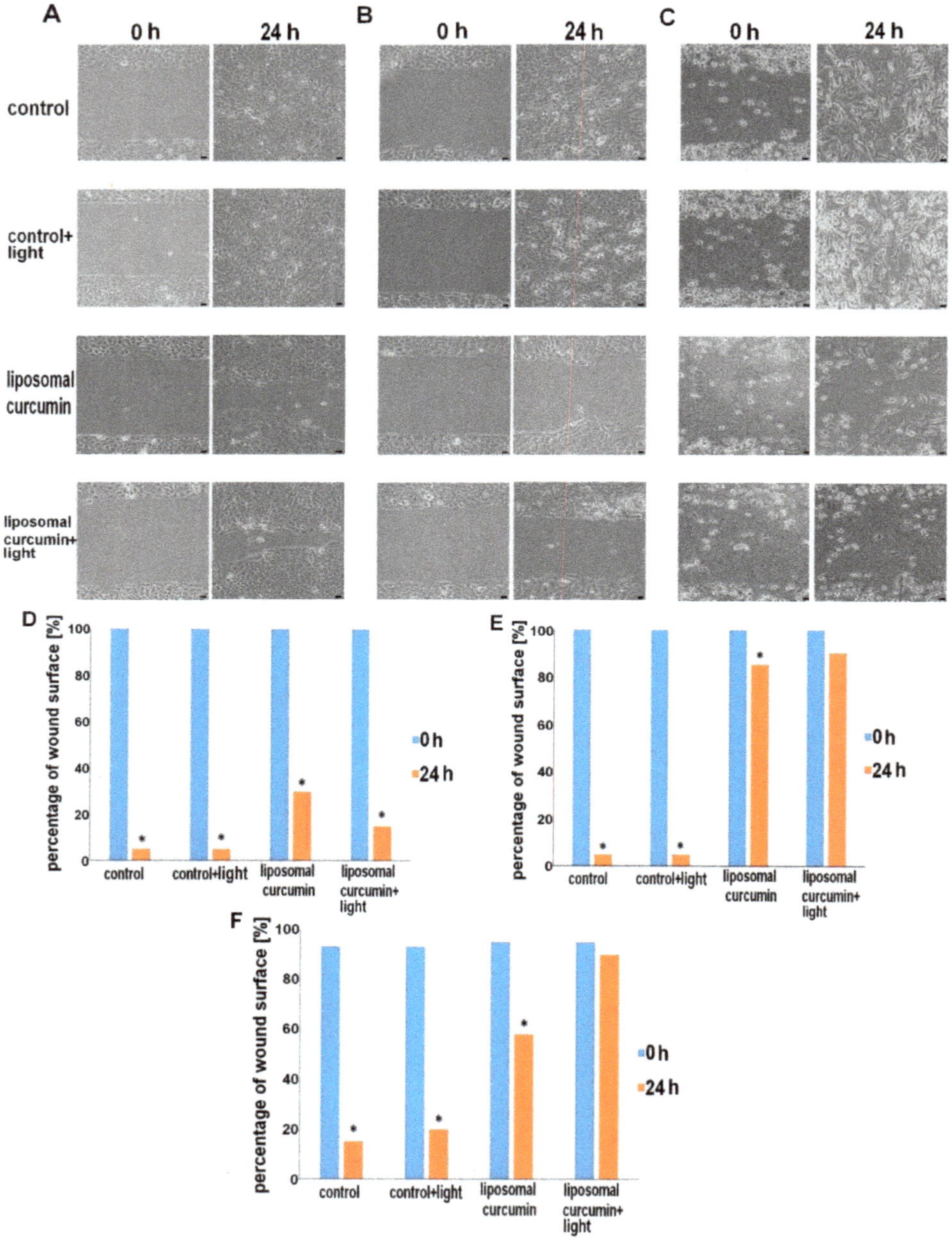

Figure 3. Wound-healing assay in time point 0 h and 24 h of (**A**) HaCaT, (**B**) SCC-25, and (**C**) MUG-Mel2 cell line. Representative images show that, after 24 h, the scrap in control cells is minimal compared to cancer cells treated with liposomal curcumin in dose 10 µM with blue light (2.5 J/cm^2). In treated cells, in the HaCaT control cell line, the scrap is smaller than in the other two cancer cells (**D**–**F**). (**D**) Quantification of cell migration for HaCaT, SCC-25, and MUG-Mel2 cells. Results are presented as the percentage of the wound surface. The initial wound area is expressed as 100% at 0 h. Results represent the mean from three different experiments. Scale bar = 50 µm. * $p < 0.05$.

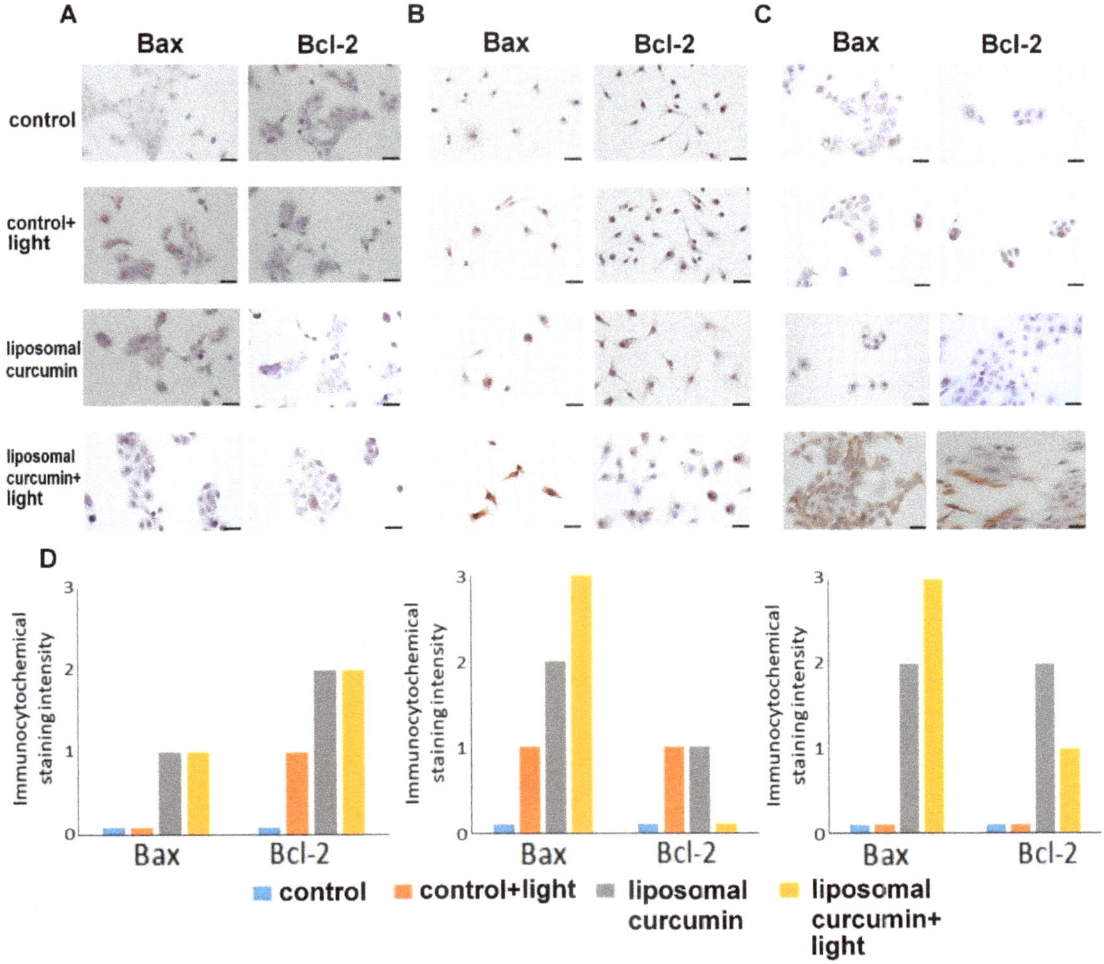

Figure 4. Presenting immunocytochemical staining of the chosen apoptosis-related proteins Bax and Bcl-2 in HaCaT (**A**), MUG-Mel2 (**B**), and SCC-25 (**C**) cells lines in four conditions: control, control with blue light only, liposomal curcumin in dose 10 μM, and liposomal curcumin in dose 10 μM with a low dose of blue light (2.5 J/cm^2). (**D**) Results of immunocytochemical analysis of HaCaT, MUG-Mel2, and the SCC-25 cell line and the chosen apoptosis-related protein expression calculated by the immunoreactivity score. Abbreviations: 0-no staining, 1-weak staining, 2-moderate staining, and 3-strong staining. Scale bar = 50 μm.

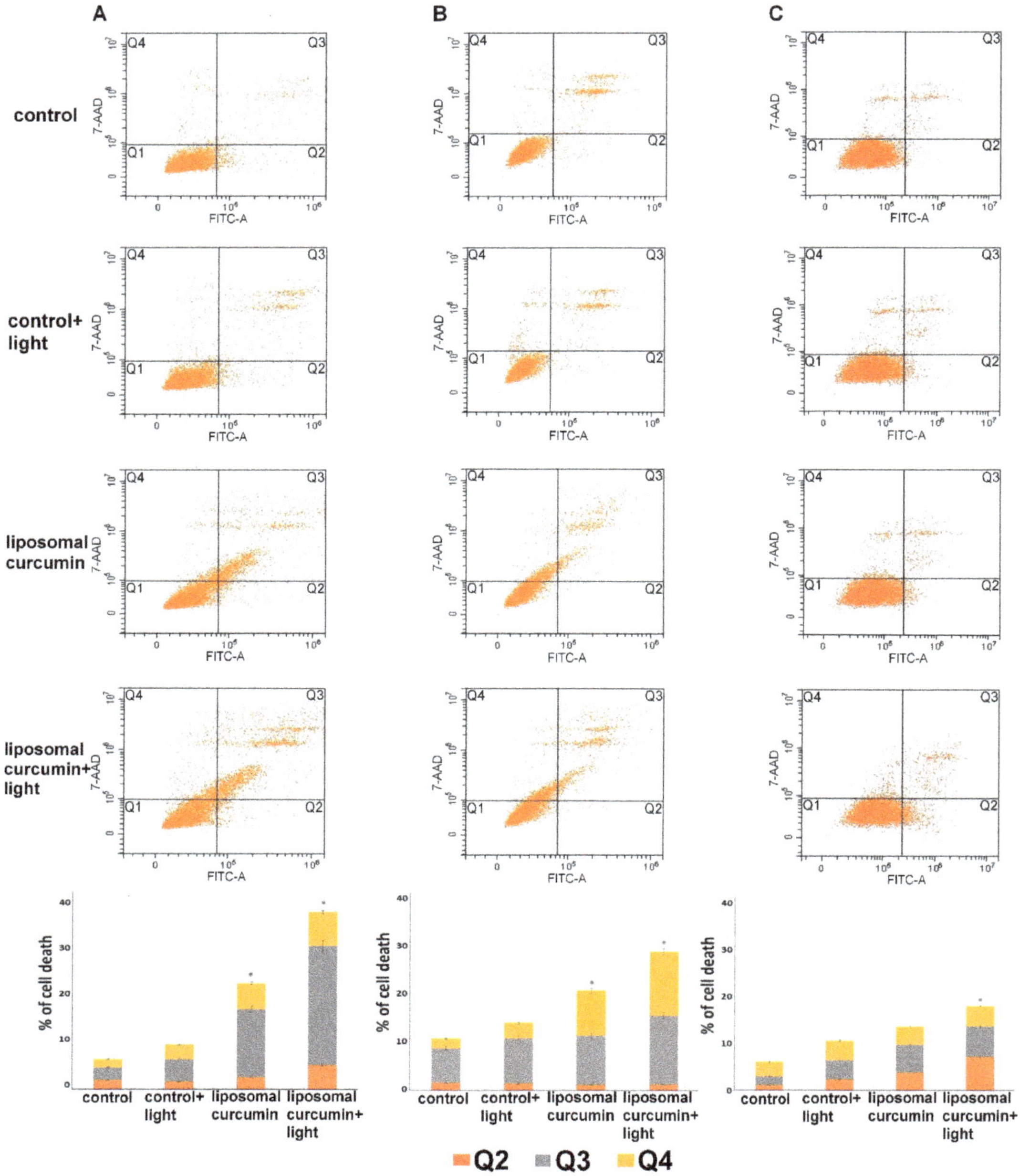

Figure 5. Effect of photodynamic therapy (PDT) with liposomal curcumin in dose 10 μM with blue light (2.5 J/cm2) on SCC-25 (**A**), MUG-Mel2 (**B**), and HaCaT (**C**) cells. Dot plots present alive-Q1, early apoptotic-Q2, late apoptotic-Q3, and dead-Q4 cells. After therapy, the cells were stained using Annexin-FICT /7-AAD Kit and were measured by flow cytometry. Bars represents the quantitative percentage of total apoptotic cells (early + late apoptosis) and necrotic cells in SCC-25, MUG-Mel2, and HaCaT cells. Results represent the mean from three different experiments. * $p < 0.05$.

3. Discussion

In the past, various approaches were undertaken in order to increase the efficacy of photodynamic therapy [32,33]. The studies included the application of chemically functionalized PS [34–36] as well as liposomal derivatives of photosensitizers for both in vitro and in vivo studies. Several studies showed an advantage of the latter modality over the routine way of photosensitizer delivery to targeted cells [37,38]. Curcumin, which revealed promising effects in PDT, can act as a direct photosensitizer exhibiting cytotoxic properties in various types of tumours, including skin cancers [12,39,40].

Although curcumin can be applied in a pure form and then sensitized with light at the proper wavelength, its liposomal formulation was proposed as the more effective strategy in killing the malignant cells [41,42]. Free curcumin is characterized by low water solubility and poor bioavailability. It is rapidly metabolized or degraded in the cell culture media or after oral administration. In contrast, nano-capsules in which the compound is confined into phospholipid bilayers dismiss the significant drawbacks and promote increased absorption of curcumin into the cells [43].

In the present study, the effectiveness of curcumin loaded in PEGylated, cholesterol-free formulation based upon hydrogenated soya PC liposomes has been investigated on three skin cell lines: melanoma MUG-Mel2, squamous cell carcinoma SCC-25, and immortalized keratinocytes HaCaT cells. In previous studies, the previously described formulation of liposomes was evaluated on the pancreatic cell line and in human plasma [31]. The results indicated that this formulation presented the best parameters of the hydrophobic drug incorporation by improved bioavailability, increased stability, and cytotoxicity. In this article, the MTT results revealed statistically significant phototoxicity of this liposomal formulation of 10 μM curcumin compared to the free substance before and after photodynamic therapy. In the case of a free substance, it interacts with the cellular outer membrane, while liposomes are quickly internalized and enter the cell through the endosomal route, which increases its bioavailability and, thus, results in more potent cytotoxic effects [43,44]. Acquired results are in accordance with the observations and conclusions of Vetha et al. and Ambreen et al. on different cancer cell lines [41,45]. Although the effect was evident for both malignant cell lines, normal HaCaT keratinocytes were slightly resistant to the therapy. These spontaneously immortalized human keratinocytes from adult skin have been used as a model cell line to study normal keratinocyte functions in different studies [46]. Additionally, HaCaT cells maintained in a culture medium without the calcium display normal morphogenesis and expression of the cellular membrane markers as keratinocytes isolated from adult skin [47]. Based on gleaned, different experimental results, conclusions emerge that immortalized HaCaT keratinocytes are less susceptible to photosensitization with curcumin than MUG-Mel2 and SCC-25 malignant cells in terms of phototoxicity. These observations are following the results of Popovic et al. [48]. The authors found that 3 μM hypercin-mediated-PDT is completely refractory to keratinocytes. Moreover, they indicated a different response toward a natural plant derivative compound-PDT in each skin cell type. On the other hand, Szlasa et al. [12] presented the increased cytotoxic impact of the free curcumin-mediated photodynamic therapy on the keratinocytes. However, according to the cell line and light dose used in their studies described in the methods paragraph, the authors used normal human epidermal keratinocytes (HEK) and 6 J/cm^2 to irradiate cells in their experiments. These differences in the cells' response to curcumin irradiated with blue light may be considered due to the distinct vulnerability of cell lines to the cell-stress induction and different PDT protocols.

The present study results showed that liposomal formulation of a compound considered a potent photosensitizer can also enhance the effectiveness of liposomal curcumin-mediated-PDT by increasing the apoptosis ratio validated by flow cytometry and the production of pro-apoptotic factors, e.g., Bax protein. At the same time, the proposed therapy decreases the production of anti-apoptotic proteins, which is, in this case, Bcl-2. The significantly increased strong Bax expression was observed in both cancer cell lines, whereas, in HaCaT cells, Bax expression was lower in the sample treated with liposomal curcumin irradiated with the light. A flow cytometry assay confirmed this effect. Cells were stained with Annexin V-FITC and propidium iodide to detect early and late apoptosis and dead cells after treatment. It has been noticed that the late apoptosis in SCC-25 and MUG-Mel2 cells was increased after 24 h from the proposed therapy. Interestingly, SCC-25 cells revealed apoptosis as a leading cause of cell death, while MUG-Mel2 cells showed both types of cell death as a possible mechanism. The above finding remains in concordance with the results of other authors and shows that, in hydrophobic photosensitizers, an increase of photodynamic efficacy could be achieved by trapping them in liposomes [41,45].

The presented observations also point toward a possible mechanism of action of curcumin in PDT via an apoptotic pathway. Cells in all three examined groups showed necrosis, which is routinely observed after the PDT [7,49].

As a result of the different proliferative and migration capabilities of examined skin cell lines, a designed treatment on migration potency by a wound healing assay has been evaluated. A further examination confirms a decreased motility of melanoma and squamous cell carcinoma cell lines compared to normal keratinocytes after liposomal curcumin only and liposomal curcumin following irradiation, which is consistent with Szlasa et al. examination of the wound [12]. Normal cells nearly filled the wound (15% remaining) by 24 h, whereas the wound in malignant cells remained unfilled after 24 h. According to Ambreen et al., it is evident that liposomal curcumin-PDT reduces cancer cell migration and contributes to malignant cell metastasis inhibition.

Conducted investigations indicate the promising role of curcumin encapsulated in hydrogenated soy phosphatidylcholine liposomes in enhancing the photokilling effect on melanoma and squamous skin cancer cells following blue light PDT. Additionally, a minimal phototoxic reaction was observed in normal, human, immortalized keratinocytes with the same curcumin dose after irradiation. In conclusion, further experiments on the specific, cellular functional differences between the skin cells and in vivo testing will help confirm the effectiveness of nanocurcumin as a photosensitizer in PDT.

4. Materials and Methods

4.1. Cell Culture

Melanoma MUG-Mel2 (DSMZ, Germany) cells were cultured in RPMI 1640 cell culture medium, SCC-25-tongue squamous carcinoma (DSMZ, Braunschweig, Germany) cells in DMEM-F12, and HaCaT human epidermal keratinocytes (CLS, Eppelheim, Germany) were cultured in DMEM (Dulbecco's Modified Eagle Medium) without calcium to maintain normal morphogenesis and expression of the cellular membrane markers. To prepare a full cell culture media, 10% FBS, 1% glutamine, and 1% antibiotics were added to the bottle. Culture reagents were bought from Gibco (Thermo Fisher Scientific Inc., Waltham, MA, USA). Cells were maintained at 37 °C and 5% CO_2 in a humidified atmosphere. For experiments, cells from the 3rd to the 10th passages were used.

4.2. Preparation of Curcumin-Loaded Liposomes and Curcumin in DMSO

Curcumin-loaded liposomes of the composition HSPC/DSPE-PEG2000 9.5:0.5 mol/mol were formulated using the extrusion technique. Hydrogenated soy phosphatidylcholine (Phospholipon 90H, HSPC), 1,2-distearol-sn-glycero-phosphoethanolamin-N-(poly[ethylene glycol]2000) (DSPE-PEG2000) were purchased from Lipoid GmbH (Ludwigshafen, Germany). In brief, lipids and curcumin were dissolved in chloroform or methanol to obtain stock solutions at 10 and 5 mg/mL, respectively. Curcumin (2 mg) was mixed together with 40 mg of lipid in a borosilicate glass tube. Solvents were removed from the sample via evaporation under a stream of nitrogen gas and the resultant lipid film was dissolved in a mixture of cyclohexane and methanol (99:1, v/v). The sample was frozen in liquid nitrogen and freeze-dried for 8 h at a low pressure using a Savant Modulyo apparatus (Thermo Fisher Scientific, Waltham, CA, USA). The lipid film was hydrated by the addition of 1.5 mL of 150 mM NaCl at 64 °C, in a water bath, with gentle mixing. The liposomal suspension was finally sonicated in a water bath sonicator for 8 min at 64 °C. The newly-formed multilamellar vesicles (MLVs) were extruded 10 times through Nucleopore polycarbonate filters (Whatman, Maidstone, UK) with pore sizes of 400 and 100 nm, respectively, using a Thermobarrel Extruder (10 mL Lipex extruder, Northern Lipids, Canada) to obtain large uni-lamellar vesicles (LUVs). The extruder was maintained at 64 °C throughout the liposome extrusion procedure.

The curcumin: (1E, 6E)-1,7-bis-(4-hydroxy-3-methoxyphenyl)-1,6- heptadiene-3,5-dione (LKT Laboratories, Inc., St. Paul, MN, USA) was diffused in dimethyl sulfoxide (DMSO, suitable for hybridoma, Sigma Aldrich, Germany) to make 25 mM stock of the drug. Afterward, a decent amount of stock was compounded with a cell culture medium to achieve the composite's appropriate concentration. The DMSO amount in the final solute used to perform incubation did not surpass 0.01% and it was affirmed that the peak amount did not statistically influence the cells.

4.3. Determination of Incorporation Efficiency and Characterization of Curcumin-Loaded Liposomes

Non-incorporated drug-crystals were separated from the curcumin-loaded liposomes during the liposome extrusion procedure (only curcumin-loaded liposomes can pass through Nucleopore polycarbonate filters). Additionally, the samples were centrifuged and then collected to ensure the absence of any free curcumin liposome samples. In total, 50 µL were taken before extrusion (initial) and after centrifugation. The lipid concentration was determined by the ammonium ferrothicyanate assay on a Varian Cary1 50 UV-Vis Spectrophotometer (Varian, Ltd., Victoria, Australia). The concentration of curcumin in the liposomes was determined photometrically at λ = 425 nm on the same spectrophotometer after the curcumin-loaded liposomes were dissolved in methanol. Curcumin encapsulation efficiency was 95 ± 1.6%. The size of the liposomes was 102 nm ± 2.3 and the polydispersity index was very low (0.051).

4.4. Curcumin-Mediated PDT Experimental Protocol

Cells were incubated with free or encapsulated curcumin (5, 10 µM) for 4 h according to Szlasa et al. [12] and Ambreen et al. [45] observations in FBS-free culture medium. Then the wells were washed twice with DPBS, fresh medium was added, and irradiation was performed using a halogen lamp (Penta Lamps, Teclas, Lugano, Switzerland) with the radiation power consistency set to 20 mW/cm^2. The cells were irradiated for 2 min (2.5 J/cm^2). The blue light (380−500 nm) was chosen to achieve the photodynamic effect (the light absorption peak of curcumin of 410 nm). Cells involved in curcumin and PDT treatment were protected from light at all times. After 24 h from irradiation, experiments were conducted according to the protocols.

4.5. Cell Viability Assay

The MTT assay is a colorimetric assay used to measure cellular metabolic activity to indicate cell viability, proliferation, and cytotoxicity. In the MTT assay, living cells transform yellow tetrazolium salt MTT into purple formazan crystals. This process is possible because living cells have an enzyme-mitochondrial dehydrogenase, which causes this change.

Cells were seeded at 8 × 10^4 in 96-well culture plates and cultured as mentioned in the experiment description with curcumin and liposomal curcumin for 4 h in the dark. Different doses of curcumin and liposomal curcumin were experimentally established for the next experiments on MUG-Mel2, SCC-25, and HaCaT to obtain IC50. The MTT assay was performed after 24 h from irradiation. The MTT solution was added to the wells in a final concentration of 1 mg/mL for 3 h. Next, formazan dye was solubilized with 50 µL DMSO for 15 min. Absorbance was measured at 490 nm in BioTek Well-plate Reader (Winooski, VT, USA). The control group absorbance was 100%, whereas treated samples' cell viability was counted using the formula: % = (A of experimental wells/A of the control wells) × 100%. After preliminary studies with different curcumin and liposomal-derivative doses (1, 2, 5, 10 μM) for the MTT assay, curcumin and liposomal curcumin was chosen in doses of 5 and 10 μM.

4.6. Wound-Healing Assay

A wound-healing assay was used to inquire cells' interactions and cell migration. According to the manufacturer's instructions, a wound-healing assay was made with the Culture-Insert 2 Well in μ-Dish 35 mm (Ibidi, Germany). The cells were seeded to achieve the monolayer in both parts of the insert. Following liposomal curcumin mediated PDT, the inserts were removed, the culture medium was exchanged, and the cells were cultured until about 100% confluency was reached in control cells. Control samples were without treatment at all. The photographs were taken after removal of the inserts at a time point 0 h and after 24 h of incubation by using a light microscope with a 10× magnifying objective (Olympus IX73 with a camera and CellSens Programme, Hamburg, Germany).

4.7. Flow Cytometry-Apoptosis Assay

Cells were drawn from each of the wells and transferred to Eppendorf tubes. Afterward, cells were centrifuged with PBS washing (7 min, 20 °C, 1000× *g*). The supernatant was gently removed and 1 mL of the Binding Buffer per 1 × 10^6 cells was added. For the next step, 4 μL AAD-7 and 8 μL FITC was added to each sample, according to the manufacturer's instruction. Eppendorf tubes were vortexed and incubated without the light for 15 min at room temperature. After incubation time, samples were analyzed with a flow cytometer using the FICT channel for Annexin 5 and PC5.5 channel for AAD-7 (Cytoflex, Beckman Coulter Life Sciences, Indianapolis, IN, USA). Negative samples were prepared without the staining and samples stained with one fluorochrome were used for compensation.

4.8. Immunocytochemistry (ICC) Staining for Apoptosis Detection

Cells were fixed with 4% paraformaldehyde for 10 min at room temperature, and then rinsed 2 × 5 min with PBS. Next, cells were blocked with endogenous peroxidase for 10 min using Peroxidase Blocking Reagent and rinsed with PBS 2 × 4 min. Non-specific proteins were blocked by Protein Block Serum-Free Ready to Use for 1 h. Following serum excess removal, anti-Bax and anti-Bcl-2 primary antibodies (Sigma-Aldrich) in dilution 1:200 were added on the slides for overnight incubation. Afterward, primary antibodies were rinsed with PBS for 2 × 4 min. A secondary rabbit antibody (Abcam, UK) in dilution 1:500 was added for 1 h at room temperature. After incubation time, cells were rinsed with PBS for 2 × 4 min and DAB Substrate in Chromogen Solution was added for 2–5 min until the light brown color was achieved. Cells were rinsed with distilled water for 2 × 4 min, and then hematoxylin was used for 1–2 min to stain cell nuclei. Next, cells were rinsed with tap water 2 × 5 min. The Fluoromount™ Aqueous Mounting Medium (Sigma Aldrich) was added onto the slides, and, the following day, the photographs were taken under the microscope (Olympus BX34 with camera DP74 and CellSens Programme, Hamburg, Germany). All ICC reagents were purchased from DAKO, Agilent (Glostrup, Denmark).

4.9. Statistical Analysis

All experiments were performed in triplicates and the values are presented as a mean ± standard deviation. Analysis between the groups was conducted using the non-parametric test Kruskal-Wallis for abnormal distributed data. a p-value below 0.05 was considered significant. PQStat Programme, version 1.8.2 (PQStat Software, Poland) was used for the calculations.

5. Conclusions

In conclusion, natural plant derivative-curcumin encapsulated in liposomes has been confirmed as a viable photosensitizer in PDT of skin cancer cell lines.

Improved bioavailability and increased stability revealed potent anti-cancer activity in squamous cell carcinoma and melanoma cell lines. The encapsulated compound preferentially accumulated in malignant skin cells. Contrarily, it showed decreased phototoxicity in normal skin keratinocytes HaCaT cells after PDT treatment. These results collectively support liposomal curcumin as a potential photosensitizer in developing natural-based photosensitizers that improve photodynamic therapy safety and efficacy. Thus, additional in vitro and in vivo studies on different normal and cancer cells are essential to confirm this less toxic natural plant derivative PS in the PDT approach.

Author Contributions: Conceptualization, M.W. and P.Z. Methodology, M.W. and M.N. Validation, M.N., K.W., and A.L. Formal analysis, M.W. Investigation, M.W., M.N., A.L., I.J., and K.S. Writing—original draft preparation, M.W., M.N., P.Z., K.W., A.L., A.G., and J.G. Writing—review and editing, M.W. Supervision, M.W., J.G., and P.Z. Project administration, M.W. Funding acquisition, M.W. All authors have read and agreed to the published version of the manuscript.

Funding: This research was funded by the Wroclaw Medical University, grant number SUB.A010.20.13.

Institutional Review Board Statement: Not applicable.

Informed Consent Statement: Not applicable.

Data Availability Statement: Data are contained within the article.

Conflicts of Interest: The authors declare no conflict of interest.

Sample Availability: Samples of the compounds (curcumin) are available from the authors.

References

1. Leiter, U.; Keim, U.; Garbe, C. Epidemiology of skin cancer: Update. In *Advances in Experimental Medicine and Biology*; Springer: Berlin, Germany, 2020; Volume 1268, pp. 123–139.
2. Cullen, J.K.; Simmons, J.L.; Parsons, P.G.; Boyle, G.M. Topical treatments for skin cancer. *Adv. Drug Deliv. Rev.* **2020**, *153*, 54–64. [CrossRef]
3. Nehal, K.S.; Bichakjian, C.K. Update on Keratinocyte Carcinomas. *N. Engl. J. Med.* **2018**, *379*, 363–374. [CrossRef]
4. Ma, Y.; Qu, S.; Xu, L.; Lu, H.; Li, B. An in vitro study of the effect of 5-ALA-mediated photodynamic therapy on oral squamous cell carcinoma. *BMC Oral Health* **2020**, *20*, 258. [CrossRef]
5. Akasov, R.A.; Sholina, N.V.; Khochenkov, D.A.; Alova, A.V.; Gorelkin, P.V.; Erofeev, A.S.; Generalova, A.N.; Khaydukov, E.V. Photodynamic therapy of melanoma by blue-light photoactivation of flavin mononucleotide. *Sci. Rep.* **2019**, *9*. [CrossRef]
6. Allegra, A.; Pioggia, G.; Tonacci, A.; Musolino, C.; Gangemi, S. Oxidative stress and photodynamic therapy of skin cancers: Mechanisms, challenges and promising developments. *Antioxidants* **2020**, *9*, 448. [CrossRef] [PubMed]
7. Bacellar, I.O.L.; Tsubone, T.M.; Pavani, C.; Baptista, M.S. Photodynamic efficiency: From molecular photochemistry to cell death. *Int. J. Mol. Sci.* **2015**, *16*, 20523–20559. [CrossRef] [PubMed]
8. Kessel, D.; Oleinick, N.L. Cell Death Pathways Associated with Photodynamic Therapy: An Update. *Photochem. Photobiol.* **2018**, *94*, 213–218. [CrossRef] [PubMed]
9. Donohoe, C.; Senge, M.O.; Arnaut, L.G.; Gomes-da-Silva, L.C. Cell death in photodynamic therapy: From oxidative stress to anti-tumor immunity. *Biochim. Biophys. Acta Rev. Cancer* **2019**, *1872*, 188308. [CrossRef] [PubMed]
10. Simões, J.C.S.; Sarpaki, S.; Papadimitroulas, P.; Therrien, B.; Loudos, G. Conjugated Photosensitizers for Imaging and PDT in Cancer Research. *J. Med. Chem.* **2020**, *63*, 14119–14150. [CrossRef]
11. He, Y.; Yue, Y.; Zheng, X.; Zhang, K.; Chen, S.; Du, Z. Curcumin, inflammation, and chronic diseases: How are they linked? *Molecules* **2015**, *20*, 9183–9213. [CrossRef] [PubMed]
12. Szlasa, W.; Supplitt, S.; Drąg-Zalesińska, M.; Przystupski, D.; Kotowski, K.; Szewczyk, A.; Kasperkiewicz, P.; Saczko, J.; Kulbacka, J. Effects of curcumin based PDT on the viability and the organization of actin in melanotic (A375) and amelanotic melanoma (C32)– in vitro studies. *Biomed. Pharmacother.* **2020**, *132*. [CrossRef]
13. Mansoori, B.; Mohammadi, A.; Amin Doustvandi, M.; Mohammadnejad, F.; Kamari, F.; Gjerstorff, M.F.; Baradaran, B.; Hamblin, M.R. Photodynamic therapy for cancer: Role of natural products. *Photodiagn. Photodyn. Ther.* **2019**, *26*, 395–404. [CrossRef]
14. Nistico, S.; Tamburi, F.; Bennardo, L.; Dastoli, S.; Schipani, G.; Caro, G.; Fortuna, M.C.; Rossi, A. Treatment of telogen effluvium using a dietary supplement containing Boswellia serrata, Curcuma longa, and Vitis vinifera: Results of an observational study. *Dermatol. Ther.* **2019**, *32*, e12842. [CrossRef]
15. Giordano, A.; Tommonaro, G. Curcumin and cancer. *Nutrients* **2019**, *11*, 2376. [CrossRef] [PubMed]
16. Patel, S.S.; Acharya, A.; Ray, R.S.; Agrawal, R.; Raghuwanshi, R.; Jain, P. Cellular and molecular mechanisms of curcumin in prevention and treatment of disease. *Crit. Rev. Food Sci. Nutr.* **2020**, *60*, 887–939. [CrossRef]

17. Chen, Y.; Wu, Q.; Zhang, Z.; Yuan, L.; Liu, X.; Zhou, L. Preparation of curcumin-loaded liposomes and evaluation of their skin permeation and pharmacodynamics. *Molecules* **2012**, *17*, 5972. [CrossRef]
18. Rata, D.M.; Cadinoiu, A.N.; Atanase, L.I.; Popa, M.; Mihai, C.T.; Solcan, C.; Ochiuz, L.; Vochita, G. Topical formulations containing aptamer-functionalized nanocapsules loaded with 5-fluorouracil-An innovative concept for the skin cancer therapy. *Mater. Sci. Eng. C* **2021**, *119*. [CrossRef]
19. Cadinoiu, A.N.; Rata, D.M.; Atanase, L.I.; Daraba, O.M.; Gherghel, D.; Vochita, G.; Popa, M. Aptamer-functionalized liposomes as a potential treatment for basal cell carcinoma. *Polymers (Basel)* **2019**, *11*, 1515. [CrossRef]
20. Rizvi, I.; Obaid, G.; Bano, S.; Hasan, T.; Kessel, D. Photodynamic therapy: Promoting in vitro efficacy of photodynamic therapy by liposomal formulations of a photosensitizing agent. *Lasers Surg. Med.* **2018**. [CrossRef] [PubMed]
21. Kiesslich, T.; Berlanda, J.; Plaetzer, K.; Krammer, B.; Berr, F. Comparative characterization of the efficiency and cellular pharmacokinetics of Foscan®- and Foslip®-based photodynamic treatment in human biliary tract cancer cell lines. *Photochem. Photobiol. Sci.* **2007**, *6*, 619–627. [CrossRef] [PubMed]
22. Ghalandarlaki, N.; Alizadeh, A.M.; Ashkani-Esfahani, S. Nanotechnology-applied curcumin for different diseases therapy. *Biomed. Res. Int.* **2014**, *2014*, 1–23. [CrossRef] [PubMed]
23. Castro, K.A.D.F.; Costa, L.D.; Prandini, J.A.; Biazzotto, J.C.; Tomé, A.C.; Hamblin, M.R.; da Graça, P.M.S.; Neves, M.; Faustino, M.A.F.; da Silva, R.S. The Photosensitizing Efficacy of Micelles Containing a Porphyrinic Photosensitizer and KI against Resistant Melanoma Cells. *Chem. A Eur. J.* **2021**. [CrossRef]
24. Fan, Y.; Zhang, Q. Development of liposomal formulations: From concept to clinical investigations. *Asian J. Pharm. Sci.* **2013**. [CrossRef]
25. Mufamadi, M.S.; Pillay, V.; Choonara, Y.E.; Du Toit, L.C.; Modi, G.; Naidoo, D.; Ndesendo, V.M.K. A Review on Composite Liposomal Technologies for Specialized Drug Delivery. *J. Drug Deliv.* **2011**. [CrossRef]
26. Lammers, T.; Kiessling, F.; Hennink, W.E.; Storm, G. Drug targeting to tumors: Principles, pitfalls and (pre-) clinical progress. *J. Control. Release* **2012**, *161*, 175–187. [CrossRef]
27. Nichols, J.W.; Bae, Y.H. Odyssey of a cancer nanoparticle: From injection site to site of action. *Nano Today* **2012**, *7*, 606–618. [CrossRef]
28. Yallapu, M.M.; Jaggi, M.; Chauhan, S.C. Curcumin nanoformulations: A future nanomedicine for cancer. *Drug Discov. Today* **2012**, *17*, 71–80. [CrossRef]
29. Naksuriya, O.; Okonogi, S.; Schiffelers, R.M.; Hennink, W.E. Curcumin nanoformulations: A review of pharmaceutical properties and preclinical studies and clinical data related to cancer treatment. *Biomaterials* **2014**, *35*, 3365–3383. [CrossRef] [PubMed]
30. Siviero, A.; Gallo, E.; Maggini, V.; Gori, L.; Mugelli, A.; Firenzuoli, F.; Vannacci, A. Curcumin, a golden spice with a low bioavailability. *J. Herb. Med.* **2015**, *5*, 57–70. [CrossRef]
31. Mahmud, M.; Piwoni, A.; Filiczak, N.; Janicka, M.; Gubernator, J. Long-circulating curcumin-loaded liposome formulations with high incorporation efficiency, stability and anticancer activity towards pancreatic adenocarcinoma cell lines in vitro. *PLoS ONE* **2016**, *11*, e0167787. [CrossRef] [PubMed]
32. Larue, L.; Myrzakhmetov, B.; Ben-Mihoub, A.; Moussaron, A.; Thomas, N.; Arnoux, P.; Baros, F.; Vanderesse, R.; Acherar, S.; Frochot, C. Fighting hypoxia to improve PDT. *Pharmaceuticals* **2019**, *12*, 163. [CrossRef]
33. Kwiatkowski, S.; Knap, B.; Przystupski, D.; Saczko, J.; Kędzierska, E.; Knap-Czop, K.; Kotlińska, J.; Michel, O.; Kotowski, K.; Kulbacka, J. Photodynamic therapy–mechanisms, photosensitizers and combinations. *Biomed. Pharmacother.* **2018**, *106*, 1098–1107. [CrossRef] [PubMed]
34. Zhang, L.J.; O'Shea, D.; Zhang, C.Y.; Yan, Y.J.; Wang, L.; Chen, Z.L. Evaluation of a bacteriochlorin-based photosensitizer's anti-tumor effect in vitro and in vivo. *J. Cancer Res. Clin. Oncol.* **2015**, *141*, 1921–1930. [CrossRef] [PubMed]
35. Narsireddy, A.; Vijayashree, K.; Adimoolam, M.G.; Manorama, S.V.; Rao, N.M. Photosensitizer and peptide-conjugated PAMAM dendrimer for targeted in vivo photodynamic therapy. *Int. J. Nanomed.* **2015**, *10*, 6865–6878.
36. Chen, J.J.; Hong, G.; Gao, L.J.; Liu, T.J.; Cao, W.J. In vitro and in vivo antitumor activity of a novel porphyrin-based photosensitizer for photodynamic therapy. *J. Cancer Res. Clin. Oncol.* **2015**, *141*, 1553–1561. [CrossRef] [PubMed]
37. Ghosh, S.; Carter, K.A.; Lovell, J.F. Liposomal formulations of photosensitizers. *Biomaterials* **2019**, *218*, 119341. [CrossRef] [PubMed]
38. Yakavets, I.; Millard, M.; Lamy, L.; Francois, A.; Scheglmann, D.; Wiehe, A.; Lassalle, H.P.; Zorin, V.; Bezdetnaya, L. Matryoshka-type liposomes offer the improved delivery of temoporfin to tumor spheroids. *Cancers (Basel)* **2019**, *11*, 1366. [CrossRef] [PubMed]
39. He, G.; Mu, T.; Yuan, Y.; Yang, W.; Zhang, Y.; Chen, Q.; Bian, M.; Pan, Y.; Xiang, Q.; Chen, Z.; et al. Effects of notch signaling pathway in cervical cancer by curcumin mediated photodynamic therapy and its possible mechanisms in vitro and in vivo. *J. Cancer* **2019**. [CrossRef]
40. Basniwal, R.K.; Khosla, R.; Jain, N. Improving the anticancer activity of curcumin using nanocurcumin dispersion in water. *Nutr. Cancer* **2014**. [CrossRef]
41. Swami Vetha, B.S.; Oh, P.S.; Kim, S.H.; Jeong, H.J. Curcuminoids encapsulated liposome nanoparticles as a blue light emitting diode induced photodynamic therapeutic system for cancer treatment. *J. Photochem. Photobiol. B Biol.* **2020**, *205*. [CrossRef]
42. Duse, L.; Pinnapireddy, S.R.; Strehlow, B.; Jedelská, J.; Bakowsky, U. Low level LED photodynamic therapy using curcumin loaded tetraether liposomes. *Eur. J. Pharm. Biopharm.* **2018**, *126*, 233–241. [CrossRef] [PubMed]

43. Nasery, M.M.; Abadi, B.; Poormoghadam, D.; Zarrabi, A.; Keyhanvar, P.; Khanbabaei, H.; Ashrafizadeh, M.; Mohammadinejad, R.; Tavakol, S.; Sethi, G. Curcumin delivery mediated by bio-based nanoparticles: A review. *Molecules* **2020**, *25*, 689. [CrossRef] [PubMed]
44. Chang, M.; Wu, M.; Li, H. Antitumor activities of novel glycyrrhetinic acid-modified curcumin-loaded cationic liposomes in vitro and in H22 tumor-bearing mice. *Drug Deliv.* **2018**. [CrossRef] [PubMed]
45. Ambreen, G.; Duse, L.; Tariq, I.; Ali, U.; Ali, S.; Pinnapireddy, S.R.; Bette, M.; Bakowsky, U.; Mandic, R. Sensitivity of papilloma virus-associated cell lines to photodynamic therapy with curcumin-loaded liposomes. *Cancers (Basel)* **2020**, *12*, 3278. [CrossRef]
46. Colombo, I.; Sangiovanni, E.; Maggio, R.; Mattozzi, C.; Zava, S.; Corbett, Y.; Fumagalli, M.; Carlino, C.; Corsetto, P.A.; Scaccabarozzi, D.; et al. HaCaT Cells as a Reliable in Vitro Differentiation Model to Dissect the Inflammatory/Repair Response of Human Keratinocytes. *Mediat. Inflamm.* **2017**. [CrossRef]
47. Micallef, L.; Belaubre, F.; Pinon, A.; Jayat-Vignoles, C.; Delage, C.; Charveron, M.; Simon, A. Effects of extracellular calcium on the growth-differentiation switch in immortalized keratinocyte HaCaT cells compared with normal human keratinocytes. *Exp. Dermatol.* **2009**. [CrossRef]
48. Popovic, A.; Wiggins, T.; Davids, L.M. Differential susceptibility of primary cultured human skin cells to hypericin PDT in an in vitro model. *J. Photochem. Photobiol. B Biol.* **2015**. [CrossRef]
49. Rocha, L.B.; Soares, H.T.; Mendes, M.I.P.; Cabrita, A.; Schaberle, F.A.; Arnaut, L.G. Necrosis Depth and Photodynamic Threshold Dose with Redaporfin-PDT. *Photochem. Photobiol.* **2020**, *96*, 692–698. [CrossRef]

pharmaceuticals

MDPI

Review

Photodynamic Therapy Combined with Antibiotics or Antifungals against Microorganisms That Cause Skin and Soft Tissue Infections: A Planktonic and Biofilm Approach to Overcome Resistances

Vanesa Pérez-Laguna [1,*], Isabel García-Luque [2], Sofía Ballesta [2], Antonio Rezusta [3,4,5,†] and Yolanda Gilaberte [3,6,†]

1 Max Planck Institute for Evolutionary Biology, 24306 Plön, Germany
2 Department of Microbiology, School of Medicine, University of Sevilla and Instituto de Biomedicina de Sevilla (IBIS), 41009 Sevilla, Spain; igarcial@us.es (I.G.-L.); mudarra@us.es (S.B.)
3 Aragon Health Research Institute (IIS Aragón), 50009 Zaragoza, Spain; arezusta@unizar.es (A.R.); ygilaberte@salud.aragon.es (Y.G.)
4 Department of Microbiology, Miguel Servet University Hospital, 50009 Zaragoza, Spain
5 Department of Microbiology, Preventive Medicine and Public Health, University of Zaragoza, 50001 Zaragoza, Spain
6 Department of Dermatology, Miguel Servet University Hospital, 50009 Zaragoza, Spain
* Correspondence: perezlaguna@evolbio.mpg.de
† These authors contributed equally to this work.

Citation: Pérez-Laguna, V.; García-Luque, I.; Ballesta, S.; Rezusta, A.; Gilaberte, Y. Photodynamic Therapy Combined with Antibiotics or Antifungals against Microorganisms That Cause Skin and Soft Tissue Infections: A Planktonic and Biofilm Approach to Overcome Resistances. *Pharmaceuticals* **2021**, *14*, 603. https://doi.org/10.3390/ph14070603

Academic Editor: Serge Mordon

Received: 15 May 2021
Accepted: 16 June 2021
Published: 23 June 2021

Abstract: The present review covers combination approaches of antimicrobial photodynamic therapy (aPDT) plus antibiotics or antifungals to attack bacteria and fungi in vitro (both planktonic and biofilm forms) focused on those microorganisms that cause infections in skin and soft tissues. The combination can prevent failure in the fight against these microorganisms: antimicrobial drugs can increase the susceptibility of microorganisms to aPDT and prevent the possibility of regrowth of those that were not inactivated during the irradiation; meanwhile, aPDT is effective regardless of the resistance pattern of the strain and their use does not contribute to the selection of antimicrobial resistance. Additive or synergistic antimicrobial effects in vitro are evaluated and the best combinations are presented. The use of combined treatment of aPDT with antimicrobials could help overcome the difficulty of fighting high level of resistance microorganisms and, as it is a multi-target approach, it could make the selection of resistant microorganisms more difficult.

Keywords: photoinactivation; resistance; antimicrobial agents; SSTI; MDR

1. Introduction

1.1. The Problem of Skin and Soft Tissue Infections

Skin and soft tissue infections (SSTIs) are defined as clinical entities of variable presentation, etiology, and severity that involve microbial invasion of the layers of the skin and underlying soft tissues. Related to clinical manifestations of the wide range of pathologies they represent, the minimum diagnostic criteria are erythema, edema, warmth, and pain or tenderness. The affected area may also become dysfunctional depending on the severity of the infection, and, much more relevantly, patient comorbidity can easily transform a normally mild infection into a rapidly advancing threat to life [1–3]. Complicated forms of SSTI (cSSTI) may need, apart from antibiotic or antifungal treatment, surgical procedures, or have significant underlying co-morbidities such as diabetes, systemic immunosuppression, or neurological diseases [4–6].

Their treatment has become more challenging because of the increase in the frequency and severity of infections mainly due to the ageing of the general population, the increased number of critical and immunocompromised patients, and because of the emergence of resistance to many of the antimicrobial agents commonly used to treat SSTIs in the past [7–9].

1.2. Antimicrobial Resistance in Skin and Soft Tissue Infections Causal Agents

Antimicrobial resistance (AMR) makes treatments more tedious and it has adverse consequences such as prolonged hospitalization, increased medical expenses, overburdened public health system, and even increased mortality rates [10].

The main causative agent of SSTI is *Staphylococcus aureus*, which is one of the bacteria most frequently involved in the problem of AMR. Other causal agents of a high percentage of SSTI, also involved in AMR problem, are *Enterobacteriaceae* and *Pseudomonas aeruginosa* (e.g., 45.9% of hospital-acquired SSTI in North America are caused by *S. aureus*, 10.8% by *P. aeruginosa*, and 8.2% by *Enterococcus* spp.) [1]. In fact, in 2017 the World Health Organization (WHO) published the first list of bacteria for which new antibiotics were urgently needed [11] (Table 1). Carbapenem-resistant or third generation cephalosporin-resistant *Enterobacteriaceae* (e.g., *Escherichia coli*) and carbapenem-resistant *P. aeruginosa* are classified in this list within the most dangerous class (Priority 1: critical) and methicillin-resistant or vancomycin intermediate and resistant *S. aureus* is classified as "Priority 2: high".

Table 1. Global priority list of antibiotic-resistant bacteria (Adapted from: https://www.who.int/medicines/publications/WHO-PPL-Short_Summary_25Feb-ET_NM_WHO.pdf?ua=1 (accessed on 20 June 2021)).

Priority 1: CRITICAL
-*Acinetobacter baumannii*, carbapenem-resistant -*Pseudomonas aeruginosa*, carbapenem-resistant -Enterobacteriaceae (*Klebsiella pneumonia, Escherichia coli, Enterobacter* spp., *Serratia* spp., *Proteus* spp., and *Providencia* spp., *Morganella* spp.), carbapenem-resistant, 3rd generation cephalosporin-resistant
Priority 2: HIGH
-*Enterococcus faecium*, vancomycin-resistant -*Staphylococcus aureus*, methicillin-resistant, vancomycin intermediate and resistant -*Helicobacter pylori*, clarithromycin-resistant -*Campylobacter*, fluoroquinolone-resistant -*Salmonella* spp., fluoroquinolone-resistant -*Neisseria gonorrhoeae*, 3rd generation cephalosporin-resistant, fluoroquinolone-resistant
Priority 3: MEDIUM
-*Streptococcus pneumoniae*, penicillin-non-susceptible -*Haemophilus influenzae*, ampicillin-resistant -*Shigella* spp., fluoroquinolone-resistant

Carbapenem-resistant *Acinetobacter baumannii* completes the list of bacteria classified as "Priority 1". *A. baumannii*-associated SSTI is an emerging infection in patients who experience trauma; although it causes few cases, usually they are severe [12].

Regarding atypical mycobacteria, which also cause SSTI on certain occasions, they are highly resistant to antibacterial treatments due to the special characteristics of their wall (the mycobacterial cell has four main layers: (i) the plasma membrane, (ii) the peptidoglycan–arabinogalactan complex, (iii) an asymmetrical outer membrane covalently linked with mycolic acids, and (iv) an outermost capsule). They are a globally established priority for which innovative new treatments are urgently needed according to the WHO [11].

On the other hand, the recent emergence of fungi that are resistant to more than one class of antifungal drug is a serious concern, especially because currently only three primary classes of agents are used to treat invasive fungal infections: (1) Azoles (Fluconazole, voriconazole, posaconazole), (2) Echinocandins (Caspofungin, micafungin, anidulafungin) and (3) Polyne (Amphotericin B). *Candida glabrata*, *Candida krusei* and *Candida auris* are species with intrinsic or high rates of resistance against the first, *Cryptococcus* spp. and

Fusarium spp. against the second, and *Candida auris* and *Aspergillus terreus* against the last [13].

With regard to yeasts that cause SSTI, *Candida* spp. stand out as causative agents. They are implicated in the AMR problem due to the increasingly frequent existence of fluconazole resistant strains [13,14].

Finally, dermatophytes and non-dermatophytes molds cause cutaneous and subcutaneous infections than are often chronic and recalcitrant. Invasive infections are rare but occur especially in immunocompromised and debilitated individuals. Its treatment is a great challenge especially in these patients. [15,16].

1.3. Antimicrobial Photodynamic Therapy Combined with Antibiotics or Antifungals to Treat SSTIs

Antimicrobial photodynamic therapy (aPDT) is based on the use of photosensitizer molecules (PS) that are activated by light in the presence of oxygen. Reactive oxygen species (ROS) are generated in the photodynamic reaction resulting in a toxic effect for microorganisms that is capable of destroying them (Figure 1). Hence, aPDT has been proposed as an alternative treatment for SSTIs [17,18].

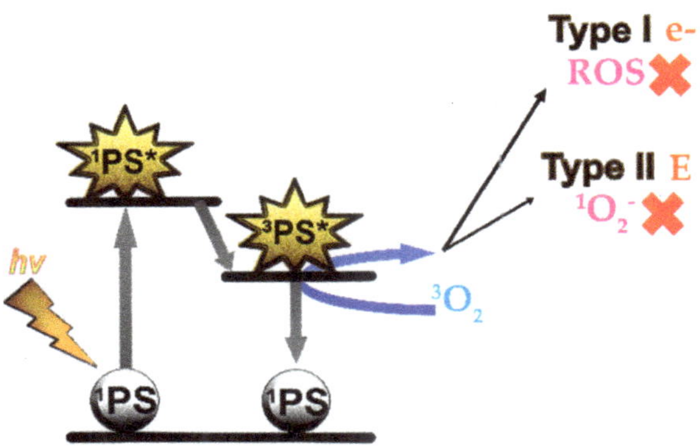

Figure 1. Mechanism of action of antimicrobial photodynamic therapy: photosensitizer molecules (PS) in ground state (^{1}PS) are activated by light (*hv*) of a wavelength that matches the absorption wavelength of the molecule. PS reaches an excited state, first singlet (^{1}PS*), very unstable, and rapidly triplet (^{3}PS*). This reacts with the oxygen by transferring electrons (Type I) or energy (Type II) in its reversion to the ground state (^{1}PS*). Type I reaction is characterized by the formation of reactive oxygen species (ROS) and Type II by the formation of singlet oxygen ($^1O_2^-$). These molecules are highly reactive and are capable of damaging cells resulting in a toxic effect for microorganisms (Adapted with permission from ref. [18], 2018, Pérez-Laguna et al.).

Among the advantages of the aPDT is the photoinactivation of the microorganisms independently of their pattern of AMR, without its use contributing to the selection of drug-resistant strains, and its broad spectrum of activity; therefore it does not require a precise microbial diagnosis and it is very useful for infections caused by several microorganisms [19,20]. By contrast, the most import limitation is the possibility of microbial regrowth by those who have not been inactivated during irradiation [17,18,21]. The combination of aPDT with antibiotic or antifungal treatment is a promising approach to the fight against infectious diseases due to aPDT's ability to increase the susceptibility of microorganisms to antimicrobial drugs and minimize the possibility of the regrowth after aPDT [22–24].

1.4. *Objective*

We present a review of articles related to in vitro activities of aPDT in combination with antibiotics or antifungals, focusing on microorganisms that cause SSTIs. The aim is to combine the information that exists about each microorganism (gather the bases of knowledge). This review serves as a starting point for new combined treatment research (expand and apply knowledge).

The questions that are intended to be answered are: (1) Which microorganisms that cause SSTIs have been exposed in vitro to combinations of aPDT and antimicrobials? (2) Which methodologies have been used? (3) Are there an additive or synergistic antimicrobial effect? (4) Which are the best combinations clinically and microbiologically?

2. Methodology

2.1. *Eligibility Criteria*

We have considered in vitro studies that used antimicrobial agents (antibiotics or antifungals) plus aPDT against microorganisms that cause SSTIs. The specific requirements for inclusion of the studies were (1) in vitro studies in planktonic state or forming biofilm and ex vivo studies on the skin or mucous membrane; (2) aimed to inactivate identified bacteria, or fungi that cause SSTIs; (3) used antibiotics or antifungals as a fundamental part of the treatment; (4) used aPDT as a fundamental part of the treatment; and (5) published in indexed journals and written in English or Spanish.

2.2. *Study Selection, Data Collection Process, and Characteristics*

The keywords used for the search in Pubmed and Embase library databases were: photodynamic therapy, PDT, antimicrobial photodynamic therapy, aPDT, photodynamic antimicrobial chemotherapy, PACT, photoinactivation, photodynamic inactivation, PDI, combination, combined treatment, antimicrobial agents, antibiotics, and antifungals. The last search was carried out on 1 June 2021 and no time limits in the past were applied. Research papers that meet the eligibility criteria (Section 2.1) were included. A huge number of studies contained the keywords; nevertheless, after applying the eligibility criteria, the number was drastically reduced to a total of 33.

The data recapitulated were: (1) etiological agent of SSTIs; (2) type of study: in vitro (planktonic or forming biofilm) and ex vivo; (3) antimicrobial methodology: antibiotics or antifungals used and their application and concentration; (4) aPDT methodology: PS used and parameters of irradiation (source type, wavelength, intensity, and fluence); (5) effect of combined treatment against microorganism.

The 33 included studies were grouped depending on the etiological agent of SSTIs. The structure used is as follows:

-Gram-positive bacteria: *Staphylococcus* spp. (Section 3.1: Section 3.1.1. S. aureus and Section 3.1.2. *Staphylococcus epidermidis* and *Staphylococcus hemolytic)*

-Mycobacteria: *Mycobacterium fortuitum* (Section 3.2.)

-Gram-negative bacteria: *E. coli, P. aeruginosa*, and *A. baumannii* (Section 3.3, Section 3.4, Section 3.5 respectively)

-Fungal infectious agents: Yeast (Section 3.6. *Candida* spp.), dermatophytes, and molds (Section 3.7).

3. Results of Studies on In Vitro aPDT Combined with Antimicrobial Agents against Infectious Microorganism of Skin and Soft Tissues

3.1. *Staphylococcus* spp.

3.1.1. Staphylococcus aureus

S. aureus, frequently involved in the problem of AMR, is the main causative agent of SSTI and represents one of the most important pathogens involved in cSSTI [2,25,26]. For this reason, this is the bacteria with more studies trying the combination of different antibiotics and PS published so far. Table 5 summarizes the studies concerning in vitro aPDT plus other antimicrobial treatments against *S. aureus*.

Porphycene and Porphyrin Studies

To our knowledge, one of the most recent articles about aPDT and antimicrobials' combination was published by Nieves et al. They report the synthesis of a new porphycene namely 2-aminothiazolo[4,5-c]-2,7,12,17-tetrakis(methoxyethyl)porphycene (ATAZTMPo)-gentamicin conjugate. It outperforms the antimicrobial effect against *S. aureus* and *E. coli* of the two components delivered separately. This novel photoantimicrobial agent may be used to enhance the therapeutic index of gentamicin, broaden the spectrum of pathogens against which it is effective, and reduce its side effects [27].

The combination of gentamicin with porphyrins also seems to be effective against *S. aureus* [26,28].

5-aminolevulinic acid (5-ALA), a pre-drug that, once metabolized by proliferating bacteria, is converted into the natural PS protoporphyrin IX (PpIX), combined with a low concentration of gentamycin (2 μg/mL) improved qualitative and quantitative antibacterial effects against *S. aureus* biofilms. The proposed explanation was that photoactivation generated ROS which damages or kills the cells, while gentamicin, even at low doses, completes the eradication. The mechanism of action of gentamicin is based on its capacity to bind tightly to the 30S subunit of the bacterial ribosome impairing protein synthesis and the bacterial cells die. The fundamental requirement for gentamicin to act as an antibiotic is the need to penetrate the membrane of the bacterial cell, and aPDT can possibly damage the membrane thus increasing antibiotic penetration [28].

Deuteroporphyrin-aPDT combined with different antibiotics included gentamicin, vancomycin, rifampicin, fusidic acid, and oxacillin did not detect synergistic effect except for oxacillin. They concluded that only the combination of deuteroporphyrin-aPDT plus oxacillin had potential for aPDT to improve traditional antibiotic treatment with cell wall synthesis inhibitors [29].

Regarding the combination of porphyrins with other antibiotics different from gentamycin, when ALA-aPDT was combined with netilmicin, vancomycin or cefaclor against *S. aureus* biofilms, synergistic bactericidal effect was observed Destruction occurred predominantly in the upper layer of the biofilm, and in a strain-dependent manner. That is why they suggested that a drug sensitivity test should be performed in advance [30].

Meso-tetrakis(*N*-methyl-4-pyridyl) porphine tetra-tosylate (*TMP*), a cationic porphyrin combined with vancomycin, was highly effective against *S. aureus* biofilms. It seems that the combination blocks cell wall synthesis, and the damaged biofilms may be more susceptible to host defenses which could be useful for biofilms adhering to medical implant surfaces [31].

In contrast, the combination of other cationic porphyrin, 5.10,15,20-tetrakis[4-(3-*N,N*-dimethylaminopropoxy) phenyl] porphyrin (TAPP)-aPDT with the cell wall inhibiting antibiotics ceftriaxone or vancomycin did not significantly increase the aPDT effect on planktonic *S. aureus*. However, the bactericidal effect of TAPP-aPDT was additive with the protein synthesis inhibitors chloramphenicol and tobramycin against *S. aureus* and *E. coli*, and was synergistic against methicillin resistant *S. aureus* (MRSA) and *S. epidermidis*. It seems that vancomycin and ceftriaxone presented limited effects when combined with TAPP-aPDT; nevertheless, tobramycin and chloramphenicol reached additive effects for *S. aureus* and *E. coli* and synergy against MRSA and *S. epidermidis*, even when sub-minimum inhibitory concentration (MIC) levels of TAPP and antibiotics were used. The results suggest that even sub-MIC levels of photo-activated TAPP could be used to boost the activity of waning antibiotics [32].

Tetracationic porphyrin *meso*-tetrakis(1-methylpyridinium)porphyrin (Tetra-Py$^+$-Me)-PDT combined with ampicillin significantly reduced *S aureus* survival in both states (planktonic and biofilm) [33].

Finally, the endogenous porphyrins accumulated by some microorganism, such as *E. coli* or *S. aureus*, combined with different wavelengths of light (blue, 470 nm, or red, 625 nm) in the presence of ciprofloxacin were more effective than the antibiotic alone [34].

Overall, there are multiple promising combinations of antibiotics and porphyrins-PDT which could increase the efficacy in vitro of conventional antibiotics against *S. aureus* (See Table 5).

Phenothiazine Studies

Another chemical group of PS for PDT are phenothiazinium dyes, such as methylene blue (MB) and toluidine blue O (TBO).

MB-aPDT combined with chloramphenicol showed an additive effect against planktonic *S. aureus* [32], while ciprofloxacin plus MB-aPDT was synergistic either in planktonic and in biofilm forms [35]. As usual, bacteria forming biofilms were less affected than bacteria in planktonic phase and a bigger effect was seen when ciprofloxacin was applied after aPDT rather than before or simultaneously. Kashef et al. using MB or TBO-aPDT with linezolid against biofilm resulted in a greater reduction in the viable count of bacteria than either of them separately. TBO-aPDT plus linezolid gave the greatest reduction (although the killing was only <3 log_{10} reduction in viable count for any of the strains) [36].

TBO-aPDT combined with gentamicin and incubated 45-min prior to irradiation showed positive effects against *S. aureus* strains in planktonic state, both in a reference strain and in a multidrug resistant (MDR) clinical isolate; they showed that TBO-aPDT treatment permeabilized the bacterial membranes, promoted gentamicin cellular accumulation and the intracellular ROS generation by the combination was much higher than that of single treatment groups [37].

Our group explored the effect of MB-aPDT, alone or in combination with gentamicin, against planktonic and biofilm *S. aureus*. The addition of gentamicin (concentrations that by themselves do not achieve any effect) caused the complete photoelimination in the case of planktonic *S. aureus*, with a lower MB-PDT dose, whereas it did not produce any change in biofilm [38]. In a previous study, it was proven that concentrations of linezolid or mupirocin which did not harm the bacteria also exert a synergistic effect when they were combined with MB, also reducing the concentration of the PS needed to photoinactivate planktonic *S. aureus* [39]. However, recently we have verified in a superficial abrasion mouse model of *S. aureus* skin infection that the addition of mupirocin to MB-aPDT treatment improved antimicrobial activity but it did not enhance wound healing [40].

Xanthene Studies

Rose Bengal (RB, a xanthene dye) has been rehearsed combined with mupirocin, linezolid, or gentamicin with a synergistic bactericidal effect in planktonic *S. aureus* [39]. Additionally, the combination of RB-aPDT plus gentamicin was evaluated against *S. aureus* biofilms. However, only high concentration of RB (64 μg/mL) and gentamicin (40 μg/mL) showed a synergistic effect against biofilms [41]. It is interesting that the combination of RB-aPDT with methicillin significantly reduced the MIC of methicillin either of MRSA or methicillin sensitive *S. aureus* [26].

3.1.2. *Staphylococcus epidermidis* and *Staphylococcus haemolyticus*

S. epidermidis and *S. haemolyticus* are part of the skin flora of humans but in specific circumstances, such as with immunocompromised or hospitalized patients with catheters or other surgical implants, they are well-known opportunistic pathogens which cause local or systemic infections. The highly antibiotic-resistant phenotype and their ability to form biofilms makes infections difficult to treat [42–45].

The study of Barra et al. has explored 5-ALA-aPDT plus gentamicin at a low concentration against clinical isolates of *S. haemolyticus*, *S. epidermidis*, and *S. aureus* biofilms; they reported a synergistic effect being *S. haemolyticus* the most sensitive to photoinactivation [28].

Other porphyrins, such as TMP-aPDT combined with vancomycin in biofilms [46] or TAPP-aPDT combined with chloramphenicol or tobramycin in planktonic cells, showed a higher efficacy against *S. epidermidis* than any of them separately. The authors hypothesized that TAPP could act in combination with lower concentrations of antibiotics providing a controlled release of the antibiotic, and in consequence improving their efficacy to maintain an antimicrobial effect [32].

In conclusion, multiple antibiotics in combination with different types of PDT based on protoporphyrins showed improved bactericidal effects against planktonic *S. epidermidis* or forming biofilms [28,32,46].

Table 5 includes a summary of all the studies of aPDT combined with antibiotics. Table 2 summarizes in detail the methodology and the results of different studies that reported the in vitro activity of aPDT plus antibiotics against *S. haemolyticus* and *S. epidermidis* (Table 2: Biofilm Studies and biofilm state, planktonic—Table 2: Planktonic Cell Studies).

Table 2. Studies combining in vitro aPDT plus other treatment against *S. haemolyticus* and *S. epidermidis* comparing the methodology and the results biofilm studies, planktonic cell studies.

Biofilm Ttudies											
Target	**PS**	**PS Concentration (μM)**	**Antibiotic**	**Antibiotic Concentration (μg/mL)**	**Source Type**	**Wavelength (nm)**	**Intensity (mw/cm²)**	**Fluence (J/cm²)**	**Inactivation Fraction (%)**	**Synergistic Observed Effect (*)**	**Reference**
S. haemolyticus clinical isolate	5-ALA	40	gentamicin	2	50-LED	630 ± 15	83	250	~70	>inactivation	[28]
S. epidermidis clinical isolate	5-ALA	40	gentamicin	2	50-LED	630 ± 15	83	250	~75	>inactivation	[28]
S. epidermidis RP62A & 5179R	TMP	10	vancomycin	200	tungsten lamp	400–800	166	300	~99.9999	>inactivation	[46]
Planktonic Cell Studies											
Target	**PS**	**PS Concentration (μM)**	**Antibiotic**	**Antibiotic Concentration (mg/L)**	**Preincubation Time (h)**	**Irradiation Time (h)**	**Source Type**	**Wave length**	**Media**	**CFU/200 μL Well**	**Log_{10} Reduction**
S. epidermidis ATCC 35984	TAPP	5	chloram phenicol	2	19	5	100 W, 120 V Sylvania white light	Broad spectrum	TSB	~10^6	~2
S. epidermidis ATCC 35984	TAPP	5	tobramycin	4.5	19	5	100 W, 120 V Sylvania white light	Broad spectrum	TSB	~10^6	~3

(*): the combination causes an increase in percentage of microbial inactivation greater than the sum of the antibacterial activity of the aPDT plus the antibiotic treatment when they are applied alone; 5-ALA: 5-aminolevulinic acid; aPDT: antimicrobial photodynamic therapy; CFU: colony forming unit; LED: light-emitting diode; PS: photosensitizer; TAPP: meso-tetra (4-aminophenyl) porphine; TMP: tetra-substituted N-methyl-pyridyl-porphine; TSB: trypticase soy broth.

3.2. *Mycobacterium fortuitum*

M. fortuitum is an atypical mycobacteria involved in skin infections very difficult to treat, thus usually causes refractory infections [47,48].

The combination of MB-aPDT with ciprofloxacin, moxifloxacin, or amikacin against a clinical isolate had in vitro synergistic antimicrobial effect (the latter also was tested in vivo to treat a model of keratitis in rabbits resulting in significantly less bacterial burden). Sublethal concentrations of antibiotic plus sublethal dosage of MB reached a mycobacterial survival reduction in the colony forming units (CFU) of at least 2 log_{10} lesser compared to the reduction achieved with antibiotics alone, although this effect became insignificant when higher doses of antibiotics were used. The maximum degree of inactivation was achieved by the combination of MB-aPDT with ciprofloxacin or moxifloxacin ($\geq$4 log_{10} reduction, with amikacin (~3 log_{10} reduction) [49]. Table 5 and Table 3 summarize the methodology and the results of this study.

Table 3. Comparison of methodology and results of an in vitro study of aPDT followed by culture with antibiotics on *Mycobacterium fortuitum*.

Strain	PS	PS Concentration (mg/mL)	Antibiotic	Antibiotic Concentration (mg/mL)	Preincubation Time (h)	Source Type	Wave length (nm)	Intensity (mw/cm^2)	Fluence (J/cm^2)	Media/Culture	CFU/200μL Well	Synergistic Observed Effect (*)	Reference
M. fortuitum clinical isolate	MB	50	amikacin	0–0.5	0 + 72 with antib	metal halogen lamp	560–780	100	100	PBS with 0.02% Tween 80 / Muller Hilton	~10^8	>inactivation (≥2 Log_{10} reduction)	[49]
			ciprofloxacin hydrochloride	0–0.06									
			moxifloxacin hydrochloride	0–0.06									

(*): the combination causes an increase in percentage of microbial inactivation greater than the sum of the antibacterial activity of the aPDT plus the antibiotic treatment when they are applied alone; aPDT: antimicrobial photodynamic therapy; MB: methylene blue; PBS: phosphate-buffered solution.

3.3. *Escherichia coli*

E. coli is the best-known bacteria of the *Enterobacteriaceae* family [50]. Even though some serotypes of *E. coli* can cause serious disease, most strains are harmless and can only act as opportunistic pathogens [51].

Table 5 includes a summary of the in vitro aPDT plus antibiotics against *E. coli* studies and Table 4 summarizes the methodology and the results of Gram-negatives studies.

Table 4. Studies on combination of in vitro aPDT plus other treatment against Gram-negative bacteria that cause SSTIs comparing the methodology and the results.

Strain	PS	Antibiotic	Phase	Source Type	Wavelength (nm)	Intensity (mw/cm^2)	Fluence (J/cm^2)	Synergy	Observed Effect (*)	Reference
E. coli ATCC 25922	ATAZTMPo	gentamicin	planktonic	LED (Sorisa Photocare)	638 ± 9 nm	17	45	yes	>inactivation	[27]
E. coli ATCC 25922	TAPP	tobramycin or chloramphenicol	planktonic	100 W, 120 V Sylvania white light	Broad spectrum	ND	ND	no	additivity	[32]
E. coli ATCC 25922	MB	ciprofloxacin	planktonic	IrradLED® biopdi, São Carlos, SP, Brazil	~660	ND	2.8 and 5.6	yes	>inactivation	[35]
E. coli ATCC 25922	MB	ciprofloxacin	biofilm	IrradLED® biopdi, São Carlos, SP, Brazil	660	ND	11.2 and 22	yes	>inactivation	[35]
E. coli ATCC 9027 and MDR clinical isolates	endogenous porphyrins	ciprofloxacin or norfloxacin n	planktonic	LED Dermaled®	~470 and ~625	ND	ND	yes	incrase in halo	[34]
E. coli	Chlorin e6	colistin, ciprofloxacin or amikacin	planktonic	diode laser, Laser Coupler 635 (Wroclaw, Poland)	635	290	120	yes	>inactivation	[51]
E. coli	Tetra-Py$^+$-Me	ampicillin or chloramphenicol	planktonic	white light lamps (13 lamps OSRAM 21 of 18 W each)	Broad spectrum 380 to 700	40	-	yes	>inactivation	[52]
E. coli, *E. aerogenes*, and *K. pneumoniae* resistant to 3rd-cephalosporins, clinical isolates	MB	ceftriaxone	planktonic	LED (Biopdi/Irrad-Led 660)	660n ± 5	25	25	no	indifference	[53]
P. aeruginosa PAO1	TMP	tobramycin	biofilm	mercury vapor lamp	Broad spectrum	-	220–240	yes	>inactivation & tobramycin MIC decreased	[54]

Table 4. *Cont.*

Strain	PS	Antibiotic	Phase	Source Type	Wavelength (nm)	Intensity (mw/cm²)	Fluence (J/cm²)	Synergy	Observed Effect (*)	Reference
P. aeruginosa PAO1	endogenous porphyrins	gentamicin	planktonic	Nd:YAG laser continuous / Pulsed-Q switched	532	106		yes	>inactivation	[55]
P. aeruginosa PAO1and others MDR and XDR	endogenous porphyrins	gentamicin, meropenem or ceftazidime	planktonic	Single-emitter diode lamp	410	15.7	50	yes	>inactivation & antibiotic MIC decreased	[56]
ATCC 27853 *P. aeruginosa* ATCC 27853	MB	ofloxacin	planktonic	LED	~637	44		yes	>inactivation	[57]
P. aeruginosa ATCC 27853	MB	gentamicin	planktonic	LED lamp (Showtec LED Par 64 Short 18 x RGB 3-in-1 LED, Highlite International B.V. Spain)	625 ± 10	7	18	yes	bactericidal effect with lower MB-PDT dose	[38]
P. aeruginosa ATCC 27853	MB	gentamicin	biofilms	LED lamp (Showtec LED Par 64 Short 18 x RGB 3-in-1 LED, Highlite International B.V. Spain)	625 ± 10	7	18	yes	bactericidal effect with lower MB-PDT dose	[38]
P. aeruginosa ATCC 10145 and 35 clinical isolates including MDR and XDR	RB	camel or pexiganan	planktonic	LED lamps (SecureMe-dia, Poland)	~514	23	60	yes	>inactivation	[58]
P. aeruginosa ATCC 25668 and sensitive and resistant clinical isolates	RB	sulfanilamide	planktonic	18 W white luminescent lamp	Broad spectrum 400–700	1.25	-	no	indifference	[26]
A. baumannii 2 XDR clinical isolates	RB	gentamycin, doxycicline, trimethoprim-sulfamethoxa zole, ciprofloxacin, imipenem, piperacillin-tazobactam, ceftazidime, ampicillin-sulbactam, colistin	planktonic	LED	515	70	300	yes	>inactivation & antibiotic MIC decreased	[59]
A. baumannii 2 XDR clinical isolates	endogenous porphyrins	gentamycin, doxycicline, trimethoprim-sulfamethoxa zole, ciprofloxacin, imipenem, piperacillin-tazobactam, ceftazidime, ampicillin-sulbactam, colistin	planktonic	LED	411	130	109.1	yes	>inactivation & antibiotic MIC decreased	[59]

(*) > inactivation: the combination causes an increase in the percentage of microbial inactivation greater than the sum of the antibacterial activity of the aPDT plus the antibiotic treatment when they are applied alone; additivity the combination causes an increase in percentage of microbial inactivation equal to the sum of the antibacterial activity of the aPDT plus the antibiotic treatment when they are applied alone: Indifference was defined as no change from the most active antibiotic treatment; aPDT: antimicrobial photodynamic therapy; ATAZTMPo; 2-aminothiazolo[4,5-c]-2,7,12,17-tetrakis(methoxyethyl)porphycene; RB: rose bengal; LED: light-emitting diode; MB: methylene blue; MDR: multidrug resistant; MIC: minimum inhibitory concentration; ND: no data; PS: photosensitizer; TAPP: meso-tetra (4-aminophenyl) porphine; TBO: toluidine blue O; Tetra-Py$^+$-Me: 5,10,15,20-tetrakis(1-methylpyridinium-4-yl)porphyrin tetra-iodide; TMP: meso-tetra (*N*-methyl-4-pyridyl) porphine tetra tosylate; XDR: extensively-drug resistant.

Dastgheyb et al., Ronqui et al., and Nieves et al. investigated not only strains of *E. coli* but also *S. aureus*. As has been previously commented, results are similar in both species [27,32,35].

3.3.1. Porphycene Study

Nieves et al. report the synthesis of a new porphycene, ATAZTMPo -gentamicin conjugate, that is able to reduce 8 $\log_{10}$ of surviving planktonic *E. coli* while it did not exert any effect in darkness. The gentamicin concentration used was close to the reported MIC (3.8 µg/mL and 4 µg/mL respectively). Nevertheless, MIC values were determined at inoculum sizes 100–1000-fold, more diluted than those used in this study, and the antimicrobial effect usually was evaluated after 24 h of incubation with the antibiotic, while only 1 h of preirradiation incubation was allowed in these experiments; therefore, it is not surprising the lack of gentamicin toxicity in darkness [27].

3.3.2. Phenothiazine and Porphyrin Studies

Among Dastgheyb et al. and Ronqui et al. studies, the best results against planktonic *E. coli* were obtained by the later using MB-aPDT followed by ciprofloxacin. The former evaluated TAPP-aPDT associated with tobramycin or chloramphenicol achieving only additive antibacterial effect [32]. The later achieved a reduction of approximately 7 logs of *E. coli* in planktonic and biofilm using MB-aPDT followed by ciprofloxacin (while the reduction against *S. aureus* biofilms was 5.4 logs). The most remarkable is that the synergistic effect of aPDT plus ciprofloxacin overcame the resistance of biofilm to aPDT. On the other hand, when the antibiotic is applied after aPDT, lower concentrations than the MIC of ciprofloxacin could be used, since the first sub-MIC led to bacterial reduction of both *S. aureus* and *E. coli* in planktonic state. They hypothesize that this combination achieves a higher effect in biofilm than in planktonic, and higher in *E. coli* than in *S. aureus*, because MB-aPDT works worse on *E. coli* biofilm and the effect of the addition of antibiotic therapy may be more evident [35]. Pereira et al. studied the effect of ciprofloxacin and norfloxacin against planktonic *E. coli*, including MDR clinical isolates, when the strains were irradiated, and they observed because of the endogenous porphyrins an increase in the inhibition halo diameter, indicating that these combinations were synergistic [34]. Other combinations, such as ampicillin or chloramphenicol with aPDT using 5,10,15,20-tetrakis(1-methylpyridinium-4-yl) porphyrin tetra-iodide, achieved a greater increase in MDR *E. coli* killing, especially with ampicillin, at sub-inhibitory and inhibitory concentrations [52]. On the other hand, Costa et al. in a comparative study of the effect of MB-aPDT alone or in combination with ceftriaxone against planktonic clinical isolates of Gram-negative bacteria (*E. coli*, *Klebsiella aerogenes*—formerly *Enterobacter aerogenes*—and *Klebsiella pneumoniae* resistant to third-generation cephalosporins) concluded that the combination did not cause an effect on bacterial viability greater than that MB-PDT itself [53].

3.3.3. Chlorophyll Study

The combination of ciprofloxacin, amikacin, or colistin with aPDT using chlorin e6 improves the antibacterial activity of antibiotics against planktonic *E. coli*, being the best combination with ciprofloxacin [51].

3.4. *Pseudomonas aeruginosa*

P. aeruginosa is an opportunistic human pathogen especially causing infections in chronic ulcers and burns. Table 5 includes a summary of the in vitro aPDT plus antibiotics against *P. aeruginosa* studies and more information on the different methodologies can be consulted in Table 4.

3.4.1. Porphyrin Studies

Sequential treatments of TMP-aPDT followed by tobramycin against *P. aeruginosa* biofilms showed a synergistic effect. The survival bacteria decreased and biofilms pre-treated with TMP-aPDT were substantially more sensitive to tobramycin than untreated biofilms [54].

Different studies employed endogenous porphyrins as the PS. The combination of green light (532 nm) and gentamicin resulted in an antibacterial effect against planktonic *P. aeruginosa* [55]. On the other hand, the combination of sublethal doses of blue light (410 nm) and antibiotics (gentamicin, meropenem, or ceftazidime) reduced the MIC against some planktonic *P. aeruginosa* strains that displayed MDR and extensive drug resistance profiles [56].

3.4.2. Phenothiazine Studies

Another study evaluated the in vitro phototoxicity of MB plus ofloxacin against sensitive or resistant *P. aeruginosa* strains that tolerated ofloxacin. MB-aPDT in combination with antibiotic significantly reduced the viability of *P. aeruginosa* strains compared to either one in monotherapy [57]. Recently, our group showed how gentamicin enhances MB-aPDT-induced bacterial photinactivation of planktonic and biofilm *P. aeruginosa*; an in vitro synergistic effect against both states of *P. aeruginosa* was found, therefore the authors hypothesized that this combination could be useful to manage difficult-to-treat SSTI caused by *P. aeruginosa* [38].

3.4.3. Xanthene Studies

The use of RB-aPDT combined with CAMEL or pexiganan (antimicrobial peptides) achieves a total elimination of different *P. aeruginosa* clinical isolates, including MDR and XDR strains, in contrast to none or partial reduction when the treatments are applied separately. In addition, they demonstrated that the combination is safe, without harmful effects in human keratinocytes [58]. On other hand, Ilizirov et al. studied the combination of RB-aPDT with sulfanilamide against sensitive and resistant *P. aeruginosa* clinical isolates reporting the absence of additional effect. They hypothesize that it could be because of the very low rates of bacterial envelope damage by RB at sub-MIC, which was not sufficient for enhancing the action of sulfanilamide and they suggest that maybe is better to choose antibiotics which affect the same cell target sites in order to achieve amplification of their activity [26].

3.5. *Acinetobacter baumannii*

A. baumannii is a threatening human pathogen with outstanding capability to acquire AMR [11] although it rarely causes skin infection SSTIs.

Different antibiotics were combined with RB-aPDT or only blue light to excite endogenous PS against extremely drug resistant *A. baumannii* clinical isolates. The final result was the effective eradication. The bactericidal effect of antibiotics was enhanced with sublethal aPDT or blue light addition applied after them. Moreover, they measured the production of ROS and claimed that its increase with the combined treatment could explain the synergistic activity observed. Table 5 includes a summary of this study (and more information can be seen in Table 3) that stands out for covering all antibiotic categories as well as all antimicrobial mechanisms of action and the wide variety of methods used to test synergy (diffusion assays—disk diffusion and E-test, serial dilution methodology—checkerboard assay, and CFU counting- and time kill curves method—postantibiotic effect) [59].

3.6. *Candida* spp.

Candida spp. cause skin and oral and genital mucosa infections and *C. albicans* is the most relevant one [60].

Table 5 includes a summary of the different combinations studied using in vitro aPDT plus antifungals against yeasts that cause SSTIs; specific details of the methodology of these studies are provided in Table 6.

The effect of MB-aPDT plus fluconazole was evaluated against fluconazole-resistant *C. albicans*, *Candida glabrata*, and *Candida krusei*. A synergistic combination effect against the strains of *C. albicans* and *C. glabrata*, but not against *C. krusei*, was found. The effect of MB-aPDT alone against *C. krusei* was not statistically significant compared to the effect of its combination with fluconazole [61].

Snell et al. showed that fluconazole did not enhance *C. albicans* killing induced by aPDT using TMP. However, miconazole improved the fungicidal activity of aPDT using either TMP or MB [62].

3.7. Dermatophytes and Moulds

Table 5 includes a summary of the combination treatment studies using in vitro aPDT plus other compounds against dermatophytes and molds that cause SSTIs and an extended version can be consulted in Table 6.

Trichophyton rubrum causes of athlete's foot, fungal infection of nail and ringworm, worldwide athlete's foot, onychomycosis, jock itch, and ringworm. Morton et al. reported that clotrimazole combined with RB-aPDT had the potential to reduce the MIC of the antifungal drug against spores of *T. rubrum* clinical isolate. This occurred when pretreatment with the antifungal (sublethal dose) was followed by RB-aPDT. When the order of combination was changed, no reduction in the antifungal MIC was observed [63].

Fonsecaea monophora and *Fonsecaea pedrosoi* are the main causative agents of chromoblastomycosis in Southern China. It is a chronic skin and subcutaneous fungal infection with low cure and high relapse rates. Hu et al. (2015) treated five refractory and complex cases of chromoblastomycosis with 5-ALA-aPDT combined with oral antifungal drugs. The isolates were evaluated for susceptibility to terbinafine, itraconazole, and voriconazole and 5-ALA-aPDT in vitro, revealing sensitivity to the antifungals, with 5-ALA-aPDT altering the cell wall and increasing ROS production. The results showed that there was an unclear synergistic effect of itraconazole plus 5-ALA-aPDT [64]. Altogether, they conclude that 5-ALA-aPDT combined with oral antifungal drugs is a promising method for the treatment of refractory and complex cases of chromoblastomycosis. This idea was also previously defended by the same authors in a previous clinical case of chromoblastomycosis caused by *F. monophora* that was successfully treated with terbinafine plus 5-ALA-aPDT. This study was also supported by in vitro experiments where they showed the good outcome of 5-ALA-aPDT applied for the inhibition of *F. monophora* [65].

Exophiala spp. is an ubiquitous fungal species commonly found in soil and plants which causes chromoblastomycosis [66]. On the other hand, *Fusarium solani* and *Fusarium oxysporum* are responsible for approximately 60% and 20% of the cases of fusariosis, respectively, which is the second most common mold infection in humans after aspergillosis [67]. Gao et al. investigated both planktonic suspensions and biofilms of *E. dermatitidis* and *Fusarium* spp. with MB-aPDT combined with standard antifungal treatments (itraconazole, voriconazole, posaconazole, and amphotericin) achieving encouraging results. The pretreatment with MB-aPDT made them more susceptible to antifungals either in planktonic cultures or in biofilms. Therefore, the combination may help to enhance the antifungal susceptibility to overcome problems with drug resistance issues, and has the potential to reduce antifungal drug dosages decreasing their toxicity [15]. According to the authors, this may be due to an increased membrane permeability caused by aPDT, as suggested previously for *C. albicans* [68].

Table 5. In vitro antimicrobial photodynamic therapy plus other treatment studies on infectious microorganism of skin and soft tissues grouped by photosensitizing family.

S. aureus						
PS Group	**PS**	**Antibiotic**	**Phase**	**Synergy**	**Observed Effect (*)**	**Reference**
Porphycene and Porphyrin	ATAZTMPo	gentamicin	planktonic	yes	>inactivation	[27]
	5-ALA	gentamicin	biofilm	yes	>inactivation	[28]
	DP	gentamicin, vancomycin, rifampin, fusidic acid	planktonic	no	additivity	[29]
	DP	oxacillin	planktonic	yes	oxacillin MIC decreased	[29]
	5-ALA	netilmicin, cefaclor, vancomycin	biofilm	yes	>inactivation	[30]
	TMP	vancomycin	biofilm	yes	>inactivation	[31]
	TAPP	vancomycin, ceftriaxone	planktonic	no	indifference	[32]
	TAPP	chloramphenicol, tobramycin	planktonic	yes-no	>inactivation MRSA; additive MSSA	[32]
	Tetra-Py+-Me	ampicillin	planktonic	yes	faster bactericidal effect	[33]
	Tetra-Py+-Me	ampicillin	pork skin (ex vivo)	yes	>inactivation	[33]
	(endogenous)	ciprofloxacin, norfloxacin	planktonic	yes	>inactivation	[34]
Phenothiazines	MB	chloramphenicol	planktonic	no	additivity	[32]
	MB	ciprofloxacin	planktonic	yes	>inactivation	[35]
	MB	ciprofloxacin	biofilm	yes	>inactivation	[35]
	TBO	linezolid	biofilm	yes	>inactivation	[35]
	MB	linezolid	biofilm	no	indifference	[35]
	TBO	gentamicin	planktonic	yes	>inactivation	[37]
	MB	gentamicin	planktonic	yes	bactericidal effect with lower MB-PDT dose	[38]
	MB	gentamicin	biofilm	no	no significant > inactivation	[38]
	MB	linezolid, mupirocin	planktonic	yes	bactericidal effect with lower MB-PDT dose	[39]
Xanthenes	RB	linezolid, mupirocin	planktonic	yes	bactericidal effect with lower RB-PDT dose	[39]
	RB	gentamicin	planktonic	yes	bactericidal effect with lower RB-PDT dose	[41]
	RB	gentamicin	biofilm	yes	>inactivation	[41]
	RB	methicillin	planktonic	yes	methicillin MIC decreased	[26]

Table 5. *Cont.*

S. haemolyticus						
PS Group	**PS**	**Antibiotic**	**Phase**	**Synergy**	**Observed Effect (*)**	**Reference**
Porphycene and Porphyrin	5-ALA	gentamicin	biofilm	yes	>inactivation	[28]
S. epidermidis						
PS group	**PS**	**Antibiotic**	**Phase**	**Synergy**	**Observed effect (*)**	**Reference**
Porphycene and Porphyrin	5-ALA	gentamicin	biofilm	yes	>inactivation	[28]
	TMP	vancomycin	biofilm	yes	>inactivation	[31]
	TAPP	chloramphenicol, tobramycin	planktonic	yes	>inactivation	[32]
M. fortuitum						
PS Group	**PS**	**Antibiotic**	**Phase**	**Synergy**	**Observed Effect (*)**	**Reference**
Phenothiazines	MB	ciprofloxacin, moxifloxacin or amikacin	planktonic	yes	>inactivation	[49]
E. coli						
PS Group	**PS**	**Antibiotic**	**Phase**	**Synergy**	**Observed Effect (*)**	**Reference**
Porphycene andPorphyrin	ATAZTMPo	gentamicin	planktonic	yes	>inactivation	[27]
	TAPP	tobramycin or chloramphenicol	planktonic	no	additivity	[32]
	endogenous porphyrins	ciprofloxacin or norfloxacin n	planktonic	yes	incrase in halo	[34]
	Tetra-Py$^+$-Me	ampicillin or chloramphenicol	planktonic	yes	>inactivation	[52]
Phenothiazines	MB	ciprofloxacin	planktonic	yes	>inactivation	[35]
	MB	ciprofloxacin	biofilm	yes	>inactivation	[35]
	MB	ceftriaxone	planktonic	no	indifference	[52]
Chlorophylls	Chlorin e6	colistin, ciprofloxacin or amikacin	planktonic	yes	>inactivation	[51]

Table 5. *Cont.*

P. aeruginosa						
PS Group	**PS**	**Antibiotic**	**Phase**	**Synergy**	**Observed Effect (*)**	**Reference**
Porphycene and Porphyrin	TMP	tobramycin	biofilm	yes	>inactivation & tobramycin MIC decreased	[54]
	endogenous porphyrins	gentamicin	planktonic	yes	>inactivation	[55]
	endogenous porphyrins	gentamicin, meropenem or ceftazidime	planktonic	yes	>inactivation & antibiotic MIC decreased	[56]
Phenothiazines	MB	ofloxacin	planktonic	yes	>inactivation	[57]
	MB	gentamicin	planktonic	yes	bactericidal effect with lower MB-PDT dose	[38]
	MB	gentamicin	biofilm	yes	bactericidal effect with lower MB-PDT dose	[38]
Xanthenes	RB	camel or pexiganan	planktonic	yes	>inactivation	[58]
	RB	sulfanilamide	planktonic	no	indifference	[26]
A. baumannii						
PS Group	**PS**	**Antibiotic**	**Phase**	**Synergy**	**Observed Effect (*)**	**Reference**
Xanthenes	RB	gentamycin, doxycicline, trimethoprim-sulfamethoxazole, ciprofloxacin, imipenem, piperacillin-tazobactam, ceftazidime, ampicillin-sulbactam, colistin	planktonic	yes	>inactivation & antibiotic MIC decreased	[59]
Porphycene and Porphyrin	endogenous porphyrins	gentamycin, doxycicline, trimethoprim-sulfamethoxazole, ciprofloxacin, imipenem, piperacillin-tazobactam, ceftazidime, ampicillin-sulbactam, colistin	planktonic	yes	>inactivation & antibiotic MIC decreased	[59]
***Candida* spp.**						
PS Group	**PS**	**Antibiotic**	**Phase**	**Synergy**	**Observed Effect (*)**	**Reference**
Phenothiazines	MB	fluconazole	planktonic	yes/no	≥inactivation	[61]
	MB	miconazole	planktonic	yes	>inactivation	[62]
Porphycene and Porphyrin	TMP	miconazole	planktonic	yes	>inactivation	[62]
	TMP	fluconazole	planktonic	no	indifference	[62]

Table 5. *Cont.*

Dermatophytes and moulds						
PS Group	**PS**	**Antibiotic**	**Phase**	**Synergy**	**Observed Effect (*)**	**Reference**
Xanthenes	RB	clotrimazole	Planktonic (in vitro: spores)	yes	≥inactivation	[63]
Porphycene and Porphyrin	5-ALA	itraconazole	planktonic	yes	≥inactivation	[64]
Phenothiazines	MB	itraconazole, voriconazole, posaconazole, amphotericin	planktonic and biofilms	yes	MIC decreased	[15]
	MB	itraconazole	planktonic	yes	>inactivation	[69]

(*) > inactivation: the combination causes an increase in the percentage of microbial inactivation, greater than the sum of the antibacterial activity of the aPDT plus the antibiotic treatment when they are applied alone. Additivity: the combination causes an increase in the percentage of microbial inactivation equal to the sum of the antibacterial activity of the aPDT plus the antibiotic treatment when they are applied alone. Indifference was defined as no change from the most active treatment. 5-ALA = 5-aminolevulinic acid; ATAZTMPo = 2-aminothiazolo[4,5-c]-2,7,12,17-tetrakis(methoxyethyl)porphycene; DP = deuteroporphyrin; MB = methylene blue; MIC = minimum inhibitory concentration; MRSA = methicillin resistant *S. aureus*; MSSA = methicillin sensitive *S. aureus*; RB = rose bengal; PS = photosensitizer; TAPP = 5,10,15,20-tetrakis [4-(3-*N*,*N*-dimethylaminopropoxy) phenyl] porphyrin; TBO = toluidine blue O; TAPP = meso-tetra (4-aminophenyl) porphine; TMP = tetra-substituted *N*-methyl-pyridyl-porphine; Tetra-Py$^+$-Me = 5,10,15,20-tetrakis (1-methylpyridinium-4-yl) porphyrin tetra-iodide.

Table 6. Studies on combination of in vitro aPDT plus other treatment against fungus that cause SSTIs comparing the methodology and the results.

Strain	PS	Antifungal	Phase	Source Type	Wavelength (nm)	Intensity (mw/cm^2)	Fluence (J/cm^2)	Synergy	Observed Effect (*)	Reference
fluconazole-resistant *C. albicans* and *C. glabratai*	MB	fluconazole	planktonic	InGaAlP LED	nd	200	-	yes	>inactivation	[61]
fluconazole-resistant *C. krusei*	MB	fluconazole	planktonic	InGaAlP LED	nd	200	-	no	indifference	[61]
C. albicans SC5314	TMP	miconazole	planktonic	broadband visible light (Sylvania GRO-LUX, 15 W, part no. F15T8/GRO)	575–700	4	1	yes	>inactivation	[62]
C. albicans SC5314	TMP	fluconazole	planktonic	broadband visible light (Sylvania GRO-LUX, 15 W, part no. F15T8/GRO)	575–700	4	1	no	indifference	[62]
C. albicans SC5314	MB	miconazole	planktonic	broadband visible light (Sylvania GRO-LUX, 15 W, part no. F15T8/GRO)	575–700	4	7.2	yes	>inactivation	[62]
T. rubrum clinical isolate	RB	clotrimazole	Planktonic (in vitro: spores)	LED	530	13.4	12	yes	≥inactivation	[63]
F. monophora clinical isolates	5-ALA	itraconazole	planktonic	Zeiss KL 2500 LED	635	36.8	10	yes	≥inactivation	[64]
E. dermatitidis, *F. solani*, *F. oxysporum* clinical isolates	MB	itraconazole, voriconazole, posaconazole, amphotericin	planktonic and biofilms	LED	635 ± 10	100	12-24	yes	MIC decreased	[15]
S. globosa 5 clinical isolates	MB	itraconazole	planktonic	LED	640 ± 10	22.2	40	yes	>inactivation	[69]

(*) > inactivation: the combination causes an increase in percentage of microbial inactivation greater than the sum of the antibacterial activity of the aPDT plus the antibiotic treatment when they are applied alone. = inactivation: the combination has no effect on the percentage of inactivation; 5-ALA: 5-aminolevulinic acid; aPDT: antimicrobial photodynamic therapy; nd: no data; RB: rose bengal; LED: light-emitting diode; MB: methylene blue; PS: photosensitizer; TMP: meso-tetra (*N*-methyl-4-pyridyl) porphine tetra tosylate.

Sporothrix globosa is an etiological agent of sporotrichosis whose most common clinical manifestation is cutaneous and subcutaneous nodular lesions with lymphangitis involvement. Currently, AMR and complications are the major concerns, especially in patients who have liver disorders, children, and pregnant women. Li et al. compared the efficacy in the inactivation of *S. globosa* of MB-aPDT alone or combined with itraconazole in planktonic culture and in a murine model. The combined treatment offered better results in terms of inactivation percentage and improvement of the lesion size. Therefore, they conclude that MB-aPDT could be an effective adjuvant therapy for resistant infections caused by *Sporothrix* spp. [69]. In fact, our group treated a patient with cutaneous sporotrichosis using intralesional 1% MB-aPDT in combination with intermittent low doses of itraconazole obtained complete microbiological and clinical response [70].

4. Summary of Evidence and Limitations

In the main, the combination of antibiotics or antifungals with aPDT against in vitro SSTI-etiological agents seems to be beneficial. Combined therapy is more effective than individual treatments alone and often the effects are greater than additivity, i.e., there is synergy. Among the effects reported, the following stand out: (i) the increase in percentage of microbial inactivation; or (ii) the same inactivation percentage is achieved using lower doses of antimicrobials.

It is remarkable that in some cases, drug sensitivity of resistant strains can be restored by combining antibiotics/antifungals with aPDT [15,26,59].

The highlight combinations and the best treatment protocols supported by the existing evidence of in vitro studies on combined aPDT therapies against SSTI-causing agents are included in Table 7. Nevertheless, there is not enough evidence to establish the best combination against each causal agent of SSIT according to this review. The number of studies is limited, and the methodologies used are varied, making direct comparison difficult. In addition, they mostly report the effect on inactivation, but the mechanism of action remains unknown.

This review provides additional and updated information to the one published by Wozniak and Grinholc in 2018, and it is complementary to the review focused on in vivo studies published by our group in 2019 [23,24]. All three types of review agree on the promising approach of combining both therapies and the need to expand knowledge in this line.

The coating of surfaces such as catheters with antimicrobial drugs and aPDT are extremely effective and virtually overcome any resistance build-up. This is more complex on the skin and soft tissues because more variables become part of the process, especially with fungi. However, there is sufficient evidence to support this combined treatment strategy and to lay the foundation for this SSIT treatment approach [18].

Among the obstacles to the incorporation of aPDT as part of the SSIT-treatment, the need to require materials such as lamps for exciting the PS and the need to dedicate more time, because of the irradiation time and because often more than one session of aPDT is required are highlighted. However, it does not require much more clinician specialization, various studies have proven that lamps do not have to be especially expensive and specific, and sources of light with a wide irradiation spectrum or even daylight of radiation can be used effectively [18,23]. The methodology and evolution of these treatments need to be reported to the scientific community to continue expanding knowledge and increasingly implement this combination therapeutic option.

Table 7. The highlight combinations and the best treatment protocols supported by the existing evidence of in vitro studies on combined aPDT therapies against SSTI-causing agents.

✓	The combination of different antibiotics with aPDT in general improves the efficacy against in vitro Gram-positive bacteria.
✓	Different antibiotics combined with aPDT using porphyrins, phenothiaziniums or RB have synergistic effects in vitro against Gram-positive bacteria being the combination with protoporphyrin the most studied.
✓	The combination of aPDT with gentamicin has been extensively tested in both Gram-positive and negative bacteria, reporting positive effects in all cases.
✓	The combination of MB-aPDT with sublethal concentrations of antibiotics seems to be a good option against mycobacteria.
✓	MB-aPDT combined with ciprofloxacin is the best option in vitro against *E. coli.*
✓	Endogenous porphyrins or MB based-aPDT used in combination with antibiotics is a promising option against in vitro *P. aeruginosa* regardless of its antibiotic resistance pattern.
✓	The administration of aPDT before antifungals seems to enhance their in vitro antimicrobial effect, especially against yeast and dermatophytes.

aPDT: antimicrobial photodynamic therapy; RB: rose bengal; MB: methylene blue.

5. Conclusions

aPTD combined with antimicrobial agents is promising for the management of microorganisms that cause SSTI. It can help to fight them and to overcome AMR.

Author Contributions: Conceptualization and methodology, V.P.-L., Y.G. and A.R.; writing—original draft preparation, V.P.-L.; writing—review and editing V.P.-L., I.G.-L., S.B., Y.G. and A.R.; supervision, Y.G. and A.R.; funding acquisition, I.G.-L., S.B., Y.G. and A.R. All authors have read and agreed to the published version of the manuscript.

Funding: Y.G. And A.R. were founded by Aragon Health Research Institute (IIS Aragón), A.R. was funded by Research Groups of Government of Aragón (Spain) B10_17R Infectious Diseases of Difficult Diagnosis and Treatment. I.G.-L. and S.B. were funded by Plan Nacional de I+D+i and Instituto de Salud Carlos III, Ministerio de Ciencia e Innovación, Spanish Network for Research in Infectious Diseases (REIPI; RD16/0016/0001)-co-financed by European Development Regional Fund "A way to achieve Europe", ERDF.

Institutional Review Board Statement: Not applicable.

Informed Consent Statement: Not applicable.

Data Availability Statement: No new data were created or analyzed in this study. Data sharing is not applicable to this article.

Acknowledgments: The authors thank the research groups of IIS Aragon GIIS-023 and GIS096 for their support.

Conflicts of Interest: The authors declare no conflict of interest.

References

1. Ki, V.; Rotstein, C. Bacterial skin and soft tissue infections in adults: A review of their epidemiology, pathogenesis, diagnosis, treatment and site of care. *Can. J. Infect. Dis. Med. Microbiol.* **2008**, *19*, 173–184. [CrossRef]
2. Russo, A.; Concia, E.; Cristini, F.; De Rosa, F.G.; Esposito, S.; Menichetti, F., Petrosillo, N.; Tumbarello, M.; Venditti, M.; Viale, P.; et al. Current and future trends in antibiotic therapy of acute bacterial skin and skin-structure infections. *Clin. Microbiol. Infect.* **2016**, *22* (Suppl. 2), S27–S36. [CrossRef]
3. Sartelli, M.; Guirao, X.; Hardcastle, T.C.; Kluger, Y.; Boermeester, M.A.; Raşa, K.; Ansaloni, L.; Coccolini, F.; Montravers, P.; Abu-Zidan, F.M.; et al. 2018 WSES/SIS-E consensus conference: Recommendations for the management of skin and soft-tissue infections. *World J. Emerg. Surg.* **2018**, *13*, 58. [CrossRef] [PubMed]
4. Ibrahim, F.; Khan, T.; Pujalte, G.G.A. Bacterial Skin Infections. *Prim. Care* **2015**, *42*, 485–499. [CrossRef] [PubMed]
5. Moffarah, A.S.; Mohajer, M.A.; Hurwitz, B.L.; Armstrong, D.G. Skin and soft tissue infections. *Microbiol. Spectr.* **2016**, *4*, 4. [CrossRef]
6. Polk, C.; Sampson, M.M.; Roshdy, D.; Davidson, L.E. Skin and Soft Tissue Infections in Patients with Diabetes Mellitus. *Infect. Dis. Clin. N. Am.* **2021**, *35*, 183–197. [CrossRef] [PubMed]
7. Esposito, S.; Noviello, S.; de Caro, F.; Boccia, G. New insights into classification, epidemiology and microbiology of SSTIs, including diabetic foot infections. *Infez. Med.* **2018**, *26*, 3–14. [PubMed]
8. Leong, H.N.; Kurup, A.; Tan, M.Y.; Kwa, A.L.H.; Liau, K.H.; Wilcox, M.H. Management of complicated skin and soft tissue infections with a special focus on the role of newer antibiotics. *Infect. Drug Resist.* **2018**, *11*, 1959–1974. [CrossRef] [PubMed]
9. Stevens, D.L.; Bisno, A.L.; Chambers, H.F.; Dellinger, E.P.; Goldstein, E.J.; Gorbach, S.L.; Hirschmann, J.V.; Kaplan, S.L.; Montoya, J.G.; Wade, J.C.; et al. Practice Guidelines for the Diagnosis and Management of Skin and Soft Tissue Infections: 2014 Update by the Infectious Diseases Society of America. *Clin. Infect. Dis.* **2014**, *59*, e10–e52. [CrossRef]
10. Dadgostar, P. Antimicrobial Resistance: Implications and Costs. *Infect. Drug Resist.* **2019**, *12*, 3903–3910. [CrossRef]
11. Tacconelli, E. *Magrani Global Priority List of Antibiotic-Resistant Bacteria to Guide Research, Discovery, and Development of New Antibiotics*; WHO: Geneva, Switzerland, 2017, Available online: https://www.who.int/medicines/publications/WHO-PPL-Short_Summary_25Feb-ET_NM_WHO.pdf?ua=1 (accessed on 20 June 2021).
12. Sebeny, P.J.; Riddle, M.S.; Petersen, K. *Acinetobacter baumannii* skin and soft-tissue infection associated with war trauma. *Clin. Infect. Dis.* **2008**, *47*, 444–449. [CrossRef]
13. Berman, J.; Krysan, D.J. Drug resistance and tolerance in fungi. *Nat. Rev. Microbiol.* **2020**, *18*, 319–331 [CrossRef] [PubMed]
14. Pappas, P.G.; Rex, J.H.; Sobel, J.D.; Filler, S.G.; Dismukes, W.E.; Walsh, T.J.; Edwards, J.E. Infectious Diseases Society of America Guidelines for treatment of candidiasis. *Clin. Infect. Dis.* **2004**, *38*, 161–189. [CrossRef]
15. Gao, L.; Jiang, S.; Sun, Y.; Deng, M.; Wu, Q.; Li, M.; Zeng, T. Evaluation of the Effects of Photodynamic Therapy Alone and Combined with Standard Antifungal Therapy on Planktonic Cells and Biofilms of Fusarium spp. and Exophiala spp. *Front. Microbiol.* **2016**, *7*, 617. [CrossRef]
16. Tainwala, R.; Sharma, Y. Pathogenesis of Dermatophytoses. *Indian J. Dermatol.* **2011**, *56*, 259–261. [CrossRef] [PubMed]
17. Kharkwal, G.B.; Sharma, S.K.; Huang, Y.-Y.; Dai, T.; Hamblin, M.R. Photodynamic therapy for infections: Clinical applications. *Lasers Surg. Med.* **2011**, *43*, 755–767. [CrossRef]
18. Pérez-Laguna, V.; García-Malinis, A.J.; Aspiroz, C.; Rezusta, A.; Gilaberte, Y. Antimicrobial effects of photodynamic therapy: An overview. *G. Ital. Dermatol. Venereol.* **2018**. [CrossRef] [PubMed]
19. Cieplik, F.; Deng, D.; Crielaard, W.; Buchalla, W.; Hellwig, E.; Al-Ahmad, A.; Maisch, T. Antimicrobial photodynamic therapy—What we know and what we don't. *Crit. Rev. Microbiol.* **2018**, *44*, 571–589. [CrossRef] [PubMed]
20. O'Riordan, K.; Akilov, O.E.; Hasan, T. The potential for photodynamic therapy in the treatment of localized infections. *Photodiagnosis Photodyn. Ther.* **2005**, *2*, 247–262. [CrossRef]
21. Dai, T.; Huang, Y.-Y.; Hamblin, M.R. Photodynamic therapy for localized infections–state of the art. *Photodiagnosis Photodyn. Ther.* **2009**, *6*, 170–188. [CrossRef]
22. Nisnevitch, M.; Valkov, A.; Nakonechny, F.; Gutterman, M.; Nitzan, Y. Antibiotics Combined with Photosensitizers: A Novel Approach to Antibacterial Treatment. In *Antibiotic Therapy: New Developments*; Turner, A., Hall, J., Eds.; Nova Science Inc.: New York, NY, USA, 2013; pp. 63–88, ISBN 978-1-62808-171-8.
23. Pérez-Laguna, V.; Gilaberte, Y.; Millán-Lou, M.I.; Agut, M.; Nonell, S.; Rezusta, A.; Hamblin, M.R. A combination of photodynamic therapy and antimicrobial compounds to treat skin and mucosal infections: A systematic review. *Photochem. Photobiol. Sci.* **2019**, *18*, 1020–1029. [CrossRef] [PubMed]
24. Wozniak, A.; Grinholc, M. Combined Antimicrobial Activity of Photodynamic Inactivation and Antimicrobials-State of the Art. *Front. Microbiol.* **2018**, *9*, 930. [CrossRef] [PubMed]
25. Bassetti, M.; Baguneid, M.; Bouza, E.; Dryden, M.; Nathwani, D.; Wilcox, M. European perspective and update on the management of complicated skin and soft tissue infections due to methicillin-resistant Staphylococcus aureus after more than 10 years of experience with linezolid. *Clin. Microbiol. Infect.* **2014**, *20* (Suppl. 4), 3–18. [CrossRef] [PubMed]
26. Ilizirov, Y.; Formanovsky, A.; Mikhura, I.; Paitan, Y.; Nakonechny, F.; Nisnevitch, M. Effect of Photodynamic Antibacterial Chemotherapy Combined with Antibiotics on Gram-Positive and Gram-Negative Bacteria. *Molecules* **2018**, *23*, 3152. [CrossRef] [PubMed]

27. Nieves, I.; Hally, C.; Viappiani, C.; Agut, M.; Nonell, S. A porphycene-gentamicin conjugate for enhanced photodynamic inactivation of bacteria. *Bioorganic Chem.* **2020**, *97*, 103661. [CrossRef]
28. Barra, F.; Roscetto, E.; Soriano, A.A.; Vollaro, A.; Postiglione, I.; Pierantoni, G.M.; Palumbo, G.; Catania, M.R. Photodynamic and Antibiotic Therapy in Combination to Fight Biofilms and Resistant Surface Bacterial Infections. *Int. J. Mol. Sci.* **2015**, *16*, 20417–20430. [CrossRef] [PubMed]
29. Iluz, N.; Maor, Y.; Keller, N.; Malik, Z. The synergistic effect of PDT and oxacillin on clinical isolates of Staphylococcus aureus. *Lasers Surg. Med.* **2018**, *50*, 535–551. [CrossRef]
30. Zhang, Q.-Z.; Zhao, K.-Q.; Wu, Y.; Li, X.-H.; Yang, C.; Guo, L.-M.; Liu, C.-H.; Qu, D.; Zheng, C.-Q. 5-aminolevulinic acid-mediated photodynamic therapy and its strain-dependent combined effect with antibiotics on Staphylococcus aureus biofilm. *PLoS ONE* **2017**, *12*, e0174627. [CrossRef]
31. Di Poto, A.; Sbarra, M.S.; Provenza, G.; Visai, L.; Speziale, P. The effect of photodynamic treatment combined with antibiotic action or host defence mechanisms on Staphylococcus aureus biofilms. *Biomaterials* **2009**, *30*, 3158–3166. [CrossRef]
32. Dastgheyb, S.S.; Eckmann, D.M.; Composto, R.J.; Hickok, N.J. Photo-activated porphyrin in combination with antibiotics: Therapies against Staphylococci. *J. Photochem. Photobiol. B* **2013**, *129*, 27–35. [CrossRef]
33. Branco, T.M.; Valério, N.C.; Jesus, V.I.R.; Dias, C.J.; Neves, M.G.P.M.S.; Faustino, M.A.F.; Almeida, A. Single and combined effects of photodynamic therapy and antibiotics to inactivate Staphylococcus aureus on skin. *Photodiagnosis Photodyn. Ther.* **2018**, *21*, 285–293. [CrossRef] [PubMed]
34. Pereira, N.L.F.; Aquino, P.E.A.; Júnior, J.G.A.S.; Cristo, J.S.; Vieira Filho, M.A.; Moura, F.F.; Ferreira, N.M.N.; Silva, M.K.N.; Nascimento, E.M.; Correia, F.M.A.; et al. Antibacterial activity and antibiotic modulating potential of the essential oil obtained from Eugenia jambolana in association with led lights. *J. Photochem. Photobiol. B* **2017**, *174*, 144–149. [CrossRef]
35. Ronqui, M.R.; de Aguiar Coletti, T.M.S.F.; de Freitas, L.M.; Miranda, E.T.; Fontana, C.R. Synergistic antimicrobial effect of photodynamic therapy and ciprofloxacin. *J. Photochem. Photobiol. B* **2016**, *158*, 122–129. [CrossRef]
36. Kashef, N.; Akbarizare, M.; Razzaghi, M.R. In vitro Activity of Linezolid in Combination with Photodynamic Inactivation Against Staphylococcus aureus Biofilms. *Avicenna J. Med. Biotechnol.* **2017**, *9*, 44–48. [PubMed]
37. Liu, S.; Mai, B.; Jia, M.; Lin, D.; Zhang, J.; Liu, Q.; Wang, P. Synergistic antimicrobial effects of photodynamic antimicrobial chemotherapy and gentamicin on Staphylococcus aureus and multidrug-resistant Staphylococcus aureus. *Photodiagn. Photodyn. Ther.* **2020**, *30*, 101703. [CrossRef] [PubMed]
38. Pérez-Laguna, V.; García-Luque, I.; Ballesta, S.; Pérez-Artiaga, L.; Lampaya-Pérez, V.; Rezusta, A.; Gilaberte, Y. Photodynamic therapy using methylene blue, combined or not with gentamicin, against Staphylococcus aureus and Pseudomonas aeruginosa. *Photodiagn. Photodyn. Ther.* **2020**, *31*, 101810. [CrossRef] [PubMed]
39. Pérez-Laguna, V.; Pérez-Artiaga, L.; Lampaya-Pérez, V.; García-Luque, I.; Ballesta, S.; Nonell, S.; Paz-Cristobal, M.P.; Gilaberte, Y.; Rezusta, A. Bactericidal Effect of Photodynamic Therapy, Alone or in Combination with Mupirocin or Linezolid, on Staphylococcus aureus. *Front. Microbiol.* **2017**, *8*, 1002. [CrossRef]
40. Pérez, M.; Robres, P.; Moreno, B.; Bolea, R.; Verde, M.T.; Pérez-Laguna, V.; Aspiroz, C.; Gilaberte, Y.; Rezusta, A. Comparison of Antibacterial Activity and Wound Healing in a Superficial Abrasion Mouse Model of Staphylococcus aureus Skin Infection Using Photodynamic Therapy Based on Methylene Blue or Mupirocin or Both. *Front. Med.* **2021**, *8*. [CrossRef]
41. Pérez-Laguna, V.; García-Luque, I.; Ballesta, S.; Pérez-Artiaga, L.; Lampaya-Pérez, V.; Samper, S.; Soria-Lozano, P.; Rezusta, A.; Gilaberte, Y. Antimicrobial photodynamic activity of Rose Bengal, alone or in combination with Gentamicin, against planktonic and biofilm Staphylococcus aureus. *Photodiagn. Photodyn. Ther.* **2018**, *21*, 211–216. [CrossRef] [PubMed]
42. de Allori, M.C.G.; Jure, M.A.; Romero, C.; de Castillo, M.E.C. Antimicrobial resistance and production of biofilms in clinical isolates of coagulase-negative Staphylococcus strains. *Biol. Pharm. Bull.* **2006**, *29*, 1592–1596. [CrossRef]
43. Fey, P.D.; Olson, M.E. Current concepts in biofilm formation of Staphylococcus epidermidis. *Future Microbiol.* **2010**, *5*, 917–933. [CrossRef]
44. Foster, T. Staphylococcus. In *Medical Microbiology*, 4th ed.; Baron, S., Ed.; University of Texas Medical Branch at Galveston: Galveston, TX, USA, 1996. Available online: http://www.ncbi.nlm.nih.gov/books/NBK8448/ (accessed on 10 July 2018).
45. Viale, P.; Stefani, S. Vascular catheter-associated infections: A microbiological and therapeutic update. *J. Chemother. Florence Italy* **2006**, *18*, 235–249. [CrossRef]
46. Sbarra, M.S.; Arciola, C.R.; Di Poto, A.; Saino, E.; Rohde, H.; Speziale, P.; Visai, L. The photodynamic effect of tetra-substituted N-methyl-pyridyl-porphine combined with the action of vancomycin or host defense mechanisms disrupts Staphylococcus epidermidis biofilms. *Int. J. Artif. Organs* **2009**, *32*, 574–583. [CrossRef] [PubMed]
47. Gong, N.; Tan, Y.; Li, M.; Lu, W.; Lei, X. ALA-PDT combined with antibiotics for the treatment of multiple skin abscesses caused by Mycobacterium fortuitum. *Photodiagn. Photodyn. Ther.* **2016**, *15*, 70–72. [CrossRef] [PubMed]
48. Sun, K.; Yang, H.; Huang, X.; Gong, N.; Qin, Q.; Lu, W.; Lei, X. ALA-PDT combined with antibiotics for the treatment of atypical mycobacterial skin infections: Outcomes and safety. *Photodiagnosis Photodyn. Ther.* **2017**, *19*, 274–277. [CrossRef] [PubMed]
49. Shih, M.-H.; Huang, F.-C. Effects of photodynamic therapy on rapidly growing nontuberculous mycobacteria keratitis. *Investig. Ophthalmol. Vis. Sci.* **2011**, *52*, 223–229. [CrossRef]
50. Boucher, H.W.; Talbot, G.H.; Bradley, J.S.; Edwards, J.E.; Gilbert, D.; Rice, L.B.; Scheld, M.; Spellberg, B.; Bartlett, J. Bad bugs, no drugs: No ESKAPE! An update from the Infectious Diseases Society of America. *Clin. Infect. Dis.* **2009**, *48*, 1–12. [CrossRef] [PubMed]

51. Tichaczek-Goska, D.; Wojnicz, D.; Symonowicz, K.; Ziółkowski, P.; Hendrich, A.B. Photodynamic enhancement of the activity of antibiotics used in urinary tract infections. *Lasers Med. Sci.* **2019**. [CrossRef] [PubMed]
52. Almeida, J.; Tomé, J.P.C.; Neves, M.G.P.M.S.; Tomé, A.C.; Cavaleiro, J.A.S., Cunha, Â.; Costa, L.; Faustino, M.A.F.; Almeida, A. Photodynamic inactivation of multidrug-resistant bacteria in hospital wastewaters: Influence of residual antibiotics. *Photochem. Photobiol. Sci.* **2014**, *13*, 626–633. [CrossRef]
53. Magacho, C.C.; Pinto, J.G.; Souza, B.M.N.; Pereira, A.H.C.; Ferreira-Strixino, J. Comparison of photodynamic therapy with methylene blue associated with ceftriaxone in gram-negative bacteria; an in vitro study. *Photodiagn. Photodyn. Ther.* **2020**, *30*, 101691. [CrossRef]
54. Collins, T.L.; Markus, E.A.; Hassett, D.J.; Robinson, J.B. The effect of a cationic porphyrin on Pseudomonas aeruginosa biofilms. *Curr. Microbiol.* **2010**, *61*, 411–416. [CrossRef] [PubMed]
55. Reznick, Y.; Banin, E.; Lipovsky, A.; Lubart, R.; Polak, P.; Zalevsky, Z. The synergistic effect of visible light and gentamycin on Pseudomona aeruginosa microorganisms. *J. Vis. Exp. JoVE* **2013**, e4370. [CrossRef] [PubMed]
56. Fila, G.; Kawiak, A.; Grinholc, M.S. Blue light treatment of Pseudomonas aeruginosa: Strong bactericidal activity, synergism with antibiotics and inactivation of virulence factors. *Virulence* **2017**, *8*, 938–958. [CrossRef] [PubMed]
57. Oppezzo, O.J.; Forte Giacobone, A.F. Lethal Effect of Photodynamic Treatment on Persister Bacteria. *Photochem. Photobiol.* **2018**, *94*, 186–189. [CrossRef] [PubMed]
58. Nakonieczna, J.; Wolnikowska, K.; Ogonowska, P.; Neubauer, D.; Bernat, A.; Kamysz, W. Rose Bengal-Mediated Photoinactivation of Multidrug Resistant Pseudomonas aeruginosa Is Enhanced in the Presence of Antimicrobial Peptides. *Front. Microbiol.* **2018**, *9*. [CrossRef] [PubMed]
59. Wozniak, A.; Rapacka-Zdonczyk, A.; Mutters, N.T.; Grinholc, M. Antimicrobials Are a Photodynamic Inactivation Adjuvant for the Eradication of Extensively Drug-Resistant Acinetobacter baumannii. *Front. Microbiol.* **2019**, *10*, 229. [CrossRef]
60. Kashem, S.W.; Kaplan, D.H. Skin Immunity to Candida albicans. *Trends Immunol.* **2016**, *37*, 440–450. [CrossRef]
61. Lyon, J.P.; Carvalho, C.R.; Rezende, R.R.; Lima, C.J.; Santos, F.V.; Moreira, L.M. Synergism between fluconazole and methylene blue-photodynamic therapy against fluconazole-resistant Candida strains. *Indian J. Med. Microbiol.* **2016**, *34*, 506–508. [CrossRef]
62. Snell, S.B.; Foster, T.H.; Haidaris, C.G. Miconazole induces fungistasis and increases killing of Candida albicans subjected to photodynamic therapy. *Photochem. Photobiol.* **2012**, *88*, 596–603. [CrossRef]
63. Morton, C.O.; Chau, M.; Stack, C. In vitro combination therapy using low dose clotrimazole and photodynamic therapy leads to enhanced killing of the dermatophyte Trichophyton rubrum. *BMC Microbiol.* **2014**, *14*, 261. [CrossRef]
64. Hu, Y.; Qi, X.; Sun, H.; Lu, Y.; Hu, Y.; Chen, X.; Liu, K.; Yang, Y.; Mao, Z.; Wu, Z.; et al. Photodynamic therapy combined with antifungal drugs against chromoblastomycosis and the effect of ALA-PDT on Fonsecaea in vitro. *PLoS Negl. Trop. Dis.* **2019**, *13*, e0007849. [CrossRef] [PubMed]
65. Hu, Y.; Huang, X.; Lu, S.; Hamblin, M.R.; Mylonakis, E.; Zhang, J.; Xi, L. Photodynamic therapy combined with terbinafine against chromoblastomycosis and the effect of PDT on Fonsecaea monophora in vitro. *Mycopathologia* **2015**, *179*, 103–109. [CrossRef] [PubMed]
66. Yang, Y.; Hu, Y.; Zhang, J.; Li, X.; Lu, C.; Liang, Y.; Xi, L. A refractory case of chromoblastomycosis due to Fonsecaea monophora with improvement by photodynamic therapy. *Med. Mycol.* **2012**, *50*, 649–653. [CrossRef] [PubMed]
67. Guarro, J. Fusariosis, a complex infection caused by a high diversity of fungal species refractory to treatment. *Eur. J. Clin. Microbiol. Infect. Dis.* **2013**, *32*, 1491–1500. [CrossRef]
68. Giroldo, L.M.; Felipe, M.P.; de Oliveira, M.A.; Munin, E.; Alves, L.P.; Costa, M.S. Photodynamic antimicrobial chemotherapy (PACT) with methylene blue increases membrane permeability in Candida albicans. *Lasers Med. Sci.* **2009**, *24*, 109–112. [CrossRef]
69. Li, J.; Zhu, M.; An, L.; Chen, F.; Zhang, X. Fungicidal efficacy of photodynamic therapy using methylene blue against Sporothrix globosa in vitro and in vivo. *Eur. J. Dermatol. EJD* **2019**, *29*, 160–166. [CrossRef]
70. Gilaberte, Y.; Aspiroz, C.; Alejandre, M.C.; Andres-Ciriano, E.; Fortuño, B.; Charlez, L.; Revillo, M J.; Hamblin, M.R.; Rezusta, A. Cutaneous Sporotrichosis Treated with Photodynamic Therapy: An in Vitro and in Vivo Study. *Photomed. Laser Surg.* **2014**, *32*, 54–57. [CrossRef]

pharmaceuticals

Article

Effectiveness of Riboflavin and Rose Bengal Photosensitizer Modified Adhesive Resin for Orthodontic Bonding

Ali Alqerban [1,2]

[1] Department of Preventive Dental Sciences, College of Dentistry, Dar al Uloom University, Riyadh 45142, Saudi Arabia; a.alqerban@dau.edu.sa

[2] Department of Preventive Dental Sciences, College of Dentistry, Prince Sattam Bin Abdulaziz University, Al-Kharj 11942, Saudi Arabia

Abstract: This study aimed to evaluate the effect of riboflavin (RF) and Rose Bengal (RB) photosensitizer modified adhesive resin on the degree of conversion (DC), and antimicrobial capacity after bonded to tooth surface. Different concentrations of RB and RF were prepared by homogenization method. An ultraviolet light source A (UVA) (375 nm wavelength, 3 mW/cm^2 power) was used for 30 min irradiation. FTIR was performed for control and test adhesives to analyze the DC. Antibacterial testing was performed using the MTT assay. Metal brackets were bonded using the modified adhesives and subjected for SEM examination. The surfaces of teeth and metal brackets were examined at ×10 magnification for assessing adhesive remnant index (ARI) after PDT, 24 h and thermocycling. For DC, control group, 0.1% RB and RF after PDT showed the highest value. SEM imaging indicated lowest growth of *Streptococcus mutans* over 0.5% of RB-PDT and RF-PDT as compared to the control group. The MTT assay outcomes reported that the activity of *S. mutans* substantially decreased with the addition of a high amount of either RB or RF ($p < 0.01$). Mean ARI scores showed a significant difference between all groups. This study concluded that 0.1% of either RB or RF after PDT can be used for bonding orthodontic brackets to the tooth surface with substantial antibacterial properties.

Keywords: cross-linking agents; fluorescent dyes; oxidation reduction; photochemistry; photosensitizing agents; protein conformation; Rose Bengal

Citation: Alqerban, A. Effectiveness of Riboflavin and Rose Bengal Photosensitizer Modified Adhesive Resin for Orthodontic Bonding. *Pharmaceuticals* **2021**, *14*, 48. https://doi.org/10.3390/ph14010048

Received: 17 December 2020
Accepted: 1 January 2021
Published: 10 January 2021

Publisher's Note: MDPI stays neutral with regard to jurisdictional claims in published maps and institutional affiliations.

1. Introduction

With the advancement of dentistry and patient knowledge, orthodontic therapy has gained popularity amongst the general population. The therapy is done to provide treatment for misaligned teeth that involves the use of orthodontic wires and brackets. The brackets used can either be metal or ceramic in origin [1,2]. Although orthodontic treatment assists in creating skeletal and dental improvement, this treatment modality is also responsible for creating an environment that favors bacterial growth [3–5].

The factors responsible for the increased development of bacterial biofilm include the application of orthodontic appliances close to the soft tissue structures, malaligned teeth, and improper oral hygiene maintenance [6]. The abundance of bacterial biofilm can lead to the development of oral health problems that include plaque-induced gingivitis, white spot lesions, and oral malodor (bad breath) [7–10]. The most common treatment modality used to tackle this problem addresses the repeated application of scalers (manual or ultrasonic). In adjunct to this, the patients are also advised to follow strict oral hygiene protocols [11].

To improve oral health care and patient satisfaction, dentists and scientists have been working together to establish treatment protocols that benefit on a larger scale. One such protocol that is being readily used in dentistry as an established therapeutic regimen is a photodynamic therapy (PDT) [12–17]. This treatment involves three main components, which include laser light (specific wavelength), nascent or free oxygen, and photosensitizer (PS). The procedure for this treatment is initiated by the application of laser light on the dye molecules. These dye molecules absorb the coherent light and transform a dormant singlet state to an excited triplet state. The excited molecules then react with the free molecular oxygen to form a highly toxic product called reactive oxygen species (ROS). These species further facilitate an irreversible oxidative movement in the bacterial cell composition, which ultimately leads to bacterial cell death [18–20]. Various types of photosensitizers are used to treat oral diseases including methylene blue, toluidine blue, indocyanine green, and many others. Rose Bengal (RB) and riboflavin (RF) have been extensively used for medical applications. RB and Rf both act as cross-linking agents. These photosensitizer-like agents act as inhibitors of collagen degradation, after the destructive activity of the acid-based bacteria, such as *Streptococcus mutans* (*S. mutans*) [21].

According to the authors' knowledge, no study exists in the database that evaluated RB and RF as photosensitized proteins and modified inside the orthodontic adhesive resin. Therefore, this laboratory study aimed to characterize and evaluate the effect of RB and RF photosensitized orthodontic adhesive resin on degree of conversion, and antimicrobial capacity bonded to the tooth surface.

2. Results

The FTIR spectra of the powders riboflavin and rose Bengal are presented in Figure 1A. Characteristic peaks at 1576 cm^{-1}, 1652 cm^{-1}, and 1728 cm^{-1} are noted. The spectrum at 1728 cm^{-1} indicates C=O stretching frequency of riboflavin. In the components, the aromatic C=C stretching mode of riboflavin appears at 1576 cm^{-1} and 1652 cm^{-1} (indicated with pointers). Rose-Bengal possesses important peaks at 456 cm^{-1} which indicates C–I bonding, at 952 cm^{-1} for C–C stretching vibration, and 1335 cm^{-1} and 1613 cm^{-1} for C=C stretching vibration and C=O stretching vibration, respectively. After incorporating these powders in orthodontic adhesives, the ascribed peaks were evident forming strong bonds with the resin matrix. The degree of conversion was plotted for the area on the spectrum ranging between 1608 cm^{-1} to 1640 cm^{-1} (Figure 1B).

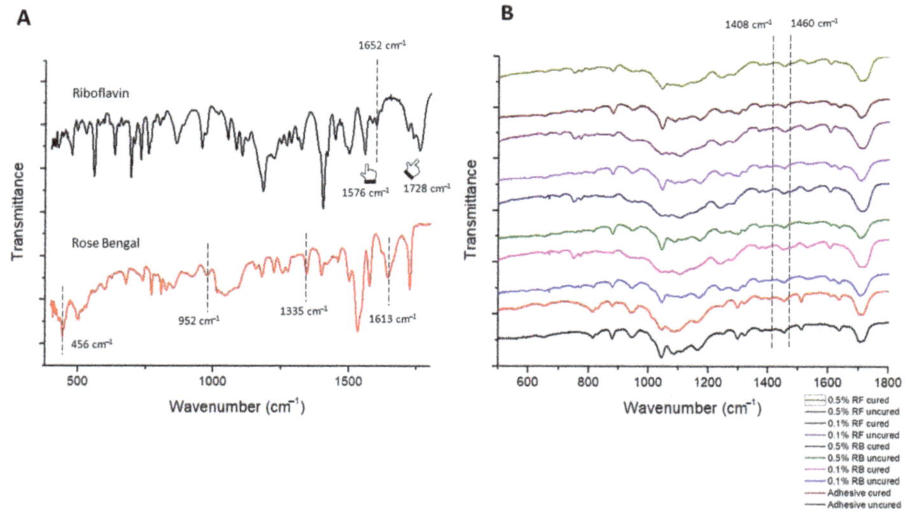

Figure 1. FTIR spectra of (**A**) riboflavin (black spectrum) and Rose Bengal (red spectrum) powder. Characteristic peaks at 1576 cm^{-1}, 1652 cm^{-1}, and 1728 cm^{-1} are noted. The spectrum at 1728 cm^{-1} indicates C=O stretching frequency of riboflavin. In the components, the aromatic C=C stretching mode of riboflavin appears at 1576 cm^{-1} and 1652 cm^{-1} (indicated with pointers). Rose-Bengal possesses important peaks at 456 cm^{-1} which indicates C–I bonding, at 952 cm^{-1} for C–C stretching vibration, and 1335 cm^{-1} and 1613 cm^{-1} for C=C stretching vibration and C=O stretching vibration, respectively. (**B**) After incorporating these powders in orthodontic adhesives, the ascribed peaks were evident forming strong bonds with the resin matrix.

For DC, the groups control, 0.1% RB and 0.1% RF after PDT showed the highest degree of conversion indicating a significant conversion of monomer from the resin followed by lowest degree of conversion of 0.5% RB and 0.5% RF adhesive after PDT (Table 1).

Table 1. Degree of conversion in percentage of all groups.

Groups	Degree of Conversion (%Mean ± SD)	Tukey ($p < 0.05$)
Transbond XT	$n = 43.5 \pm 6.6$	A
0.1% RB-PDT adhesive	$n = 45.8 \pm 7.3$	A
0.1% RF-PDT adhesive	$n = 46.2 \pm 5.8$	A
0.5% RB-PDT adhesive	$n = 41.3 \pm 4.3$	B
0.5% RF-PDT adhesive	$n = 39.7 \pm 5.1$	B

The SEM images of the bonded brackets are shown in Figure 2A–E. It is noted that all adhesive showed a considerable excellent bonding between bracket mesh and tooth surface. However, 0.5% RF-PDT indicated slightly poor bonding with defective adhesive within the mesh of the bracket (Figure 2E).

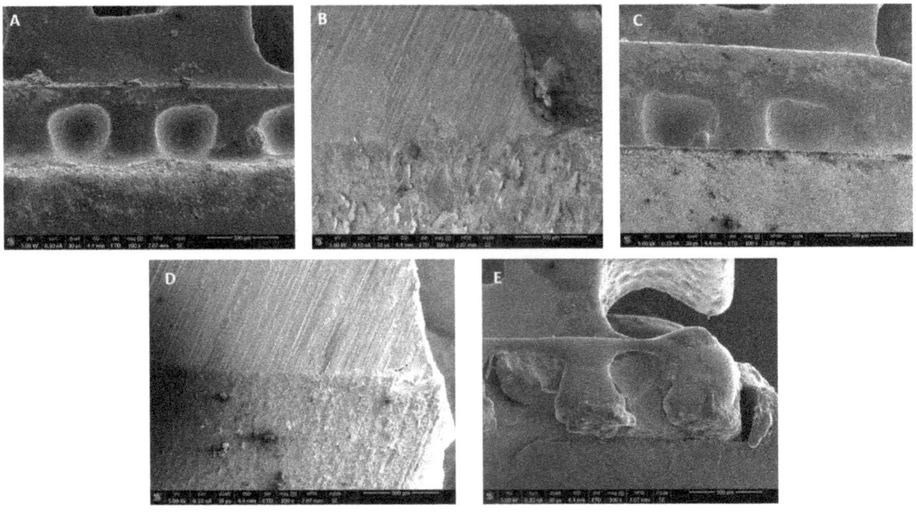

Figure 2. Representative SEM images of bracket-resin interface depicting excellent bonding of tooth surface with (**A**) Transbond XT (control); (**B**) 0.1% RB-PDT adhesive; (**C**) 0.1% RF-PDT adhesive; and (**D**) 0.5% RB-PDT adhesive; but (**E**) 0.5% RF-PDT adhesive, showed slightly poor bonding with defective adhesive within the mesh of the bracket.

SEM imaging of the experimental adhesives using 0.1% and 0.5% RB and RF indicated substantial low growth of *S. mutans* over 0.1% of RB-PDT and RF-PDT as compared to control group consisting of Transbond XT. Higher addition of RB and RF with 0.5% produced even lower growth of *S. mutans* over the modified adhesive surfaces (Figure 3A–E).

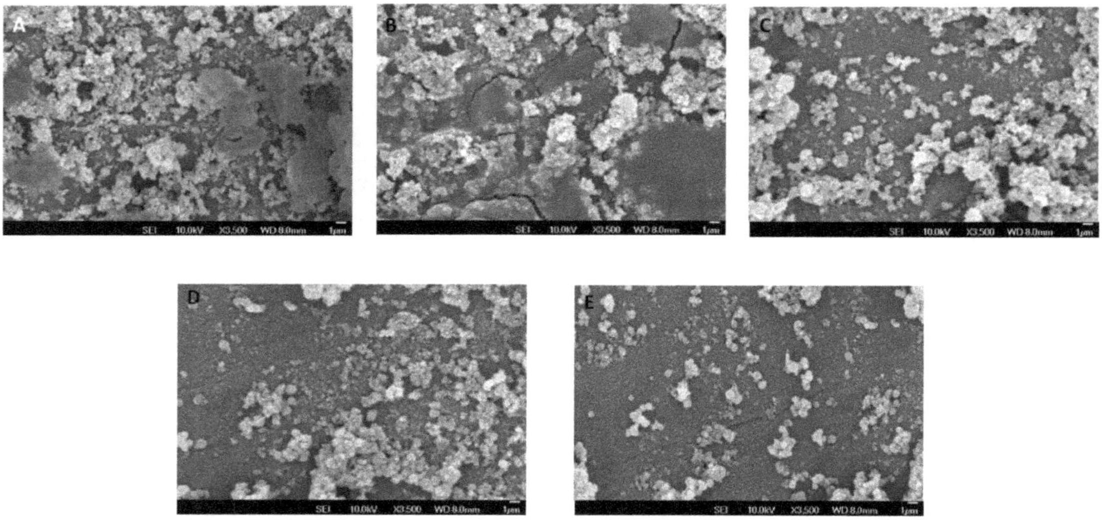

Figure 3. SEM micrographs showing growth of *Streptococcus mutans* over the surface of different modified adhesives used (**A**) Transbond XT (control); (**B**) 0.1% RB-PDT adhesive; (**C**) 0.1% RF-PDT adhesive showed low growth of *S. mutans*, Whereas (**D**) 0.5% RB-PDT adhesive and (**E**) 0.5% RF-PDT adhesive showed a substantial decrease in the microbial content of *S.mutans*.

The results of MTT assay are depicted in Figure 4. The outcomes showed that regardless of the day of assessment (Day 1 or Day 30), the metabolic activity of *S. mutans* substantially decreased with the addition of high amount of either RB or RF ($p < 0.01$). On comparison with day 1, although the metabolic activity of 0.1% RB or RF after PDT reduced to a certain extent after Day 30, 0.5% of either RB-PDT or RF-PDT showed relatively reduced viability of *S. mutans* well under 35%.

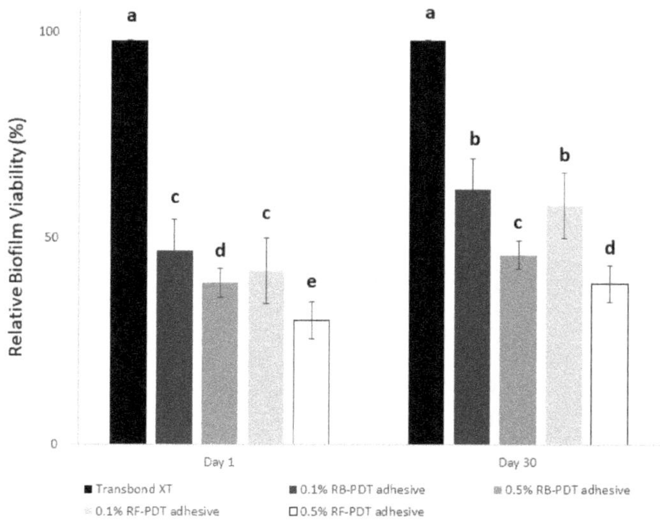

Figure 4. The relative biofilm viability expressed in percentage after Day 1 of incubation of *S. mutans* on experimental adhesives of 10 groups (5 for Day 1 groups, and 5 for Day 30 groups). The data are expressed as means and SD, groups with the dissimilar letters are statistically significant ($p < 0.05$).

Table 2 depicts the results for ARI. The results of the ARI are shown in Table 2. When ARI was compared between groups in different time points (immediately after bonding, after 24 h bonding and after thermocycling), there was a significant difference in the mean ARI scores between all five groups.

Table 2. Mean adhesive remnant index (ARI) scores for all groups immediately after photodynamic therapy (PDT), after 24 h and after thermocycling.

Groups (Time Point)	Mean ARI	*p*-Value
Transbond XT		
Immediately after	$n = 1.33 \pm 0.57$	
After 24 h	$n = 1.27 \pm 0.69$	
After thermocycling	$n = 1.24 \pm 0.71$	
0.1% RB-PDT adhesive		
Immediately after	$n = 1.47 \pm 0.65$	
After 24 h	$n = 1.42 \pm 0.71$	
After thermocycling	$n = 1.37 \pm 0.75$	
0.1% RF-PDT adhesive		
Immediately after	$n = 1.72 \pm 0.55$	
After 24 h	$n = 1.68 \pm 0.46$	$n = 0.022$
After thermocycling	$n = 1.61 \pm 0.84$	
0.5% RB-PDT adhesive		
Immediately after	$n = 1.91 \pm 0.85$	
After 24 h	$n = 1.73 \pm 0.77$	
After thermocycling	$n = 1.66 \pm 0.90$	
0.5% RF-PDT adhesive		
Immediately after	$n = 1.66 \pm 0.83$	
After 24 h	$n = 1.62 \pm 0.97$	
After thermocycling	$n = 1.58 \pm 0.74$	

3. Discussion

According to the data obtained after a comprehensive literature review, the present in-vitro experiment is said to be a novel study as it focused on orthodontic resin modification using RB and Rf with different concentrations (0.1% and 0.5%) after incorporating PDT to assess its effect on properties such as degree of conversion (DC), antimicrobial capacity and adhesive remnant index (ARI), respectively. According to the obtained results from the study, Group II and Group IV obtained better and equal results in comparison to the other included groups.

The resin–dentin interface is a very important interaction when it comes to restorative or orthodontic treatment. The durability of this interface comes into question when resistant oral bacterial influx usually harbors along with this complex interface. With increased acid accumulation, the adhesive–dentin interface is subjected to microleakage of bacterial by-products, further leading to secondary caries [22]. The secondary caries is instigated via the collagenase enzyme which is responsible for collagen degradation present in dentin [23]. During the procedure of bonding brackets, the conventional acid-etch technique also accounts for the production of an acidic environment on the tooth surface. This procedure allows the formation of micropores inside the tooth enamel by dissolving the calcium and phosphate ions [24]. So, to avoid further enamel degradation, the adhesive agent with antimicrobial properties is an absolute indication for tooth bonding. In our study, modified adhesives which included photosensitizing agents (RB, RF) were manufactured to test their effect in the presence of PDT. RB is a xanthene based fluorescent dye which has been used as a photosensitizing agent in numerous studies [25]. Being photoactive, this dye has been used to explore anti-microbial activity. Moreover, it has been used as a diagnostic tool in cancer and corneal surface defects. RF is a redox biological growth factor that belongs to the flavin group. Being a photosensitizer, it undergoes photoreactions at the nuclear levels to sensitize tumor cell destruction [26]. It has also been used for antimicrobial, antiviral and blood sensitizing applications. Furthermore, RF serves as a photoactivated cross-linking agent in corneal stiffness. Both the photosensitizers achieve intersystem conversion with the help of photo-illumination [27–29]. To achieve this, PDT was employed. PDT is based on a biochemical process that involves the use of a dye molecule in the presence of laser light of a specific wavelength and free oxygen. The exposure of laser light on the dye molecules instigates the process of conversion. This conversion is achieved by activation of dye molecules from a singlet dormant state to an excited triplet state. The excited triplet state reacts with the nascent oxygen to form reactive oxygen species (ROS). Being highly toxic, these free radical molecules play a significant role in the process of bacterial cell death [30–32]. According to previous studies, both RB and Rf are termed as photosensitive agents with excellent antimicrobial properties [33–35].

Regarding the aspect of tooth protein denaturation after thermal cycling, recent studies suggested that the photosensitizers RB and RF produce no effects. This can be supported by the fact that both photosensitizers differ in cross-linking procedures, as RF freely diffuses whereas, RB adheres tightly to the collagen matrix [36,37].

The degree of conversion (DC) is of great importance as it governs the physical and mechanical properties of the adhesive resins. It is also directly related to the conversion of monomers during the polymerization process [38]. According to the results of the present study, the DC is more prominent in control groups (0.1% RB-PDT and 0.1% RF-PDT) as compared to the test groups (0.5% RB-PDT and 0.5% RF-PDT). According to Table 1, a decrease in the levels of DC is observed after the increase in the concentration of both photosensitizers. A possible explanation for this finding could be because of the agglomeration of both the photosensitizers (RB, RF) used which could impede the curing process [39].

SEM images have been described in Figures 2 and 3 In Figure 2, the images obtained depict that a strong bond was achieved between the enamel surface of the tooth and mesh of applied brackets. However, the use of 0.5% RF-PDT displayed relatively poor bonding in comparison to the other modified adhesive resins being used. The poor bonding observed might be due to the increase in the composition of riboflavin in the modified adhesive, which may have provided a shielding effect during the curing process of the bracket on the enamel surface [40]. For Figure 3, the use of RB-PDT and RF-PDT (0.5% concentration) produced a significant reduction in the *S. mutans* count. It has been reported in studies that RF and RB undergo conversion under the effect of photo-illumination by initiating a redox reaction which in turn causes changes in the bacterial cell contents, leading to necrosis [41].

Few limitations were observed for the present study. In this research, the use of Transmission electron microscope to study the sections after preparing photosensitizers mixed with orthodontic resins would have provided more accurate blending of the materials within resin matrix. Moreover, the translation of this in-vitro study into clinical experiments with the photodynamic system inside the oral cavity could result in different outcomes. Therefore, such studies are warranted in future. In addition, the use of a single type of bracket i.e., metallic is a limitation that can be addressed in other trials. The use of ceramic brackets alongside the modified adhesives could have provided data that would have been useful in devising better treatment protocol options. A reduced sample size could be another limitation for this in-vitro study. Future studies should consider more samples and compare this system within different types of ceramics surfaces and types of teeth other than molars.

4. Materials and Methods

4.1. Materials and Chemicals

The materials used for the present in-vitro study included riboflavin and Rose Bengal and MTT assay kit ordered from Sigma Aldrich (St. Louis, MO, USA). The metallic brackets (0.022 × 0.028-in and 0.022 × 0.030-in) and orthodontic adhesive (Transbond XT) was ordered from 3M, Unitek (St. Paul, MN, USA).

4.2. Specimen Preparation

This experimental study was performed on 60 extracted human molar teeth which were acquired from patients who underwent extraction. The extracted teeth were thoroughly cleaned and rinsed with saline and later stored in 0.5% chloramine T solution at 4 °C until further use. Before establishing the criterion for inclusion and exclusion, all the teeth were examined thoroughly with the naked eye and under a stereomicroscope with ×10 magnification. Extracted molar teeth were visually inspected for any fractures, enamel loss, or caries.

4.3. Photosensitizer-Modified Experimental Orthodontic Adhesives

The 0.1 and 0.5 wt.% RB and RF modified adhesives were prepared by homogenization method, respectively. A total of 5 mg and 25 mg of the powders (RB and RF) were added in 5 mL of orthodontic resin (Transbond XT, 3M, Unitek, St. Paul, MN, USA), respectively [21]. The mixtures were given a thorough mixing in an ultra-homogenizer sonicator (SALD 2300 Shimadzu, Shimadzu Corporation, Kyoto, Japan) for 10 min. These prepared solutions were kept in the dark to avoid any occurrence of photobleaching.

4.4. Groups

The experimental groups were divided based on the orthodontic adhesive modification. The groups were divided into following 5 groups, Group I: Transbond XT (control), Group II: 0.1% RB–PDT adhesive, Group III: 0.1% RF–PDT adhesive, Group IV: 0.5% RB–PDT adhesive and Group V: 0.5% RF–PDT adhesive.

4.5. Photodynamic Therapy (PDT) Protocol

The PDT protocol for this particular research was done per the method described by Arboleda et al. [21]. The samples were fixed in a petri dish. The petri-dish was set 1 cm from the irradiation source. The total spot diameter was carefully checked to be at 8 mm. Ultraviolet light source A (UVA) of 375 nm wavelength was employed to irradiate the petri-dish. The value of irradiance was 3 mW/cm^2. Moreover, the time of exposure was 30 min.

4.6. Degree of Conversion

Fourier-transformed infrared spectroscopy (FTIR) (ThermoFischer Scientific, Waltham, MA, USA) was performed for control and test adhesives to analyze the degree of conversion (DC). For the spectral analysis, the uncured adhesive specimen was exposed to oil-free air, whereas the cured adhesive specimen was constantly kept in contact with the sensor. The absorbance peaks for the uncured specimens were measured by using FTIR spectrometer (Nicolet 6700, ThermoFisher Scientific, Waltham, MA, USA). To measure the degree of conversion, the range of spectrum was set between 400–4000 cm^2. The same analytical tests were used for the cured specimens after curing. A baseline technique was used to obtain the C=C absorbance peak/1638 cm^{-1} (unpolymerized methacrylate stretching vibration) and C–C reference peak at 1607 cm^{-1} (stretching vibration of the aromatic ring). The formula used to calculate the degree of conversion is as follows:

$$\text{Degree of conversion} = [1 - (C_{\text{aliphatic}}/C_{\text{aromatic}})/(U_{\text{aliphatic}}/U_{\text{aromatic}})] \times 100\% \quad (1)$$

4.7. Anti-Bacterial Testing

The bacterial strain of *S. mutans* was grown for two days at a temperature of 37 °C in an anaerobic environment. A total of nine specimens using both the control and modified orthodontic resin specimens (0.1% RB–PDT adhesive, 0.5% RB–PDT adhesive, 0.1% RF–PDT adhesive and 0.5% RF–PDT adhesive) were used for antimicrobial testing. To obtain the inoculation medium, the prepared medium of BHI supplementing 1% (*w*/*v*) sucrose was diluted. Each adhesive disc was placed in each well of 48 well-plate. Each well plate was carefully smeared in 1 mL of inoculation medium, before being incubated for 2 days for growth of *S. mutans*, in an anaerobic environment. The temperature was set at 37 °C. The buffer solution was used to gently rinse the *S. mutans*-coated specimens.

The viability of *S. mutans* biofilm was tested using the MTT assay. Specimen from each beam was transferred to one well of 24-well plate containing 1 mL of 3-(4,5-Dimethyl-2 thiazolyl)-2,5-diphenyl-2*H*-tetrazolium bromide (MTT; Sigma-Aldrich, St. Louis, MO, USA) solution, and incubated at 37 °C for 3 h anaerobically. After that, the MTT solution of each well was washed out and replaced by 2 mL of DMSO to dissolve the formazan. This was followed by gently shaking for 15 min and the supernatant collected from each well was estimated at 600 nm using a spectrophotometer (ThermoFischer Scientific, Waltham, MA, USA) after 24 h and 30 days.

4.8. Placement of Brackets on Enamel Surface

Metal brackets (3M, Unitek, St. Paul, MN, USA; MBT, slot 0.022 × 0.028 inch) were used to be bonded on extracted molar teeth. Transbond XT light cure primer and adhesive modified with 0.1% and 0.5% RB–PDT adhesive and 0.1% and 0.5% RF–PDT adhesive along with the control adhesive were used to bond brackets by using the conventional etch-and-rinse protocol. The etchant gel (37% phosphoric acid) was applied for 30 s on all the enamel surfaces, after being rinsed with water and dried for 20 s each. All the selected teeth in each group were subjected to their respective resin application. The process of light-curing was applied for 20 s where mesial and distal sides were each subjected to 10 s, respectively.

4.9. Scanning Electron Microscopy (SEM)

A random sample from each of the study groups was selected for the studying resin-bracket interface and bacterial culture over the discs under a scanning electron microscope (FEI, Tokyo, Japan). The process of fixation of these specimens was done by using a combination of 4% paraformaldehyde and 2% glutaraldehyde in 0.1 M sodium cacodylate (NaCac) buffer solution. The pH for these samples was maintained at 7.4 and was allowed to be stored overnight. The specimens were then subjected to 2% osmium tetraoxide before undergoing dehydration via ethanol (50 to 100%). These dried discs were then taped with double-sided copper tape. 10 kV disc images were obtained with the help of a through-lens detector (TLD) for secondary electron imaging.

4.10. Adhesive Remnant Index (ARI) Assessment

After the process of enamel debonding, the surfaces of teeth and metal brackets were examined with the help of a stereomicroscope (EMZ-TR, MEIJI, Saitama, Japan) at ×10 magnification. The ARI scoring system was employed to score the remnants of adhesive resin. The scoring system was graded as:

0—total absence of resin on the surface enamel, 1—<50% of resin remaining on the surface enamel, 2—>50% of resin remaining on the surface enamel, 3—the complete presence of resin on the surface enamel having a complete impression of the base of metallic brackets. ARI was estimated immediately after PDT, after 24 h and after thermocycling process.

4.11. Statistical Analysis

The statistical analysis for the obtained data was executed by using the SPSS software (Version 26, SPSS Inc., IBM, Armonk, NY, USA). The data was reported in mean and standard deviation (mean ± SD). The value of statistical significance for each test was set at $p < 0.05$. The one-way ANOVA with Student–Neumann–Keuls post hoc test for the assessment of differences in the degree of conversion. For the MTT assay analysis, the Friedman test with post-hoc Tukey's Kramer test was performed, whereas the Kruskal–Wallis test was employed for analyzing the ARI scores.

5. Conclusions

This study showed that a versatile RB or riboflavin modified orthodontic adhesive was developed, and the impact of addition of RB and RF after photoillumination on antimicrobial capacity and orthodontic bonding characteristics were studied. Our study suggested that 0.1% of either RB or RF after PDT can be used for bonding orthodontic brackets to the tooth surface with substantial antibacterial properties. Increased concentration of RF (0.5%) produced better antimicrobial results, but relatively poor bonding properties.

Funding: This research was funded by the Deanship of Post Graduate Studies and Dar Al Uloom University, Saudi Arabia.

Institutional Review Board Statement: The study was conducted according to the guidelines of the Declaration of Helsinki and approved by the Ethics Committee of Dar al Uloom University (protocol code REC013/2020 and date of approval: 2 November 2020).

Informed Consent Statement: Not applicable.

Data Availability Statement: Not applicable.

Conflicts of Interest: The author declares no conflict of interest.

References

1. Mei, L.; Chieng, J.; Wong, C.; Benic, G.; Farella, M. Factors affecting dental biofilm in patients wearing fixed orthodontic appliances. *Prog. Orthod.* **2017**, *18*, 4. [CrossRef] [PubMed]
2. Zhang, M.; McGrath, C.; Hägg, U. Changes in oral health-related quality of life during fixed orthodontic appliance therapy. *Am. J. Orthod. Dentofacial Orthop.* **2008**, *133*, 25–29. [CrossRef] [PubMed]
3. Pachevska, A.V.; Filimonov, Y.V.; Filimonov, V.Y.; Dudik, O.P.; Popova, O.I.; Drachuk, N.V.; Kasianenko, D.M.; Biloshitska, A.V.; Istoshyn, V.M. Clinical and laboratory assessment the levels of oral hygiene, total protein, hydrogen sulfide and nitrogen metabolites in oral fluid in the development of inflammatory complications during orthodontic treatment of children. *Wiad Lek* **2019**, *72*, 744–747. [CrossRef] [PubMed]
4. Ristic, M.; Vlahovic Svabic, M.; Sasic, M.; Zelic, O. Clinical and microbiological effects of fixed orthodontic appliances on periodontal tissues in adolescents. *Orthod. Craniofacial Res.* **2007**, *10*, 187195. [CrossRef] [PubMed]
5. Verrusio, C.; Iorio-Siciliano, V.; Blasi, A.; Leuci, S.; Adamo, D.; Nicolò, M. The effect of orthodontic treatment on periodontal tissue inflammation: A systematic review. *Quintessence Int.* **2018**, *49*, 69–77. [PubMed]
6. Panhóca, V.H.; Esteban Florez, F.L.; Corrêa, T.Q.; Paolillo, F.R.; de Souza, C.W.; Bagnato, V.S. Oral decontamination of orthodontic patients using photodynamic therapy mediated by bluelight irradiation and curcumin associated with sodium dodecyl sulfate. *Photomed. Laser Surg.* **2016**, *34*, 411–417. [CrossRef] [PubMed]
7. Martha, K.; Mezei, T.; Janos, K. A histological analysis of gingival condition associated with orthodontic treatment. *Rom. J. Morphol. Embryol.* **2013**, *54*, 823–827. [PubMed]
8. Al Nazeh, A.; Alshahrani, A.A.; Almoammar, S.; Kamran, M.A.; Togoo, R.A.; Alshahrani, I. Application of photodynamic therapy against periodontal bacteria in established gingivitis lesions in adolescent patients undergoing fixed orthodontic treatment. *Photodiag. Photodyn. Ther.* **2020**, *1*, 101904. [CrossRef]
9. Huang, J.; Li, C.Y.; Jiang, J.H. Effects of fixed orthodontic brackets on oral malodor: A systematic review and meta-analysis according to the preferred reporting items for systematic reviews and meta-analyses guidelines. *Medicine (Baltimore)* **2018**, *97*, e0233. [CrossRef]
10. Gómez, C.; Abellán, R.; Palma, J.C. Efficacy of photodynamic therapy versus ultrasonic scaler for preventing gingival inflammation and white spot lesions during orthodontic treatment. *Photodiag. Photodyn. Ther.* **2018**, *24*, 377–383. [CrossRef]
11. Julien, K.C.; Buschang, P.H.; Campbell, P.M. Prevalence of white spot lesion formation during orthodontic treatment. *Angle Orthod.* **2013**, *83*, 641–647. [CrossRef] [PubMed]
12. Akram, Z.; Raffat, M.A.; Shafqat, S.S.; Mirza, S.; Ikram, S. Clinical efficacy of photodynamic therapy as an adjunct to scaling and root planing in the treatment of chronic periodontitis among cigarette smokers: A systematic review and meta-analysis. *Photodiag. Photodyn. Ther.* **2019**, *26*, 334–341. [CrossRef] [PubMed]
13. Akram, Z.; Shafqat, S.S.; Niaz, M.O.; Raza, A.; Naseem, M. Clinical efficacy of photodynamic therapy and laser irradiation as an adjunct to open flap debridement in the treatment of chronic periodontitis: A systematic review and meta-analysis. *Photodermatol. Photoimmunol. Photomed.* **2020**, *36*, 3–13. [CrossRef] [PubMed]
14. Kamran, M.A. Clinical, microbiological and immunological outcomes with photodynamic therapy as an adjunct to full-mouth scaling in patients undergoing fixed orthodontic treatment. *Photodiag. Photodyn. Ther.* **2020**, *29*, 101585. [CrossRef]
15. Akram, Z.; Javed, F.; Hosein, M.; Al-Qahtani, M.A.; Alshehri, F.; Alzahrani, A.I.; Vohra, F. Photodynamic therapy in the treatment of symptomatic oral lichen planus: A systematic review. *Photodermatol. Photoimmunol. Photomed.* **2018**, *34*, 167–174. [CrossRef]

16. Almohareb, T.; Alhamoudi, N.; Al Deeb, M.; Bin-Shuwaish, M.S.; Mokeem, S.A.; Shafqat, S.S.; Vohra, F.; Abduljabbar, T. Clinical efficacy of photodynamic therapy as an adjunct to mechanical debridement in the treatment of per-implantitis with abscess. *Photodiagnosis Photodyn. Ther.* **2020**, *30*, 101750. [CrossRef]
17. Mirza, S.; Khan, A.A.; Al-Kheraif, A.A.; Khan, S.Z.; Shafqat, S.S. Efficacy of adjunctive p.hotodynamic therapy on the clinical periodontal, HbA1c and advanced glycation end product levels among mild to moderate chronic periodontal disease patients with type 2 diabetes mellitus: A randomized controlled clinical trial. *Photodiagnosis Photodyn Ther.* **2019**, *28*, 177–182. [CrossRef]
18. Alhamoudi, N.; Mokeem, S.; Shafqat, S.S.; Vohra, F.; Abduljabbar, T. Effectiveness of antimicrobial photodynamic therapy as an adjunct to open flap debridement in patients with aggressive periodontitis. *Photodiagnosis Photodyn Ther.* **2020**, *3*, 102075. [CrossRef]
19. Javed, F.; Salehpoor, D.; Al-Dhafeeri, T.; Yousuf, M.; Malmstrom, H.; Khan, J.; Akram, Z. Is adjunctive photodynamic therapy more effective than scaling and root planing alone in the treatment of periodontal disease in hyperglycemic patients? A systematic review. *Photodiagnosis Photodyn. Ther.* **2018**, *22*, 1–6. [CrossRef]
20. Vohra, F.; Akram, Z.; Bukhari, I.A.; Sheikh, S.A.; Javed, F. Short-term effects of adjunctive antimicrobial photodynamic therapy in obese patients with chronic periodontitis: A randomized controlled clinical trial. *Photodiagnosis Photodyn. Ther.* **2018**, *21*, 10–15. [CrossRef]
21. Arboleda, A.; Miller, D.; Cabot, F.; Taneja, M.; Aguilar, M.C.; Alawa, K.; Amescua, G.; Yoo, S.H.; Parel, J.M. Assessment of rose bengal versus riboflavin photodynamic therapy for inhibition of fungal keratitis isolates. *Am. J. Ophthalmol.* **2014**, *158*, 64–70. [CrossRef] [PubMed]
22. Cheng, L.; Zhang, K.; Zhang, N.; Melo, M.; Weir, M.; Zhou, X.; Bai, Y.; Reynolds, M.; Xu, H. Developing a new generation of antimicrobial and bioactive dental resins. *J. Dent. Res.* **2017**, *96*, 855–863. [CrossRef] [PubMed]
23. Chenicheri, S.R.U.; Ramachandran, R.; Thomas, V.; Wood, A. Insight into Oral Biofilm: Primary, Secondary and Residual Caries and Phyto-Challenged Solutions. *Open Dent. J.* **2017**, *11*, 312–333. [CrossRef] [PubMed]
24. Holzmeier, M.; Schaubmayr, M.; Dasch, W.; Hirschfelder, U. A New Generation of Self-etching Adhesives: Comparison with Traditional Acid Etch Technique. *J. Orofac Orthop* **2008**, *69*, 78–93. [CrossRef]
25. Miller, J.S. Rose bengal-sensitized photooxidation of 2-chlorophenol in water using solar simulated light. *Water Res.* **2005**, *39*, 412–422. [CrossRef]
26. Insińska-Rak, M.; Sikorski, M. Riboflavin interactions with oxygen-a survey from the photochemical perspective. *Chemistry* **2014**, *20*, 15280–15291. [CrossRef]
27. Abrahamse, H.; Hamblin, M.R. New photosensitizers for photodynamic therapy. *Biochem. J.* **2016**, *473*, 347–364. [CrossRef]
28. Costa, A.C.; Rasteiro, V.M.; Pereira, C.A.; Rossoni, R.D.; Junqueira, J.C.; Jorge, A.O. The effects of rose bengal- and erythrosine-mediated photodynamic therapy on Candida albicans. *Mycoses* **2012**, *55*, 56–63. [CrossRef]
29. Spikes, J.D.; Shen, H.R.; Kopecková, P.; Kopecek, J. Photodynamic crosslinking of proteins. III. Kinetics of the FMN- and rose bengal-sensitized photooxidation and intermolecular crosslinking of model tyrosine-containing N-(2-hydroxypropyl) methacrylamide copolymers. *Photochem Photobiol.* **1999**, *70*, 130–137. [CrossRef]
30. Akram, Z.; Abduljabbar, T.; Sauro, S.; Daood, U. Effect of photodynamic therapy and laser alone as adjunct to scaling and root planing on gingival crevicular fluid inflammatory proteins in periodontal disease: A systematic review. *Photodiagnosis Photodyn. Ther.* **2016**, *16*, 142–153. [CrossRef]
31. Kwiatkowski, S.; Knap, B.; Przystupski, D.; Saczko, J.; Kędzierska, E.; Knap-Czop, K.; Kotlińska, J.; Michel, O.; Kotowski, K.; Kulbacka, J. Photodynamic therapy—Mechanisms, photosensitizers and combinations. *Biomed. Pharmacother.* **2018**, *106*, 1098–1107. [CrossRef] [PubMed]
32. Prażmo, E.J.; Kwaśny, M.; Łapiński, M.; Mielczarek, A. Photodynamic Therapy as a Promising Method Used in the Treatment of Oral Diseases. *Adv. Clin. Exp. Med.* **2016**, *25*, 799–807. [CrossRef] [PubMed]
33. Kundu, B.; Sarkar, D.; Ray, N.; Talukdar, A. Understanding the riboflavin biosynthesis pathway for the development of antimicrobial agents. *Med. Res. Rev.* **2019**, *39*, 1338–1371. [PubMed]
34. Khan, S.; Rayis, M.; Rizvi, A.; Alam, M.M.; Rizvi, M.; Naseem, I. ROS mediated antibacterial activity of photoilluminated riboflavin: A photodynamic mechanism against nosocomial infections. *Toxicol. Rep.* **2019**, *9*, 136–142. [CrossRef]
35. Anju, V.T.; Paramanantham, P.; Lal, S.B.S.; Sharan, A.; Alsaedi, M.H.; Dawoud, T.M.S.; Asad, S.; Busi, S. Antimicrobial photdynamic activity of rose bengal conjugated multi walled carbon nanotubes against planktonic cells and biofilm of *Escherichia coli. Photodiagnosis Photodyn. Ther.* **2018**, *24*, 300–310.
36. Fuentes-Lemus, E.; Mariotti, M.; Reyes, J.; Leinisch, F.; Hägglund, P.; Silva, E.; Davies, M.J.; López-Alarcón, C. Photo-oxidation of lysozyme triggered by riboflavin is O2-dependent, occurs via mixed type 1 and type 2 pathways, and results in inactivation, site-specific damage and intra- and inter-molecular crosslinks. *Free Radic. Biol. Med.* **2020**, *152*, 61–73. [CrossRef]
37. Leinisch, F.; Mariotti, M.; Rykaer, M.; Lopez-Alarcon, C.; Hägglund, P.; Davies, M.J. Peroxyl radical- and photo-oxidation of glucose 6-phosphate dehydrogenase generates cross-links and functional changes via oxidation of tyrosine and tryptophan residues. *Free Radic. Biol. Med.* **2017**, *112*, 240–252. [CrossRef]
38. Alkhudhairy, F.; AlKheraif, A.A.; Naseem, M.; Vohra, F.; Khan, R. Degree of conversion and depth of cure of Ivocerin containing photo-polymerized resin luting cement in comparison to conventional luting agents. *Pak. J. Med. Sci.* **2018**, *34*, 253–259. [CrossRef]
39. Hashem, M. Antimicrobial capacity and physico-chemical characteristics of adhesive resin containing riboflavin after photodynamic therapy. *Photodiag. Photodyn. Ther.* **2020**. [CrossRef]

40. Daood, U.; Omar, H.; Qasim, S.; Nogueira, L.P.; Pichika, M.R.; Mak, K.K.; Steier, L.; Cky, Y.; Lin, S.L.; Fawzy, A.S. New antimicrobial and collagen crosslinking formulated dentin adhesive with improved bond durability. *J. Mech. Behav. Biomed. Mater.* **2020**, *110*, 103927. [CrossRef]
41. Khan, S.; Arif, S.H.; Naseem, I. Interaction of aminophylline with photoilluminated riboflavin leads to ROS mediated macromolecular damage and cell death in benzopyrene induced mice lung carcinoma. *Chem. Biol. Interact.* **2019**, *302*, 135–142. [CrossRef] [PubMed]

pharmaceuticals

MDPI

Article

Antimicrobial Capacity and Surface Alterations Using Photodynamic Therapy and Light Activated Disinfection on Polymer-Infiltrated Ceramic Material Contaminated with Periodontal Bacteria

Elzahraa Eldwakhly [1,2], Selma Saadaldin [3], Alhanoof Aldegheishem [2,*], Marwa Salah Mostafa [4] and Mai Soliman [2]

1 Department of Fixed Prosthodontics, Faculty of Dentistry, Cairo University, Cairo 12613, Egypt; zeldwakhly@yahoo.com
2 Clinical Dental Science Department, College of Dentistry, Princess Nourah Bint Abdulrahman University, P.O. Box 84428, Riyadh 11564, Saudi Arabia; msmustafa@pnu.edu.sa
3 Division of Prosthodontics, Schulich School of Medicine and Dentistry, Western University, London, ON N6A 3K7, Canada; selsaadaldin@gmail.com
4 Department of Medical Microbiology and Immunology, Faculty of Medicine, Cairo University, Cairo 12613, Egypt; marwa75jan@yahoo.com
* Correspondence: asaldegheishem@pnu.edu.sa

Received: 20 September 2020; Accepted: 26 October 2020; Published: 29 October 2020

Abstract: This study determined the antimicrobial efficiency of light-activated disinfection (LAD) and photodynamic therapy (PDT) on polymer-infiltrated ceramic network (PICN) material contaminated with three periodontal bacteria and explored if PDT and LAD cause PICN surface alterations. Sixty PICN discs were contaminated with *Tannerella forsythia*, *Porphyromonas gingivalis*, and *Treponema denticola* and randomly divided into five groups ($n = 12$ samples/each) according to the treatment groups: Group PDT—PDT (630 ± 10 nm diode laser) with methylene blue; Group DL—808 nm diode laser in contact mode without photosensitizer; Group MB–methylene blue without light application; Group CHX—0.12% chlorhexidine digluconate solution and; Group NT—no treatment. Each disc was then placed in tubes containing phosphate buffered saline (PBS) and vortexed for 30 s to remove the remaining bacteria from the discs. A total of 10× serial dilutions were performed followed by plating of 30 µL of suspension on Brucella agar plates. The colony forming units (CFU) were calculated after 72 h. PICN discs with the attached biofilms were used for confocal microscopy investigation for live/dead bacterial viability. A random single sample from each group was selected to study the bacterial adherence and topographical alterations on PICN discs under scanning electron microscope (SEM). The PDT group showed higher reduction for each bacterial species and total counts of bacteria assessed followed by the DL group ($p < 0.05$). When compared with MB group, the two laser groups were significantly superior ($p < 0.05$). The MB group did not show significant differences for any bacteria when compared to NT. The bacteria with the CHX group and DL groups appeared dead with few areas of surviving green stained bacteria. The PDT group showed the highest dead cell count ($p < 0.05$). PDT and DL groups indicate no significant changes on the surface compared to the sterile PICN discs on visual assessment. Photodynamic therapy produced superior periodontal bacteria reduction over the surface of PICN surface. PDT group showed higher reduction for each bacterial species and total counts of bacteria assessed followed by the DL group. Both PDT and DL treatment strategies are effective without producing surface alterations on PICN.

Keywords: photodynamic therapy; laser; ceramics; bacteria; decontamination

---

1. Introduction

Biological characteristics of indirect restorative materials is pertinent especially for fixed prostheses such as crowns and bridges. Such prostheses are in direct contact with the gingival tissue and can extend down to a certain depth inside the gingival crevice [1]. Hence, a connection establishes between the internal soft tissues with the external oral environment which creates an appropriate seal in order to form a hermetic barrier to protect periodontal soft and hard tissues from bacterial insult [2]. If bacterial penetration exceeds a limit, both hard and soft tissue destruction occurs in the form of bone resorption and soft tissue recession which consequently leads to the fixed prostheses failure and compromised esthetics [3,4].

Resin-based composites and ceramics are commonly used materials for dental restorations. However, these materials possess some limitations in terms of clinical performance either in terms of polymerization shrinkage, marginal adaptation, anatomic shape, or color match [5]. For optimizing the clinical performance of these materials, the manufacturers combined composite resins and ceramics to produce material by the name polymer-infiltrated ceramic networks (PICNs) [6,7]. Among these, polymer-infiltrated ceramic crowns (PICC) have gained more attention due to their superior mechanical and esthetic properties.

Like any other dental restorations, crowns in the oral cavity tend to create microbial plaque within the gingival crevice primarily due to improper oral hygiene, or other reasons such as inaccurate tooth preparation or ill-fitting prostheses [8]. Consequently, this leads to the development of specific oral infections called periodontal diseases, if untreated, may lead to crown failure and eventual tooth loss [8]. Periodontal inflammation is caused by the adhesion of anaerobic microorganisms such as *Porphyromonas gingivalis*, *Tannerella forsythia* and *Treponema denticola* [9]. Several therapeutic strategies are employed in order to eliminate the periodontal bacteria and reduce the severity of the infection. The most widely performed method is to disinfect the oral cavity by performing dental scaling and use of adjunctive chlorhexidine biguanide [10,11]. However, it is often difficult to eliminate the bacteria from inaccessible deep periodontal spaces between the crown and tooth junction [12]. Therefore, to surmount this limitation, other promising therapeutic strategies such as diode laser (DL) therapy and photodynamic therapy (PDT) have been introduced and being researched upon for more than three decades [13]. Light-activated disinfection is based on intensification of electromagnetic fields excited by light waves to emit well-collimated, coherent, and monochromatic laser beam [14]. On the other hand, PDT is a well-known technique which relies on laser application of specific wavelength and involves the excitation of photosensitizer dye molecule from ground singlet state to hyper triplet state in the presence of oxygen to form highly reactive singlet oxygen and other reactive oxygen species. These molecules are highly fatal and facilitates bacterial cell death [15].

To the authors knowledge, no study has been performed that assessed antimicrobial capacity against periodontal bacteria and evaluated surface alterations using PDT and DL. We aimed to evaluate the antimicrobial efficacy of PDT and DL on PICN discs contaminated with three periodontal species and investigate if the PDT and DL cause PICN surface alterations. The null hypotheses of the study were: (i) the laser therapies would not affect the microbial viability over the surface of the PICC and (ii) the laser therapies would not produce any surface alterations on the PICC.

2. Results

2.1. Antibacterial Testing

Table 1 demonstrates the reduction of each bacterial species and the total bacterial count. Intergroup comparison showed statistically significant differences with regards to the three bacterial species individually and the total bacterial count ($p < 0.05$). For total colony forming units (CFUs), all the groups showed statistically significant reduction compared with the no treatment (NT) group ($p < 0.05$). It is noted that the reduction was >98% in each group. The PDT group showed the highest reduction for each bacterial species and total counts of bacteria assessed followed by the DL group ($p < 0.05$).

When compared with MB group, the two laser groups were significantly superior ($p < 0.05$). The MB group did not show significant differences for any bacteria when compared to NT. Figure 1 shows microbial cell viability in percentage for all the bacteria assessed. It is noted that the lowest bacterial viability was seen in PDT, DL and CHX groups with no significant difference between the groups ($p > 0.05$).

Table 1. Mean and standard deviation of periodontal bacteria assessed with total counts in logarithms.

Groups	*P. gingivalis*			*T. forsythia*			*T. denticola*			Total		
	Mean	**SD**	***p*-Value**	**Mean**	**SD**	***p*-Value**	**Mean**	**SD**	***p*-Value**	**Mean**	**SD**	***p*-Value**
PDT	0.6 [a]	1.1		0.5 [a]	0.7		0.5 [a]	0.8		0.7 [a]	0.9	
DL	0.8 [a]	1.2		0.8 [a]	0.9		0.6 [a]	0.7		0.9 [a]	1.0	
MB	5.7 [b]	1.0	<0.001 *	5.6 [b]	1.3	<0.001 *	5.4 [b]	1.1	<0.001 *	6.8 [b]	1.3	<0.001 *
CHX	1.0 [a]	1.1		0.9 [a]	1.0		0.8 [a]	0.9		1.3 [a]	1.4	
NT	6.5 [b]	1.3		6.2 [b]	0.9		6.1 [b]	1.3		7.1 [b]	0.9	

Dissimilar letters indicate statistical significance between groups; * *p*-value obtained using ANOVA test.

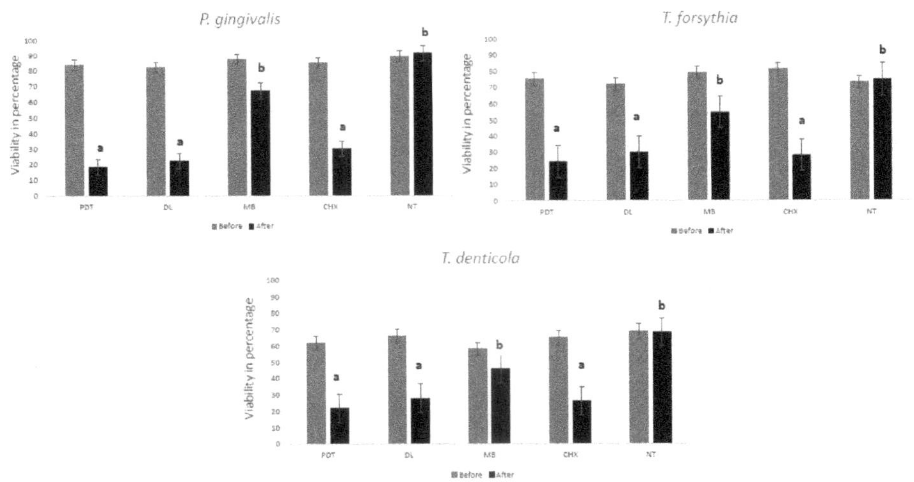

Figure 1. MTT assay showing relative percentage of *P. gingivalis*, *T. forsythia* and *T. denticola* biofilm viability with different treatment groups. Dissimilar letters indicate statistical significance between groups using ANOVA test followed by Tukey's test.

2.2. *Live/Dead Assay*

The CLSM images of bacteria cultured on PICN discs are shown in Figure 2. The dense green colonies (Figure 2a,b) with almost no area of dead bacterial cells represent the NT and MB specimens respectively. The bacteria with the CHX group (Figure 2c) and DL (Figure 2d) groups appeared dead with few areas of surviving green stained bacteria. The PDT group (Figure 2f) showed the highest dead cell count ($p < 0.05$). Figure 2e,f demonstrates pre-and post-treatment using PDT indicating the highest dead cell count and destroyed bacteria.

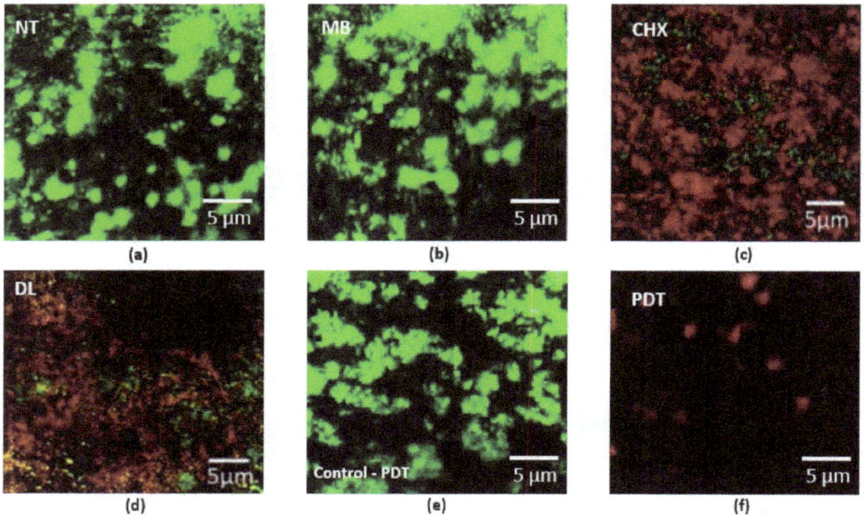

Figure 2. Representative CLSM images of periodontal bacteria grown on PICN discs are shown. (**a**) The control specimens with no treatment (NT) and (**b**) MB specimens are shown as green densely clustered colonies with almost no areas of dead cell; (**c**) represents data for CHX treated specimens. The bacteria in the CHX and (**d**) DL groups appeared dead with few areas of surviving green stained bacteria, indicating survivability; (**e**) Specimen from PDT group before decontamination (control for PDT) and (**f**) post decontamination indicating the highest dead cell count and destroyed bacteria that could be removed after treatment.

2.3. Surface Characterization

SEM investigations revealed a thick periodontal niche grown over the surface of the PICN disc (NT group) (Figure 3a). The representative SEM images taken from treatment groups are shown in Figure 3b–e.

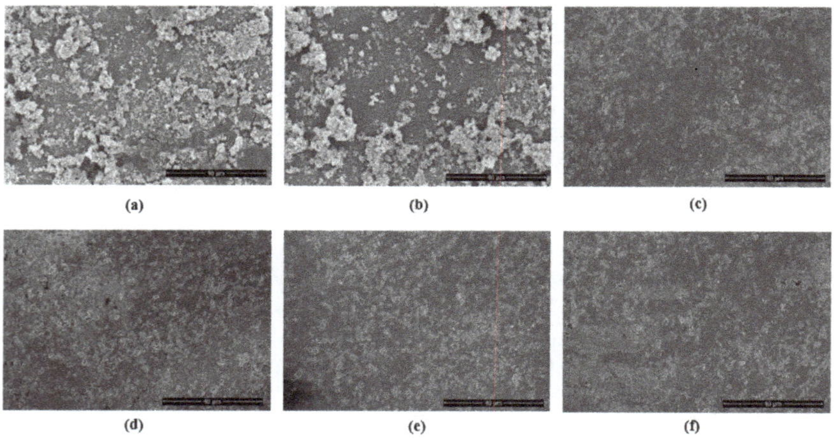

Figure 3. Representative SEM images of: (**a**) periodontal niche grown on PICN discs showing dense colonies (NT control specimen) (**b**) The specimen with MB shows no reduction in clustered colonies; (**c**) represents disc with CHX treatment; (**d**) DL therapy and; (**e**) PDT. The specimen (**f**) is shown to compare the sterile PICN disc (magnification 1500×).

DL (Figure 3d) and PDT (Figure 3e) groups indicate no significant surface alterations when compared with the image taken from the sterile PICN discs of the same type. On visual examination, they appeared to be the same as the surface of the sterile PICN disc (Figure 3e).

3. Discussion

The present study aimed to evaluate the antimicrobial efficacies of laser therapies (PDT and DL therapy) for surface decontamination of PICN and to examine if laser therapies could produce surface alterations over PICN surface. The outcomes of the present in-vitro study did not back our first null hypothesis and provide significant reduction of periodontal bacteria over the PICN surface. However, the second null hypothesis was backed in terms of laser therapies producing no surface alterations over PICN. This to our best knowledge, has never been investigated before and reports the first study in literature.

A plethora of basic and clinical research is being performed to test the efficacy of PDT in periodontics and implant dentistry [16–20]. It is proved to be a promising technique for combating periodontal and peri-implant infections around teeth and dental implants, respectively. With the growing number of dental crowns being fixed over teeth or dental implants, there is a rising need to investigate such new treatment methods for treating dental infections. It is well-known that in order to treat periodontal or peri-implant infections, it is imperative to reduce or eliminate the periodontal bacteria around the sulcus where crowns are fixed with the tooth or dental implant surface [21]. Photodynamic therapy or diode laser therapy offers maximum therapeutic outcome that is not produced by mechanical debridement or other chemicals alone such as the use of chlorhexidine [22,23]. This is true and based on the premise of how PDT works.

The mechanism of PDT is primarily based on the use of photosensitizer dye molecule that is taken up by the bacterial cell membrane that creates reactive oxygen species and other cytotoxic molecules in the presence of laser light which helps to deteriorate pathogens [15]. Our outcomes reported that PDT showed superior results as compared to other techniques reflecting how PDT could maximize the potential of antimicrobial therapy. It is important to further describe the proposed model that could be subjected to interactions by local microbiological environment [24]. The impact of PDT and LAD on various cells including microbiota, human cells and proinflammatory cytokine levels is well studied [25,26]. Future studies are warranted to test the impact of PDT and LAD on PICN on local periodontal environment including resident periodontal cells and microbiome.

The outcomes of our study represented differences among study groups and all experimental groups showed periodontal bacterial reduction compared to the group that did not undergo any treatment. In several studies performed on other materials such as titanium or zirconia surfaces, using PDT or diode laser therapy was shown to be effective in reducing bacteria [27–29]; however, the decreases were lesser than the reductions obtained in our study. Such differences may be attributed to the type of ceramic material surface being studied and according to the previous published literature, the adherence of bacteria to attach to PICN and zirconia is significantly lower as that of bacteria attached to titanium surface which is due to the significant difference in the surface free energy and surface roughness [30–32]. Therefore, we hypothesize that the periodontal bacteria did not adhered well enough on the PICN surface after 72 h of incubation period along with the rinsing of photosensitizer MB that may have caused additional detachment of the periodontal niche form PICN surface.

The present antibacterial results of PDT over PICN surface are in accordance with the previous studies conducted on titanium implant surfaces. For instance, Azizi et al. in their in-vitro study reported PDT and light activated disinfection showed high effectiveness against oral bacteria on zirconia implant surfaces [29]. Similarly, Sayar et al. [33] also reported significant reduction of a pathogenic periodontal bacteria *Aggregatibacter actinomycetemcomitans* over the titanium discs using PDT. In addition, other pre-clinical studies have investigated the impact of decontamination over other types of dental restorations [34]. For this purpose, the physical properties of dental restorations highlight important role of the impact of phototherapy against microbial niche on several types of dental biomaterials [35,36]. Furthermore, previous studies have used different laser parameters, such as different types of photosensitizers with different laser wavelengths. While these studies have indicated that PDT demonstrated the highest level of antibacterial efficacy over different types of material surfaces [37]. In relation to the laser parameters, whether PDT or DL have significant impact on surface alterations on dental restorations is still unknown [37].

Some limitation exists in the present in vitro study. The present study did not assess the cell viability and how human gingival fibroblast cells interact with the PICN material with the potential application of PDT. The oral cavity and their related tissues are rich in stem cells that are conveniently harvestable. These resident stem cells also act as anti-inflammatory and immunomodulatory elements in the local biological niche [38,39]. Such interactions with PICN material and the effects PDT and DL warrants investigation. Another important limitation exists regarding the use of conventional in vitro microbial growth/culture. Future studies should rather focus on safe and predictable in vitro culture protocols especially with those working without any additive on microbial cells and soft tissues, for instance, Bovine Serum Albumin (BSA) or Fetal Bovine Serum (FBS) to apply safely on human cells [40]. In addition, the combination of a single photosensitizer with diode laser was investigated. The including of different photosensitizer with different laser type may provide a robust comparison between laser and other non-laser groups.

With the outcomes of our study showing significant antibacterial efficacy with PDT and DL, these therapeutic strategies could be translated into clinical applications. However, the cost of the laser treatment, handling of the instrument, training/expertise with appropriate guidance with potential side effects should be taken into consideration.

4. Materials and Methods

4.1. Study Ethics and Samples

The study protocol was approved by the Faculty of Dentistry of Princess Nourah Bint Abdulrahman University under the protocol identification number: PNU-76-0001. Power analysis was performed prior to the study by keeping the alpha value at 5% and study power of 95% and 5 investigated groups [27,41]. For these incorporated parameters, it was necessary to have 12 samples/discs per group. For this study, 60 PICN discs (Vita Enamic, Vita Zahnfabrik, Bad Säckingen, Germany) were sectioned with the dimensions 10 × 10 × 3.0 mm using slow speed isomet (Isomet, Buehler, Evanston, IL, USA) under cooling water. The sectioned PICN discs were thoroughly rinsed with distilled water and further sonicated in 99% isopropanol placed in ultrasonic bath for 5 min.

4.2. Microbial Contamination of the PICN Discs

Bacterial contamination of the PICN follows the steps described in the previous study [29]. Suspensions from the three periodontal microbes [*P. gingivalis* (ATCC33277), *T. forsythia* (ATCC35405) and *T. denticola* (ATCC43037)] were prepared. All bacteria were grown separately on blood agar plates except for *P. gingivalis* which were grown on blood agar supplemented with tryptic soy broth. All bacteria were grown under anaerobic conditions. The bacterial suspension was mixed using thioglycolate in a joint suspension. A 600 nm (equivalent of 0.5 McFarland standard containing 1×10^8 CFU/mL) density was adjusted in a densitometer (Shimadzu CS 920, Tokyo, Japan). All PICN discs were placed in their full length in 2 mL Eppendorf tube (Hamburg, Germany) containing 250 µL of the bacterial suspension and incubated anaerobically for 72 h in BD GasPak™ jar (BBl, Becton Dickinson and Company, Sparks, MD, USA).

4.3. Therapies

After 72 h of incubation period in the anaerobic jars, the PICN discs were randomly divided into five groups (12 discs/group).

4.3.1. Photodynamic Therapy Group (PDT)

The contaminated PICN discs were sprayed with 150 μL of MB (Sigma Aldrich, St. Louis, MO, USA) and left undisturbed for 60 s. The discs were washed thoroughly with PBS solution and later irradiated with 630 ± 10 nm diode laser (FotoSan, CMS Dental APS, Copenhagen, Denmark) with power output 100 mW and density of 2000–4000 mW/cm^2 for 60 s. The power fluence was set at 90 J/cm^2. The distance between the diode laser tip and the specimen was maintained at 1 mm and spot area of 0.502 cm^2.

4.3.2. Diode Laser Group (DL)

In this group, the contaminated PICN were treated using an 808 nm diode laser (Lasercat 500, Medsolution, Radolfzell, Germany) in contact mode. The laser tip was positioned at 1 mm and sequentially, until the irradiation time reached 1 min per surface of the PICN section.

4.3.3. Methylene Blue Group (MB)

The PICN discs in this group were immersed in 3 mL of MB dye solution (Sigma Aldrich) of concentration 1 mg/mL for 60 s. Later on, the discs were taken out and rinsed thoroughly with PBS solution.

4.3.4. Chlorhexidine Group (CHX)

The bacterial contaminated PICN discs were immersed in 3 mL of 0.12% chlorhexidine (CHX) digluconate solution (Sigma Aldrich) and left for 2 min. The treated specimens were later thoroughly irrigated with PBS solution to remove the excess CHX solution.

4.3.5. No Treatment (NT)

The fifth group served as control specimens in which none of the decontamination technique was performed and left untreated.

4.4. Microbial Analysis

Immediately after performing therapies, each PICN discs were immersed in 600 μL PBS solution placed in 2 mL Eppendorf tube. The samples were vortexed for 30 s to detach the bacteria from the specimen surface. 100 μL from each Eppendorf tube were transferred to Mueller Hinton broth (100 μL). Subsequently, 20 μL of PBS from each Eppendorf tube was transferred to a microplate well containing 180 μL of Mueller Hinton Broth. A total of 10-fold serial dilutions were performed in 96-well plates followed by the inoculation of 30 μL of the suspension from each well and plated to Brucella agar plates. Incubation of the plates was performed for 72 h in anaerobic conditions, and CFU were counted later. Distinctive colonies were detected using MALDI Biotyper (Bruker Daltonics, Leipzig, Germany) with visual analysis.

4.5. Confocal Laser Microscopy

PICN discs with the attached biofilms were used for confocal microscopy investigation for Live/Dead Bacterial Viability. The viability of bacteria was checked using a confocal laser scanning microscope (CLSM; Fluoview FV 1000, Olympus, Tokyo, Japan). LIVE/DEAD BacLight stain (Invitrogen, Carlsbad, CA, USA) was used after mixing according to the manufacturer's instructions. The PICN specimens were incubated for 30 min in the dark, excessive stain removed and analyzed with CLSM using light emission between 500 and 550 nm with an excitation wavelength of 488 nm and ×100 objective.

4.6. Scanning Electron Microscopy

A random single sample from each group was selected to study the bacterial adherence and topographical alterations on PICN discs under scanning electron microscope (Tescan VEGA3, Tokyo, Japan). The specimens were fixed using 4% paraformaldehyde and 2% glutaraldehyde in 0.1 M sodium cacodylate buffer (NaCac) with pH 7.4 and stored overnight. Subsequently, the samples were treated with 2% osmium tetraoxide and dehydrated in a series of ethanol concentrations (60–100%). The samples were later processed for critical point drying using CO_2. The dried PICN discs were taped in double sided copper tape and sputter coated with 5 nm platinum coating. The PICN discs were imaged at 10 kV using through-lens detector (TLD) for secondary electron imaging. Representative images from each group were selected for depiction.

4.7. Statistical Analysis

Statistical data were analyzed using statistical package of SPSS (v22 IBM Corp., Armonk, NY, USA). Normality testing was performed using Kolmogorov-Smirnov test. Alpha level was set at <0.05. The differences between the groups for each periodontal bacterium assessed and overall count of bacteria were compared using ANOVA test. Multiple comparisons were applied using Tukey-Kramer method. The data for bacteria were log transformed using the following formula [29]:

$$L = \log_{10} (N + 1) \quad (1)$$

Bacterial reduction and their percentages compared to the NT group was computed using the following formula [29]:

$$1 - T/C = 100 \times (1 - T/C)\% \quad (2)$$

where T = mean value for each tested group and C = NT group.

5. Conclusions

Photodynamic therapy produced superior periodontal bacterial reduction over the surface of PICN. PDT group showed higher reduction for each bacterial species and total counts of bacteria assessed followed by the DL group. Both PDT and DL treatment strategies are effective without producing surface alterations on PICN.

Author Contributions: Conceptualization, E.E. and S.S.; methodology, A.A. and M.S.M.; software, A.A..; validation, E.E., S.S. and M.S.; formal analysis, E.E. and M.S.; investigation, E.E. and A.A.; resources, E.E., M.S., A.A. and M.S.M.; data curation, M.S.M. and M.S.; writing—original draft preparation, E.E., A.A. and M.S.M.; writing—review and editing, E.E., S.S. and M.S.; visualization, E.E. and A.A.; supervision, S.S.; project administration, E.E., S.S. and M.S.; funding acquisition, M.S. All authors have read and agreed to the published version of the manuscript.

Funding: This research was funded by the Deanship of Scientific Research at Princess Nourah bint Abdulrahman University through the Fast-track Research Funding Program.

Conflicts of Interest: The authors declare no conflict of interest.

References

1. Marcum, J.S. The effect of crown marginal depth upon gingival tissue. *J. Prosthet. Dent.* **1967**, *17*, 479–487. [CrossRef]
2. Kancyper, S.G.; Koka, S. The influence of intracrevicular crown margins on gingival health: Preliminary findings. *J. Prosthet. Dent.* **2001**, *85*, 461–465. [CrossRef]
3. Passariello, C.; Puttini, M.; Virga, A.; Gigola, P. Microbiological and host factors are involved in promoting the periodontal failure of metaloceramic crowns. *Clin. Oral Investig.* **2012**, *16*, 987–995. [PubMed]
4. Goldberg, P.V.; Higginbottom, F.L.; Wilson, T.G., Jr. Periodontal considerations in restorative and implant therapy. *Periodontology 2000* **2001**, *25*, 100–109. [PubMed]

5. Gonzaga, C.C.; Yoshimura, H.N.; Cesar, P.F.; Miranda, W.G. Subcritical crack growth in porcelains, glass-ceramics, and glass-infiltrated alumina composite for dental restorations. *J. Mater. Sci. Mater. Med.* **2009**, *20*, 1017. [PubMed]
6. Facenda, J.C.; Borba, M.; Corazza, P.H. A literature review on the new polymer-infiltrated ceramic-network material (PICN). *J. Esthet. Restor. Dent.* **2018**, *30*, 281–286. [CrossRef]
7. Leung, B.T.; Tsoi, J.K.; Matinlinna, J.P.; Pow, E.H. Comparison of mechanical properties of three machinable ceramics with an experimental fluorophlogopite glass ceramic. *J. Prosthet. Dent.* **2015**, *114*, 440–446. [PubMed]
8. Ercoli, C.; Caton, J.G. Dental prostheses and tooth-related factors. *J. Periodontol.* **2018**, *89*, S223–S236. [CrossRef]
9. Nayak, A.; Bhat, K.; Shivanaikar, S.; Pushpa, P.; Kugaji, M.; Kumbar, V. Detection of red complex organisms in chronic periodontitis by multiplex polymerase chain reaction. *J. Adv. Clin. Res. Insights* **2018**, *5*, 139–144. [CrossRef]
10. Akram, Z.; Shafqat, S.S.; Aati, S.; Kujan, O.; Fawzy, A. Clinical efficacy of probiotics in the treatment of gingivitis: A systematic review and meta-analysis. *Aust. Dent. J.* **2020**, *65*, 12–20. [CrossRef]
11. Prietto, N.R.; Martins, T.M.; dos Santos Santinoni, C.; Pola, N.M.; Ervolino, E.; Bielemann, A.M.; Leite, F.R.M. Treatment of experimental periodontitis with chlorhexidine as adjuvant to scaling and root planing. *Arch. Oral Biol.* **2020**, *110*, 104600. [CrossRef]
12. Brayer, W.K.; Mellonig, J.T.; Dunlap, R.M.; Marinak, K.W.; Carson, R.E. Scaling and root planing effectiveness: The effect of root surface access and operator experience. *J. Periodontol.* **1989**, *60*, 67–72. [CrossRef]
13. Raghavendra, M.; Koregol, A.; Bhola, S. Photodynamic therapy: A targeted therapy in periodontics. *Aust. Dent. J.* **2009**, *54*, S102–S109. [CrossRef] [PubMed]
14. Knappe, V.; Frank, F.; Rohde, E. Principles of lasers and biophotonic effects. *Photomed. Laser Surg.* **2004**, *22*, 411–417. [CrossRef] [PubMed]
15. Kwiatkowski, S.; Knap, B.; Przystupski, D.; Saczko, J.; Kędzierska, E.; Knap-Czop, K.; Kotlińska, J.; Michel, O.; Kotowski, K.; Kulbacka, J. Photodynamic therapy–mechanisms, photosensitizers and combinations. *Biomed. Pharmacother.* **2018**, *106*, 1098–1107. [CrossRef]
16. Akram, Z.; Hyder, T.; Al-Hamoudi, N.; Binshabaib, M.S.; Alharthi, S.S.; Hanif, A. Efficacy of photodynamic therapy versus antibiotics as an adjunct to scaling and root planing in the treatment of periodontitis: A systematic review and meta-analysis. *Photodiagnosis Photodyn. Ther.* **2017**, *19*, 86–92. [CrossRef] [PubMed]
17. Akram, Z.; Shafqat, S.S.; Niaz, M.O.; Raza, A.; Naseem, M. Clinical efficacy of photodynamic therapy and laser irradiation as an adjunct to open flap debridement in the treatment of chronic periodontitis: A systematic review and meta-analysis. *Photodermatol. Photoimmunol. Photomed.* **2020**, *36*, 3–13. [CrossRef] [PubMed]
18. Kellesarian, S.V.; Qayyum, F.; De Freitas, P.C.; Akram, Z.; Javed, F. Is antimicrobial photodynamic therapy a useful therapeutic protocol for oral decontamination? A systematic review and meta-analysis. *Photodiagnosis Photodyn. Ther.* **2017**, *20*, 55–61. [CrossRef]
19. Akram, Z.; Raffat, M.A.; Shafqat, S.S.; Mirza, S.; Ikram, S. Clinical efficacy of photodynamic therapy as an adjunct to scaling and root planing in the treatment of chronic periodontitis among cigarette smokers: A systematic review and meta-analysis. *Photodiagnosis Photodyn. Ther.* **2019**, *26*, 334–341. [CrossRef]
20. Sivaramakrishnan, G.; Sridharan, K. Photodynamic therapy for the treatment of peri-implant diseases: A network meta-analysis of randomized controlled trials. *Photodiagnosis Photodyn. Ther.* **2018**, *21*, 1–9. [CrossRef]
21. Mathew, M.G.; Samuel, S.; Soni, A.J.; Roopa, K.B. Evaluation of adhesion of Streptococcus mutans, plaque accumulation on zirconia and stainless steel crowns, and surrounding gingival inflammation in primary molars: Randomized controlled trial. *Clin. Oral Investig.* **2020**, *24*, 1–6. [CrossRef]
22. Karimi, M.R.; Hasani, A.; Khosroshahian, S. Efficacy of antimicrobial photodynamic therapy as an adjunctive to mechanical debridement in the treatment of peri-implant diseases: A randomized controlled clinical trial. *J. Lasers Med Sci.* **2016**, *7*, 139. [CrossRef] [PubMed]
23. Mizutani, K.; Aoki, A.; Coluzzi, D.; Yukna, R.; Wang, C.Y.; Pavlic, V.; Izumi, Y. Lasers in minimally invasive periodontal and peri-implant therapy. *Periodontology 2000* **2016**, *71*, 185–212. [CrossRef]
24. Cantore, S.; Ballini, A.; Mori, G.; Dibello, V.; Marrelli, M.; Mirgaldi, R.; De Vito, D.; Tatullo, M. Anti-plaque and antimicrobial efficiency of different oral rinses in a 3-day plaque accumulation model. *J. Biol. Regul. Homeost. Agents* **2016**, *30*, 1173–1178.
25. Akram, Z.; Abduljabbar, T.; Sauro, S.; Daood, U. Effect of photodynamic therapy and laser alone as adjunct to scaling and root planing on gingival crevicular fluid inflammatory proteins in periodontal disease: A systematic review. *Photodiagnosis Photodyn. Ther.* **2016**, *16*, 142–153. [CrossRef] [PubMed]

26. Hokari, T.; Morozumi, T.; Komatsu, Y.; Shimizu, T.; Yoshino, T.; Tanaka, M.; Tanaka, Y.; Nohno, K.; Kubota, T.; Yoshie, H. Effects of antimicrobial photodynamic therapy and local administration of minocycline on clinical, microbiological, and inflammatory markers of periodontal pockets: A pilot study. *Int. J. Dent.* **2018**, *2018*, 1748584. [CrossRef]
27. Cho, K.; Lee, S.Y.; Chang, B.-S.; Um, H.-S.; Lee, J.-K. The effect of photodynamic therapy on Aggregatibacter actinomycetemcomitans attached to surface-modified titanium. *J. Periodontal Implant Sci.* **2015**, *45*, 38–45. [CrossRef] [PubMed]
28. Prates, R.A.; Yamada Jr, A.M.; Suzuki, L.C.; Hashimoto, M.C.E.; Cai, S.; Gouw-Soares, S.; Gomes, L.; Ribeiro, M.S. Bactericidal effect of malachite green and red laser on Actinobacillus actinomycetemcomitans. *J. Photochem. Photobiol. B Biol.* **2007**, *86*, 70–76. [CrossRef]
29. Azizi, B.; Budimir, A.; Bago, I.; Mehmeti, B.; Jakovljević, S.; Kelmendi, J.; Stanko, A.P.; Gabrić, D. Antimicrobial efficacy of photodynamic therapy and light-activated disinfection on contaminated zirconia implants: An in vitro study. *Photodiagnosis Photodyn. Ther.* **2018**, *21*, 328–333. [CrossRef]
30. Al-Radha, A.S.D.; Dymock, D.; Younes, C.; O'Sullivan, D. Surface properties of titanium and zirconia dental implant materials and their effect on bacterial adhesion. *J. Dent.* **2012**, *40*, 146–153. [CrossRef]
31. Bottino, M.; Pereira, S.; Amaral, M.; Milhan, N.; Pereira, C.; Camargo, S.; Carvalho, A.; Melo, R. Streptococcus mutans Biofilm Formation and Cell Viability on Polymer-infiltrated Ceramic and Yttria-stabilized Polycrystalline Zirconium Dioxide Ceramic. *Oper. Dent.* **2019**, *44*, E271–E278. [PubMed]
32. Scarano, A.; Piattelli, M.; Caputi, S.; Favero, G.A.; Piattelli, A. Bacterial adhesion on commercially pure titanium and zirconium oxide disks: An in vivo human study. *J. Periodontol.* **2004**, *75*, 292–296. [PubMed]
33. Sayar, F.; Chiniforush, N.; Bahador, A.; Etemadi, A.; Akhondi, N.; Azimi, C. Efficacy of antimicrobial photodynamic therapy for elimination of Aggregatibacter actinomycetemcomitans biofilm on Laser-Lok titanium discs. *Photodiagnosis Photodyn. Ther.* **2019**, *27*, 462–466.
34. Marrelli, M.; Maletta, C.; Inchingolo, F.; Alfano, M.; Tatullo, M. Three-point bending tests of zirconia core/veneer ceramics for dental restorations. *Int. J. Dent.* **2013**, *2013*, 831976. [PubMed]
35. Dantas, L.C.D.M.; Silva-Neto, J.P.D.; Dantas, T.S.; Naves, L.Z.; das Neves, F.D.; da Mota, A.S. Bacterial adhesion and surface roughness for different clinical techniques for acrylic polymethyl methacrylate. *Int. J. Dent.* **2016**, *2016*, 8685796.
36. Daood, U.; Banday, N.; Akram, Z.; Tsoi, J.K.; Neelakantan, P.; Omar, H.; Abduljabbar, T.; Vohra, F.; Al-Hamoudi, N.; Fawzy, A.S. Mechanical and spectroscopic analysis of retrieved/failed dental implants. *Coatings* **2017**, *7*, 201. [CrossRef]
37. Alasqah, M.N. Antimicrobial efficacy of photodynamic therapy on dental implant surfaces: A systematic review of in vitro studies. *Photodiagnosis Photodyn. Ther.* **2019**, *25*, 349–353.
38. Spagnuolo, G.; Codispoti, B.; Marrelli, M.; Rengo, C.; Rengo, S.; Tatullo, M. Commitment of oral-derived stem cells in dental and maxillofacial applications. *Dent. J.* **2018**, *6*, 72.
39. Ballini, A.; Cantore, S.; Scacco, S.; Coletti, D.; Tatullo, M. Mesenchymal stem cells as promoters, enhancers, and playmakers of the translational regenerative medicine 2018. *Stem Cells Int.* **2018**, *2018*, 927401.
40. Marrazzo, P.; Paduano, F.; Palmieri, F.; Marrelli, M.; Tatullo, M. Highly efficient in vitro reparative behaviour of dental pulp stem cells cultured with standardised platelet lysate supplementation. *Stem Cells Int.* **2016**, *2016*, 7230987.
41. Marotti, J.; Tortamano, P.; Cai, S.; Ribeiro, M.S.; Franco, J.E.M.; de Campos, T.T. Decontamination of dental implant surfaces by means of photodynamic therapy. *Lasers Med. Sci.* **2013**, *28*, 303–309. [CrossRef] [PubMed]

Publisher's Note: MDPI stays neutral with regard to jurisdictional claims in published maps and institutional affiliations.

pharmaceuticals

Article

A Skin Cancer Prophylaxis Study in Hairless Mice Using Methylene Blue, Riboflavin, and Methyl Aminolevulinate as Photosensitizing Agents in Photodynamic Therapy

Hans Christian Wulf [1], Rami Nabil Al-Chaer [1], Martin Glud [1], Peter Alshede Philipsen [1] and Catharina Margrethe Lerche [1,2,*]

[1] Department of Dermatology, Copenhagen University Hospital, Bispebjerg, DK-2400 Copenhagen, Denmark; h.wulf@regionh.dk (H.C.W.); rami452@icloud.com (R.N.A.-C.); Martin.Glud.01@regionh.dk (M.G.); peter.alshede.philipsen@regionh.dk (P.A.P.)

[2] Department of Pharmacy, University of Copenhagen, DK-2100 Copenhagen, Denmark

* Correspondence: catharina.margrethe.lerche@regionh.dk; Tel.: +45-28207100

Abstract: The high incidence of sunlight-induced human skin cancers reveals a need for more effective photosensitizing agents. In this study, we compared the efficacy of prophylactic photodynamic therapy (PDT) when methylene blue (MB), riboflavin (RF), or methyl aminolevulinate (MAL) were used as photosensitizers. All mice in four groups of female C3.Cg/TifBomTac hairless immunocompetent mice ($N = 100$) were irradiated with three standard erythema doses of solar-simulated ultraviolet radiation (UVR) thrice weekly. Three groups received 2×2 prophylactic PDT treatments (days 45 + 52 and 90 + 97). The PDT treatments consisted of topical administration of 16% MAL, 20% MB, or 20% RF, and subsequent illumination that matched the photosensitizers' absorption spectra. Control mice received no PDT. We recorded when the first, second, and third skin tumors developed. The pattern of tumor development after MB-PDT or RF-PDT was similar to that observed in irradiated control mice ($p > 0.05$). However, the median times until the first, second, and third skin tumors developed in mice given MAL-PDT were significantly delayed, compared with control mice (256, 265, and 272 vs. 215, 222, and 230 days, respectively; $p < 0.001$). Only MAL-PDT was an effective prophylactic treatment against UVR-induced skin tumors in hairless mice.

Keywords: PDT; methyl aminolevulinate; methylene blue; riboflavin; photosensitizing agents; ultraviolet radiation; skin tumors; prophylactic treatment; hairless mice

Citation: Wulf, H.C.; Al-Chaer, R.N.; Glud, M.; Philipsen, P.A.; Lerche, C.M. A Skin Cancer Prophylaxis Study in Hairless Mice Using Methylene Blue, Riboflavin, and Methyl Aminolevulinate as Photosensitizing Agents in Photodynamic Therapy. *Pharmaceuticals* **2021**, *14*, 433. https://doi.org/10.3390/ph14050433

Academic Editors: Serge Mordon and Dimitris Tsiourvas

Received: 29 March 2021
Accepted: 29 April 2021
Published: 5 May 2021

1. Introduction

Photodynamic therapy (PDT) was discovered more than 100 years ago and has since become well established [1]. It is effective in the treatment of various human skin cancers and precancerous lesions [1–3]. PDT involves a photosensitizer that is activated by ultraviolet radiation (UVR) and/or visible light to initially form an excited singlet state, followed by transition to a long-lived excited triplet state [3]. This triplet state can undergo photochemical reactions in the presence of oxygen to form reactive oxygen species that can destroy skin cells [3]. The dual specificity of PDT relies on the accumulation of the photosensitizing agent in diseased tissue, and on localized light delivery [2]. The subcellular localization of the photosensitizing agent (e.g., in the mitochondria, lysosomes, endoplasmic reticulum, or plasma membrane) probably plays a major role in the mechanism of cell death that predominates, but other factors, such as the concentration of the photosensitizer, absorption coefficient of the photosensitizer, light dose (J/cm^2), and duration of light exposure, may also be important [1]. The molecular mechanism in PDT prophylaxis is the same as in PDT treatment, but targeting subclinical dysplatic cells instead of regular actinic keratosis or squamous cell carcinoma.

Chemical properties, such as lipophilicity, molecule size, and viscosity, affect the tissue biodistribution of photosensitizing agents [1]. Anticancer photosensitizing agents tend to be lipophilic, with little or no overall charge [1]. Many synthetic and natural non-toxic dyes have been investigated, with a view to implementing these in PDT. Methyl aminolevulinate (MAL) is a commonly used photosensitizer in PDT. However, research is ongoing to identify better alternatives. PDT with MAL is a cosmetically attractive alternative to conventional destructive treatments, including cryotherapy or surgical removal of skin cancers, such as precancerous actinic keratoses [4]. However, clinical use of PDT with MAL has been limited due to adverse side effects, the frequent recurrence of thick skin lesions, and a relapse rate of between 14 and 33% [5–11]. The treatment time is crucial in the clinic, and MAL needs to be converted to protoporphyrin IX (PpIX) in the cell before it is active. Alternative compounds could act as a photosensitizer directly, without cellular conversion.

The photosensitizing agents methylene blue (MB) and riboflavin (RF), also known as vitamin B2, are currently being assessed as potential treatments for UVR-induced skin cancer [12,13]. An RF derivate was tested for its antimicrobial and anticancer properties, and showed efficacy against malignant melanoma in an in vivo mouse model, and fewer side effects than treatments currently in use [3]. MB is a phenothiazinium dye that has been studied frequently for its antimicrobial applications, and occasionally for its anticancer activity [1].

In this study, we investigated the prophylactic effect of RF-PDT and MB-PDT, compared with MAL-PDT, in delaying the development of UVR-induced skin tumors in hairless mice.

2. Results

2.1. *Efficacy of Prophylactic PDT*

Four groups of mice were treated as shown in Figure 1. Mice treated with MB-PDT and RF-PDT showed no signs of severe discomfort during the illuminations, and no local skin reactions were observed over the following days (Figure 2). In contrast, mice treated with MAL-PDT did show discomfort during the illuminations, and erythema of the skin was observed on the day after treatment. Small wounds developed after the second MAL-PDT (day 52) and fourth MAL-PDT (day 97) treatments (Figure 2). Therefore, UVR treatments for all groups were discontinued between days 45 and 66, and between days 90 and 120.

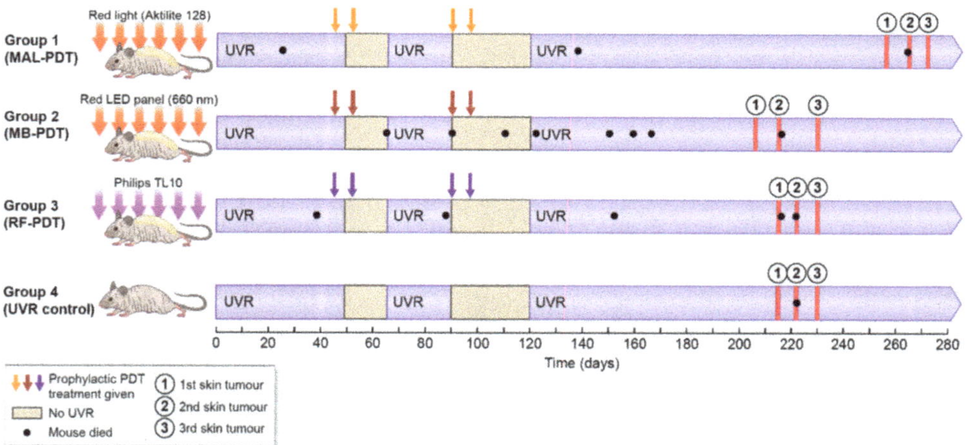

Figure 1. Study overview. Abbreviations: MAL, methyl aminolevulinate; MB, methylene blue; RF, riboflavin; PDT, photodynamic therapy; LED, light-emitting diode; UVR, ultraviolet radiation.

The weekly UVR continued until each mouse developed three 4 mm tumors or one 12 mm tumor. Histological analyses of two random tumors from different mice in each of the four groups confirmed a diagnosis of squamous cell carcinoma in all cases (Figure 3). There were no statistically significant differences in when the first tumors developed in mice from groups MB-PDT and RF-PDT, compared with mice from the UVR control group: (206 vs. 215 days, $p = 0.160$) and (215 vs. 215 days, $p = 0.394$), respectively (Figure 4 and Table 1). There was no significant difference in when the first tumor developed between mice in the MB-PDT and RF-PDT groups (206 vs. 215 days, $p = 0.491$). However, mice treated with MAL-PDT exhibited a significant delay before the first tumor developed, compared with the UVR control group (256 vs. 215 days, $p = 0.000004$; Figure 4 and Table 1). Statistically significant differences in when the first tumors developed were also found when mice from the MB-PDT and RF-PDT groups were compared with those from the MAL-PDT group: (206 vs. 256, $p = 1.6 \times 10^{-7}$) and (215 vs. 256, $p = 5.3 \times 10^{-7}$), respectively.

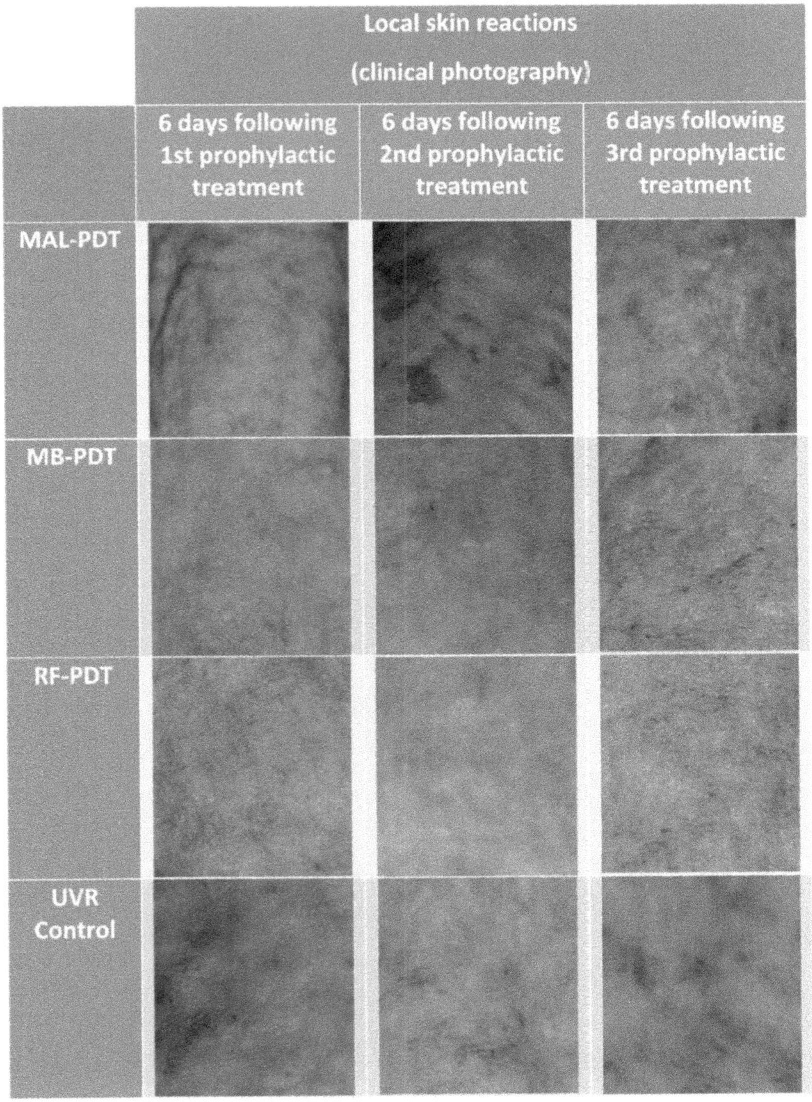

Figure 2. Clinical reactions after the prophylactic treatments. Abbreviations: MAL, methyl aminolevulinate; MB, methylene blue; RF, riboflavin; PDT, photodynamic therapy; UVR, ultraviolet radiation.

The same pattern was observed for the development of the second and third tumors. Mice in the MB-PDT, RF-PDT, and UVR control groups developed a second and third tumor at approximately the same times ($p > 0.061$) (Table 1 and Figure 4). Mice in the MAL-PDT exhibited a significant delay before the second and third tumors developed, compared with the other groups ($p \leq 0.000199$).

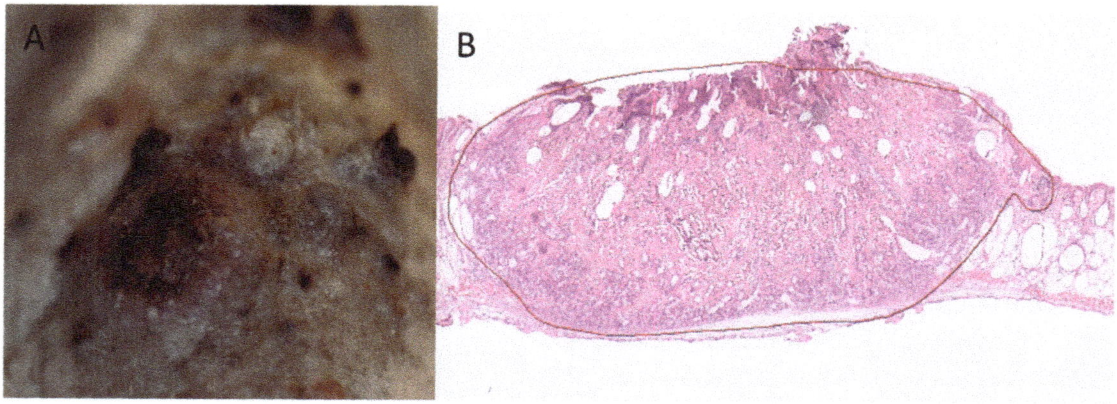

Figure 3. (**A**) Clinical photo of a representative tumor. (**B**) Tumor histology showing squamous cell carcinoma (stained with hematoxylin and eosin).

Table 1. Treatment schedule and results. Median number of days to onset of the first, second, and third tumors in 50% of the mice in each of the four groups. Interquartile ranges (25th and 75th percentiles) are shown.

Group No.	1	2	3	4
No. of Mice (*n*)	25	25	25	25
No. of Mice (*n*) Dy-ing Before End-Point	3	7	5	1
Treatment	MAL-PDT *p-value* *	MB-PDT *p-value* *	RF-PDT *p-value* *	UVR Control
Median Days to 1st Tumor	256 (237–286) 0.000004	206 (188–222) 0.160	215 (196–222) 0.394	215 (206–222)
Median Days to 2nd Tumor	265 (251–293) $3.1 \cdot 10^{-7}$	215 (215–230) 0.526	222 (206–222) 0.111	222 (215–244)
Median Days to 3rd tumor	272 (251–293) 3.6×10^{-7}	230 (215–237) 0.061	230 (215–237) 0.060	230 (222–244)

* *p*-value for the group in question is compared with the control group. Abbreviations: MAL, methyl aminolevulinate; MB, methylene blue; RF, riboflavin; PDT, photodynamic therapy; UVR, ultraviolet radiation.

No weight differences were observed among the groups ($p > 0.05$, results not shown). However, significantly more mice died before developing a first tumor in the MB-PDT group than in the UVR control group ($p = 0.048$). Three mice in the MAL-PDT group, seven mice in the MB-PDT group, five mice in the RF-PDT group, and one mouse in the UVR control group died before reaching the endpoint (Figure 1).

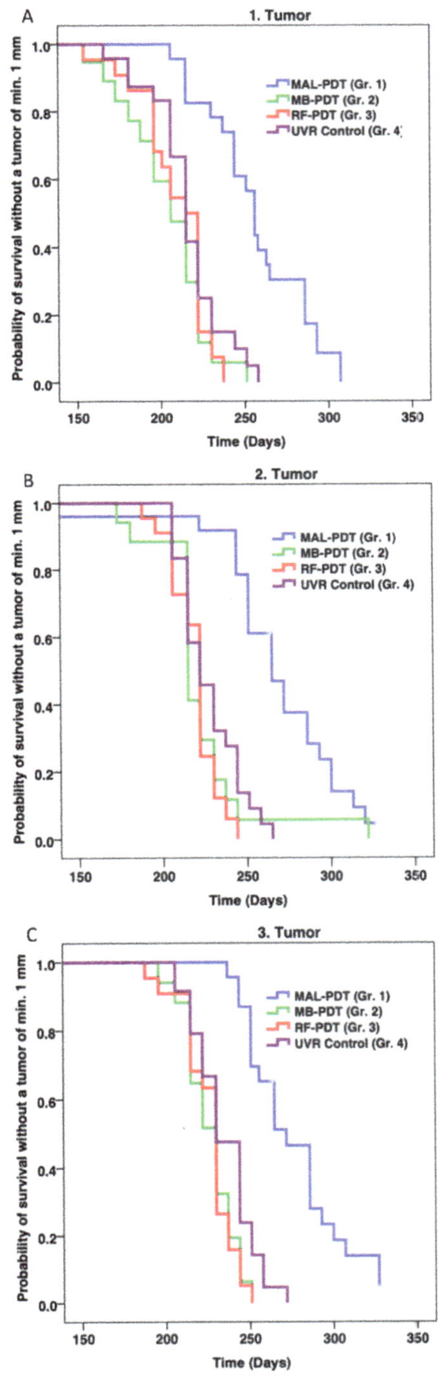

Figure 4. Kaplan–Meier plots showing the probability of survival without a first (**A**), second (**B**), and third (**C**) tumor (minimum diameter = 1 mm) as a function of time. Abbreviations: MAL, methyl aminolevulinate; MB, methylene blue; RBF, riboflavin; Gr., group.

2.2. Pigmentation

All UVR-irradiated mice developed pigmented skin (Figures 2 and 5). At days 0 and 48, there were no differences in pigmentation among the four groups ($p > 0.05$). However, after the first two PDT treatments, a significant increase in skin pigmentation was observed in mice treated with MAL-PDT (day 124: mean, 4.5 (range, 4–5)). Skin pigmentation in the MAL-PDT group was significantly greater than in the other groups at days 60 ($p < 0.0001$), 84 ($p < 0.0001$), and 124 ($p < 0.0001$).

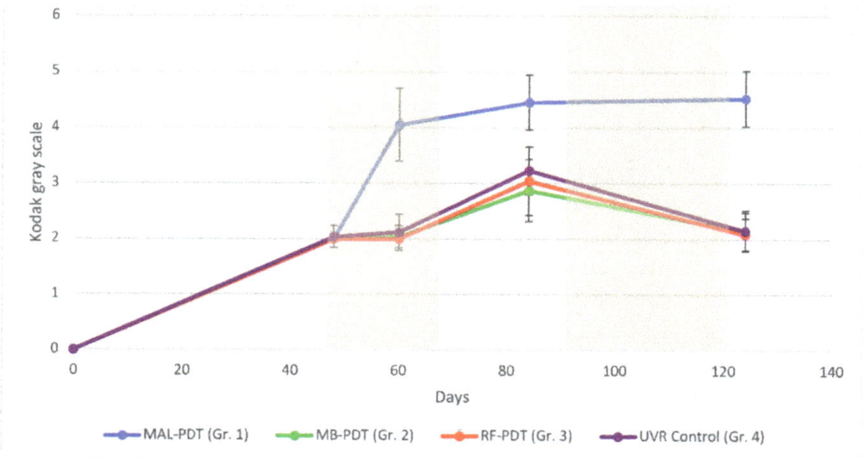

Figure 5. Pigmentation for all groups of mice measured on a Kodak gray scale (arbitrary units, au) and analyzed using one-way analysis of variance (ANOVA). The shaded areas indicate periods with no exposure to ultraviolet irradiation. Abbreviations: MAL, methyl aminolevulinate; MB, methylene blue; RF, riboflavin; Gr., group.

There were no significant differences in skin pigmentation among the MB-PDT, RF-PDT, and UVR control groups at any time ($p > 0.05$). Similar decreases in skin pigmentation were observed in the MB-PDT, RF-PDT, and UVR control groups when UVR exposure was discontinued between days 90 and 120. This decrease in pigmentation was not observed in the MAL-PDT group.

3. Discussion

The present study investigated whether prophylactic PDT, with different photosensitizing agents being applied to the skin of hairless mice, could delay UVR-induced photocarcinogenesis. We used C3.Cg/TifBomTac hairless mice, which develop pigmentation after treatment with UVR, and therefore mimic the human response to UVR better than SKH-1 mice. As anticipated, we found that chronic UVR exposure strongly induced skin cancer in the murine model.

Solar UVR consists predominantly of ultraviolet-A (UVA; >90%; 400–315 nm) and ultraviolet-B (UVB; <10%; 315–280 nm) [14]. UVR has direct effects on genomes and stimulates inflammation [14,15], and UVB is more carcinogenic than UVA [16]. In this murine model, we found that prophylactic PDT treatment with MB (MB-PDT) or RF (RF-PDT) did not delay the development of the first, second, and third tumors, compared with mice that only received UVR (UVR controls). Neither treatment showed efficacy in the prophylaxis of UVR-induced skin tumors in hairless mice. As described in the introduction, many factors are important when considering a photosensitizer for PDT. Previous research has demonstrated that MB and RF are absorbed by the skin or skin models after topical application, although neither is as lipophilic as MAL [17,18], but there are also studies showing no uptake of MB in skin without pretreatment [19,20], which could be due to the positive charge of the molecule. Ideally, light sources should exhibit maximal output at wavelengths matching the photosensitizers' maximal absorption. The wavelength of light determines the depth to which it penetrates the skin. Figure 6 shows the irradiance spectra of our light sources and the absorption spectra of the photosensitizers that were used. MAL is converted to PpIX in the cell. Blue light, red light, and daylight can all be used with MAL for PDT. Figure 6A shows that a greater dose of red light is required than blue for treatment with MAL because PpIX absorbs light more strongly in the blue region of the spectrum. However, red light penetrates the skin to a greater depth. Moreover, the absorption coefficient of the photosensitizer is an important factor, and it can be seen from Figure 6 that the absorption factors for PpIX and RF are much lower than for MB at the irradiation wavelengths, which cannot explain the poor outcome of the MB and RF treatments. Figure 6C indicates that using a light source matching the 440 nm rather than the 375 nm peak may have increased skin penetration and improved RF-PDT. Using another light source for RF-PDT with a broader absorption spectrum than TL10 could also have improved the outcome.

More mice died in the MB-PDT treated group than in the UVR control group. The reason for this significant increase in mortality is unclear, but it may be due to mice ingesting some of the photosensitizing agent applied to their skin. Although MB is poorly absorbed in the gastrointestinal tract, it may generate potentially fatal levels of serotonin at doses >5 mg/kg, resulting in anaphylactic shock [21]. We applied 720 mg/kg MB to the skin of each mouse at 2 × 2 prophylactic treatments.

The results observed in the MAL-PDT group of mice were consistent with those reported by previous studies, and confirm that tumor development can be delayed in this mouse model [2,22]. We assume that the molecular mechanism behind the effect is the same as for PDT treatment with light-mediated excitation of photosensitizer-loaded tumor cells resulting in the production of reactive oxygen species within these cells, leading to cell death. However, since the treatment is given before tumor development (day 45 and 90), the target is subclinical dysplastic cells, instead of regular actinic keratosis or squamous cell carcinoma.

The mice in the MAL-PDT group developed the most intense pigmentation, measured using the Kodak gray scale, probably due to postinflammatory hyperpigmentation. This indicates that MAL-PDT produces an intense PDT effect, as reported by previous studies [23]. The intense pigmentation may partly explain the delay in tumor development, although tumors do appear if UVR treatment is prolonged, despite the high levels of pigmentation [24,25].

Silva et al. demonstrated that application of a 1% MB solution, followed by irradiation with a diode laser for 15 min at 74 mW/cm^2 (total dose = 100 J/cm^2) could reduce tumor size. However, this treatment also reduced epithelial thickness, and keratinization occurred on the surrounding healthy skin tissue [13]. On the other hand, Khaydukov et al. demonstrated that using RF as a photosensitizer generated reactive oxygen species under UVR/blue light irradiation and killed cancer cells [12]. This phototoxicity was demonstrated using the human breast adenocarcinoma cell line SK-BR-3 [12]. We investigated the efficacy of different photosensitizing agents in prophylactic PDT treatments against UVR-induced precancerous skin changes. However, our findings are not consistent with the studies described above. Neither MB-PDT nor RF-PDT delayed the development of precancerous skin changes. It could be ideal to develop photosensitizers that are directly active without a conversion to PpIX to save time in the clinic, but, with suboptimal photosensitizer properties, MB and RF are not the best candidates as potential photosensitizers for topical use on skin cancers and precursors.

In conclusion, we found that, whereas prophylactic treatment with MB-PDT or RF-PDT was ineffective, treatment with MAL-PDT did delay the development of UVR-induced skin tumors in hairless mice.

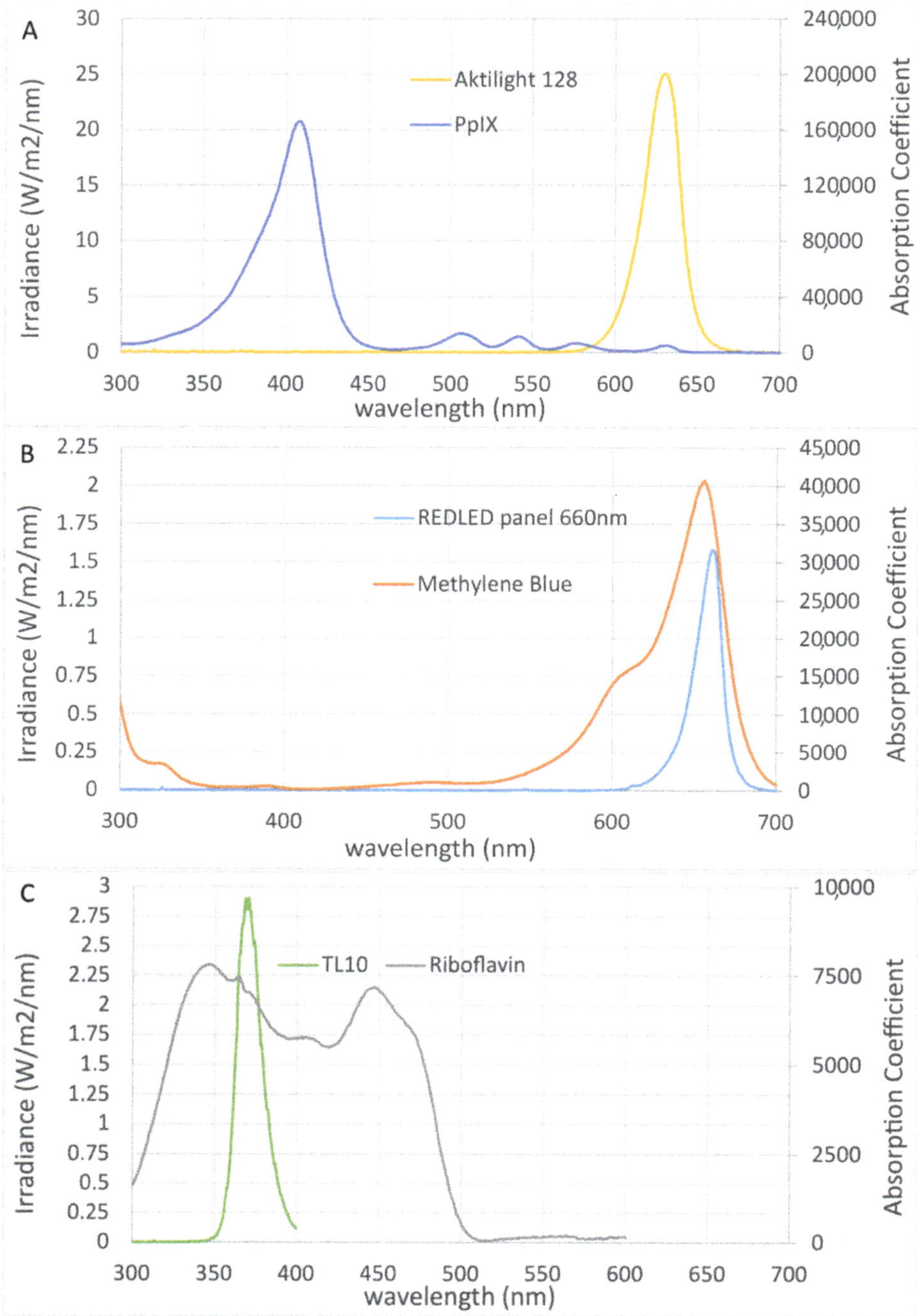

Figure 6. Absorption and irradiance spectra of the following photosensitizers and corresponding light sources: (**A**) Protoporphyrin IX (PpIX), (**B**) methylene blue, and (**C**) riboflavin. The absorption spectra shown were created using PhotochemCADTM software [26,27].

4. Materials and Methods

4.1. Animals

Four groups of 25 female C3.Cg/TifBomTac immunocompetent mice ($N = 100$) were included in the study. They were 12–16 weeks old at the beginning of our experiment. These mice were used because they are hairless and can develop UVR-induced pigmentation, mimicking the response of human skin to UVR. All mice were tattooed with consecutive numbers on the abdomen, and each group of mice was housed in a separate box where there was free access to drinking water and standard laboratory food. The mice were kept at 23–24 °C under a 12-h light/dark cycle.

4.2. Study Design

The mice were randomly divided into four groups of 25 mice. The first four tumors with a diameter of at least 1 mm were mapped separately for each mouse and followed until three of these reached a diameter of 4 mm. Mice were examined for tumors every week. The "time to the first tumor" was the number of days it took for the first 1 mm diameter tumor that later grew to 4 mm to appear. As secondary endpoints, we also recorded when the second and third tumors appeared using the same principle. Mice were euthanized when they had developed three 4 mm tumors, one 12 mm tumor, or after 365 days. After the mice were sacrificed, one tumor was taken from two randomly selected mice in each group, mounted in Tissue-Tek optimal cutting temperature compound (Sakura Finetek Europe B.V., Alphen aan den Rijn, The Netherlands), and frozen. Biopsies were sliced vertically into 10 mm thick sections for hematoxylin and eosin staining and evaluated by a Mohs surgeon. Weight and skin pigmentation were measured once every month. Pigmentation was quantified on a 20-point categorical scale of arbitrary units (au) using the Kodak gray scale.

4.3. UVR Exposure

Mice were irradiated with three standard erythema doses (SEDs) three times a week from the beginning of the study. The UVR source consisted of one UV6 tube (Waldmann, Wheeling, IL, USA) placed between five Bellarium-S SA-1-12 tubes (Wolff System, Atlanta, GA, USA). The animals were irradiated from above, in their boxes, through wire lids. Distances were adjusted every month to maintain the required doses. UVR doses were measured in SEDs using a spectroradiometer (Solatell Sola-Hazard 4D Controls Ltd., Cornwall, UK) [28]. For comparison, a Danish summer midday sun typically generates a UVR dose of 3 SEDs in 30 min [29].

4.4. Photosensitizers

Prophylactic PDT treatments were given at days 45 and 52, and again at days 90 and 97. The prophylactic treatment regimen was based on previous experience from a similar study in mice [23]. The skin was not prepared prior to the prophylactic treatments, and 100 μL of topical photosensitizer was applied to the dorsal skin of each mouse (from front legs to tail). The spectra of the illumination sources were recorded using a spectroradiometer (Jaz, Ocean Optics, FL, USA).

The photosensitizer used on group 1 mice was 16% MAL cream (Metvix®; Galderma, Lausanne, Switzerland). For MAL-PDT treatment, the animals were kept in a dark room for 3 h after the cream was applied, and any remaining cream was wiped off. The standard red light illumination (38 J/cm^2) was administered from above for approximately 9 min using a light-emitting diode (LED) lamp (Aktilite 128; Photocure ASA, Oslo, Norway; Figure 6A). The largest absorption peak for MAL is at 410 nm, with several smaller peaks between 500 and 635 nm [4].

The photosensitizer used on group 2 mice was 20% (*w*/*w*) MB (Sigma-Aldrich, St Louis, MO, USA) in a cream base (Unguentum M; Almirall, Reinbek, Germany). The illuminations commenced 0.5 h after the cream was applied. The light source was a red LED panel (A LED, Varde, Denmark), with peak intensity at 660 nm. MB absorbs light most strongly in the red spectral region, with a peak at 660 nm (Figure 6B). A light dose of 37 J/cm^2 was administered for 3 h.

The photosensitizer used on group 3 mice was 20% (*w*/*w*) RF (Sigma-Aldrich, St Louis, MO, USA) in a cream base (Unguentum M; Almirall, Reinbek, Germany). The animals were kept in a dark room for 1 h after the cream was applied. The light source was a panel of Philips TL10 tubes (Figure 6C). RF has absorption peaks in the UVA (360 nm) and blue (440 nm) spectral regions. A light dose of 37 J/cm^2 was administered for 2 h.

4.5. Statistics

We used both parametric and nonparametric statistical analyses, and reported descriptive data as medians with percentiles. The time to onset of the first, second, and third tumors was visualized in Kaplan–Meier plots. Groups were compared using log-rank tests (Mantel–Cox). Weight and skin pigmentation were analyzed by one-way analysis of variance (ANOVA), followed by Dunn's multiple comparison test for groupwise post hoc comparisons. Mortality before the first tumor was assessed among groups using chi-square tests. *P*-values less than 0.05 were considered significant. All analyses were performed using IBM SPSS 25 software for Windows (SPSS Inc., Chicago, IL, USA).

5. Conclusions

In conclusion, prophylactic treatment with MAL-PDT delayed the development of UVR-induced skin tumors in hairless mice but MB-PDT and RF-PDT was ineffective.

Author Contributions: Conceptualization, H.C.W., P.A.P. and C.M.L.; methodology, H.C.W., P.A.P. and C.M.L.; validation, M.G.; formal analysis, P.A.P., C.M.L. and M.G.; investigation, C.M.L.; resources, H.C.W. and C.M.L.; data curation, R.N.A.-C.; writing—original draft preparation, R.N.A.-C. and C.M.L.; writing—review and editing, H.C.W., R.N.A.-C., M.G., P.A.P. and C.M.L.; visualization, R.N.A.-C. and C.M.L.; supervision, H.C.W. and C.M.L.; project administration, C.M.L.; funding acquisition, C.M.L. All authors have read and agreed to the published version of the manuscript.

Funding: This research was funded by Kgl. Hofbundtmager Aage Bangs Foundation.

Institutional Review Board Statement: All animal procedures were carried out in accordance with the Guide for the Care and Use of Laboratory Animals of The Danish Animal Experiments Inspectorate. All protocols were approved by the Danish Animal Experiments Inspectorate (permits 2019-15-0201-00131 and 2014-15-0201-00096) and our institutional animal care and use committee.

Informed Consent Statement: Not applicable.

Data Availability Statement: The data presented in this study are available on request from the corresponding author.

Acknowledgments: The authors extend their gratitude to E.E. Jakob Heydenreich for recording the spectra of the illumination sources, and laboratory assistant Pia Eriksen for helping with the animals.

Conflicts of Interest: The authors declare that they have no conflict of interest.

References

1. Abrahamse, H.; Hamblin, M.R. New photossensitizers for photodynamic therapy. *Biochem. J.* **2017**, *473*, 347–364. [CrossRef]
2. Ozog, D.M.; Rkein, A.M.; Fabi, S.G.; Gold, M.H.; Goldman, M.P.; Lowe, N.J.; Martin, G.M.; Munavalli, G.S. Photodynamic Therapy: A Clinical Consensus Guide. *Dermatol. Surg.* **2016**, *42*, 804–827. [CrossRef] [PubMed]
3. Akasov, R.A.; Sholina, N.V.; Khochenkov, D.A.; Alova, A.V.; Gorelkin, P.V.; Erofeev, A.S.; Generalova, A.N.; Khaydukov, E.V. Photodynamic therapy of melanoma by blue-light photoactivation of flavin mononucleotide. *Sci. Rep.* **2019**, *9*, 9679. [CrossRef] [PubMed]
4. Morton, C.; Campbell, S.; Gupta, G.; Keohane, S.; Lear, J.; Zaki, I.; Walton, S.; Kerrouche, N.; Thomas, G.; Soto, P. Intraindividual, right-left comparison of topical methyl aminolaevulinate-photodynamic therapy and cryotherapy in subjects with actinic keratoses: A multicentre, randomized controlled study. *Br. J. Dermatol.* **2006**, *155*, 1029–1036. [CrossRef]
5. Hauschild, A.; Popp, G.; Stockfleth, E.; Meyer, K.G.; Imberger, D.; Mohr, P.; Itschert, G.; Kaufmann, R.; Neuber, K.; Frambach, Y.; et al. Effective photodynamic therapy of actinic keratoses on the head and face with a novel, self-adhesive 5-aminolaevulinic acid patch. *Exp. Dermatol.* **2009**, *18*, 116–121. [CrossRef]

6. Horn, M.; Wolf, P.; Wulf, H.C.; Warloe, T.; Fritsch, C.; Rhodes, L.E.; Kaufmann, R.; De Rie, M.; Legat, F.J.; Stender, I.M.; et al. Topical methyl aminolaevulinate photodynamic therapy in patients with basal cell carcinoma prone to complications and poor cosmetic outcome with conventional treatment. *Br. J. Dermatol.* **2003**, *149*, 1242–1249. [CrossRef] [PubMed]
7. Braathen, L.R.; Szeimies, R.M.; Basset-Seguin, N.; Bissonnette, R.; Foley, P.; Pariser, D.; Roelandts, R.; Wennberg, A.M.; Morton, C.A. Guidelines on the use of photodynamic therapy for nonmelanoma skin cancer: An international consensus. *J. Am. Acad. Dermatol.* **2007**, *56*, 125–143. [CrossRef] [PubMed]
8. Soler, A.M.; Warloe, T.; Berner, A.; Giercksky, K.E. A follow-up study of recurrence and cosmesis in completely responding superficial and nodular basal cell carcinomas treated with methyl 5-aminolaevulinate-based photodynamic therapy alone and with prior curettage. *Br. J. Dermatol.* **2001**, *145*, 467–471. [CrossRef]
9. Szeimies, R.M.; Ibbotson, S.; Murrell, D.F.; Rubel, D.; Frambach, Y.; De Berker, D.; Dummer, R.; Kerrouche, N.; Villemagne, H. A clinical study comparing methyl aminolevulinate photodynamic therapy and surgery in small superficial basal cell carcinoma (8–20 mm), with a 12-month follow-up. *J. Eur. Acad. Dermatol. Venereol.* **2008**, *22*, 1302–1311. [CrossRef] [PubMed]
10. Rhodes, L.E.; De Rie, M.; Enström, Y.; Groves, R.; Morken, T.; Goulden, V.; Wong, G.A.E.; Grob, J.J.; Varma, S.; Wolf, P. Photodynamic Therapy Using Topical Methyl Aminolevulinate vs Surgery for Nodular Basal Cell Carcinoma: Results of a Multicenter Randomized Prospective Trial. *Arch. Dermatol.* **2004**, *140*, 17–23. [CrossRef] [PubMed]
11. Vinciullo, C.; Elliott, T.; Francis, D.; Gebauer, K.; Spelman, L.; Nguyen, R.; Weightman, W.; Sheridan, A.; Reid, C.; Czarnecki, D.; et al. Photodynamic therapy with topical methyl aminolaevulinate for "difficult-to-treat" basal cell carcinoma. *Br. J. Dermatol.* **2005**, *152*, 765–772. [CrossRef] [PubMed]
12. Khaydukov, E.V.; Mironova, K.E.; Semchishen, V.A.; Generalova, A.N.; Nechaev, A.V.; Khochenkov, D.A.; Stepanova, E.V.; Lebedev, O.I.; Zvyagin, A.V.; Deyev, S.M.; et al. Riboflavin photoactivation by upconversion nanoparticles for cancer treatment. *Sci. Rep.* **2016**, *6*, 35103. [CrossRef] [PubMed]
13. Silva, A.P.; Neves, C.L.; Silva, E.D.A.; Portela, T.C.L.; Iunes, R.S.; Cogliati, B.; Severino, D.; Baptista, M.D.S.; Dagli, M.L.Z.; Blazquez, F.J.H.; et al. Effects of methylene blue-mediated photodynamic therapy on a mouse model of squamous cell carcinoma and normal skin. *Photodiagn. Photodyn. Ther.* **2018**, *23*, 154–164. [CrossRef] [PubMed]
14. Reis, J.S.; Corrêa, M.A.; Chung, M.C.; Dos Santos, J.L. Synthesis, antioxidant and photoprotection activities of hybrid derivatives useful to prevent skin cancer. *Bioorg. Med. Chem.* **2014**, *22*, 2733–2738. [CrossRef]
15. Kim, I.Y.; He, Y.Y. Ultraviolet radiation-induced non-melanoma skin cancer: Regulation of DNA damage repair and inflammation. *Genes Dis.* **2014**, *1*, 188–198. [CrossRef]
16. Lerche, C.M.; Philipsen, P.A.; Poulsen, T.; Wulf, H.C. Topical pimecrolimus and tacrolimus do not accelerate photocarcinogenesis in hairless mice after UVA or simulated solar radiation. *Exp. Dermatol.* **2009**, *18*, 246–251. [CrossRef]
17. Xiong, Z.M.; O'Donovan, M.; Sun, L.; Young Choi, J.; Ren, M.; Cao, K. Anti-Aging Potentials of Methylene Blue for Human Skin Longevity. *Sci. Rep.* **2017**, *7*, 2475. [CrossRef]
18. Damian, D.L.; Matthews, Y.J.; Halliday, G.M. Topical riboflavin attenuates ultraviolet B- and ultraviolet A-induced immunosuppression in humans. *Photodermatol. Photoimmunol. Photomed.* **2010**, *26*, 66–69. [CrossRef]
19. Chen, X.; Shah, D.; Kositratna, G.; Manstein, D.; Anderson, R.R.; Wu, M.X. Facilitation of transcutaneous drug delivery and vaccine immunization by a safe laser technology. *J. Control. Release* **2012**, *159*, 43–51. [CrossRef]
20. Johnson, P.G.; Gallo, S.A.; Hui, S.W.; Oseroff, A.R. A pulsed electric field enhances cutaneous delivery of methylene blue in excised full-thickness porcine skin. *J. Investig. Dermatol.* **1998**, *111*, 457–463. [CrossRef]
21. Gillman, P.K. Methylene blue implicated in potentially fatal serotonin toxicity. *Anaesthesia* **2006**, *61*, 1013–1014. [CrossRef]
22. Stender, I.M.; Bech-Thomsen, N.; Poulsen, T.; Wulf, H.C. Photodynamic Therapy with Topical δ-aminolevulinic Acid Delays UV Photocarcinogenesis in Hairless Mice. *Photochem. Photobiol.* **1997**, *66*, 493–496. [CrossRef]
23. Togsverd-Bo, K.; Lerche, C.M.; Poulsen, T.; Wulf, H.C.; Hædersdal, M. Photodynamic therapy with topical methyl- and hexylaminolevulinate for prophylaxis and treatment of UV-induced SCC in hairless mice. *Exp. Dermatol.* **2010**, *19*, e166–e172. [CrossRef]
24. Lerche, C.M.; Togsverd-Bo, K.; Philipsen, P.A.; Wulf, H.C. Impact of UVR Exposure Pattern on Squamous Cell Carcinoma-A Dose-Delivery and Dose-Response Study in Pigmented Hairless Mice. *Int. J. Mol. Sci.* **2017**, *18*, 2738. [CrossRef]
25. Del Bino, S.; Sok, J.; Bessac, E.; Bernerd, F. Relationship between skin response to ultraviolet exposure and skin color type. *Pigment Cell Res.* **2006**, *19*, 606–614. [CrossRef]
26. Taniguchi, M.; Du, H.; Lindsey, J.S. PhotochemCAD 3: Diverse Modules for Photophysical Calculations with Multiple Spectral Databases. *Photochem. Photobiol.* **2018**, *94*, 277–289. [CrossRef]
27. Taniguchi, M.; Lindsey, J.S. Database of Absorption and Fluorescence Spectra of >300 Common Compounds for use in PhotochemCAD. *Photochem. Photobiol.* **2018**, *94*, 290–327. [CrossRef]
28. Diffey, B.L.; Jansen, C.T.; Urbach, F.; Wulf, H.C.; Morison, W.L. The standard erythema dose: A new photobiological concept. *Photodermatol. Photoimmunol. Photomed.* **1997**, *13*, 64–66. [CrossRef]
29. Lerche, C.M.; Philipsen, P.A.; Wulf, H.C. UVR: Sun, lamps, pigmentation and Vitamin D. *Photochem. Photobiol. Sci.* **2017**, *16*, 291–301. [CrossRef]

pharmaceuticals

MDPI

Article

How Much Protoporphyrin IX Must Be Activated to Obtain Full Efficacy of Methyl Aminolevulinate Photodynamic Therapy? Implication for Treatment Modifications

Hans Christian Wulf *, Ida M. Heerfordt and Peter Alshede Philipsen

Department of Dermatology, Copenhagen University Hospital Bispebjerg, 2400 Copenhagen NV, Denmark; ida.marie.heerfordt@regionh.dk (I.M.H.); peter.alshede.philipsen@regionh.dk (P.A.P.)
* Correspondence: h.wulf@regionh.dk; Tel.: +45-30914591

Abstract: Photodynamic therapy (PDT) with methyl aminolevulinate (MAL) is a popular treatment for actinic keratoses (AK), and several PDT treatment modalities with similar cure rates are in use. The effect relies on the activation of protoporphyrin IX (PpIX) in premalignant cells. This study aimed to measure PpIX during each treatment modality to determine the minimal PpIX activation and shortest exposure time for optimal cure rate. In four different treatment modalities, we established the PpIX formation up to three hours after MAL application without illumination and measured the speed of PpIX photoactivation during 9 min of red light (37 J/cm^2). The level of PpIX three hours after MAL application was set to 100 PpIX units. In comparison, 85 PpIX units were formed during daylight PDT, 57 PpIX units during pulse PDT, and 52 PpIX units without any curettage prior to MAL. The activation of 50 PpIX units should, therefore, be enough to obtain a full effect on AK. Further, red light illumination may be shortened from 9 min to 1–2 min. The results indicate that PDT can be performed successfully with half the illumination time used in daylight PDT today and with one fourth of the illumination time used in classical PDT.

Keywords: actinic keratosis; illumination time; incubation time; photodynamic therapy; protoporphyrin IX

Citation: Wulf, H.C.; Heerfordt, I.M.; Philipsen, P.A. How Much Protoporphyrin IX Must Be Activated to Obtain Full Efficacy of Methyl Aminolevulinate Photodynamic Therapy? Implication for Treatment Modifications. *Pharmaceuticals* **2021**, *14*, 333. https://doi.org/10.3390/ph14040333

Academic Editor: Serge Mordon

Received: 12 March 2021
Accepted: 27 March 2021
Published: 6 April 2021

1. Introduction

Classic photodynamic therapy (PDT) is associated with pain during illumination but also with great inter-patient variation in pain severity when treating actinic keratosis (AK). Approximately 10% of all patients are unable to complete the illumination due to pain [1]. Acute pain sets in almost immediately when switching on the lamp and, due to pain, illumination is frequently disrupted after just a few minutes [2]. Despite this, anecdotes of unchanged treatment efficacy persist.

The classic method involves curettage, application of methyl aminolevulinate (MAL) cream (160 mg/g), and approximately 9 min of illumination with red diode light (Aktilite® 37 J/cm^2) three hours after MAL application [3,4].

With the intention of reducing pain and inflammation, several alternative ways of performing PDT of AK on the face and scalp have been developed, all producing similar cure rates. Most of these modalities build on replacing the short-time red light exposure with continuous activation of protoporphyrin IX (PpIX) during its formation by long-term, low-intensity light exposure. Any light absorbable in PpIX can be used, such as daylight or lamp-emitted artificial daylight [5].

Classic daylight PDT involves curettage, MAL application, continuous daylight illumination for 2 h, starting 30 min after application and ending 2.5 h after application [6].

Pulse PDT [7] involves curettage, MAL application, and removal of cream residue after 30 min. Illumination is performed with red diode light 37 J/cm^2 three hours after application; alternatively, continuous daylight exposure for 2 h, starting 30 min after application and ending 2.5 h after application, may be used [8].

Daylight PDT without curettage [9] involves MAL application to the face and scalp without pre-treatment. The continuous daylight illumination begins after 1 h and ends 3 h after MAL application.

The time course of the single components in the four PDT modalities is illustrated in Figure 1.

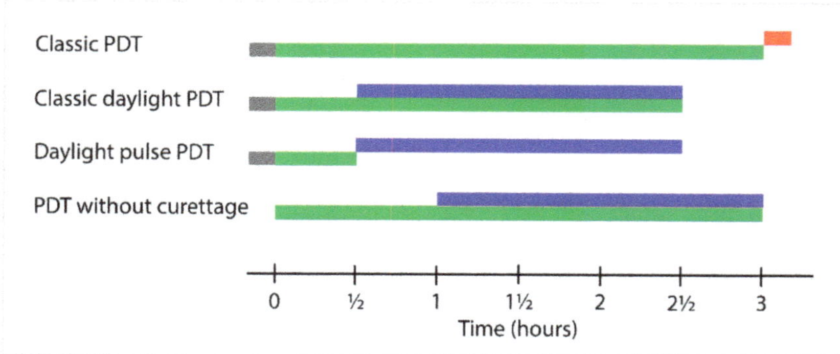

Figure 1. Time course of the single components in the four different photodynamic therapy (PDT) modalities. In most cases, skin is curettaged before MAL application (**grey color**). Methyl aminolevulinate (MAL) cream is applied at time 0. MAL incubation time (**green color**). Duration of illumination with daylight (**blue color**) or red diode light (Aktilite® 68 mW/cm^2 (**red color**)).

The efficacy of PDT builds on the formation of PpIX in the mitochondria, and its activation by appropriate light wavelengths causes cell death. When the PpIX level is kept low, pain and inflammation are lessened [2,10].

It is not clearly established how much PpIX needs to be formed and activated for optimal efficacy. As all presented methods are reported to have the same cure rate on AK of the face and scalp, the intention of this paper is to evaluate how much PpIX is formed and activated when performing classical PDT with short-term illumination 3 h after MAL application, or long-term continuous daylight illumination ending 2.5 h after MAL application. Thereby it is estimated how much shorter the MAL incubation time during PDT treatments could be. Additionally, we aim to establish whether shorter illumination time in classic PDT is likely to give a full effect when treating AK of the face and scalp.

2. Results

2.1. PpIX Formation

PpIX is an endogenous fluorophore and PpIX formation was measured by surface fluorescence. All data on PpIX formation representing the four different treatment modalities are presented in Figure 2.

(I) The amount of PpIX after 3 h in the +Cur3hourMAL group (see Methods) was normalized to 100 PpIX units, being the highest amount of PpIX measured in the study. One hundred PpIX units are the amount of PpIX available for photoactivation 3 h after MAL application before illumination with red light in classic PDT.

(II) After 2.5 h, 85 PpIX units were formed in the +Cur3hourMAL group, representing the amount of PpIX available for photoactivation during classic daylight PDT.

(III) After 2.5 h, 57 PpIX units were formed in the +Cur30MinMAL group, representing the amount of PpIX available for photoactivation during daylight pulse PDT.

(IV) When no curettage is performed, 3 h after MAL application, 52 PpIX units were formed in the -Cur3hourMAL group, representing the amount of PpIX available for photoactivation during PDT without curettage.

AK cure rate is identical in all four described treatment modalities.

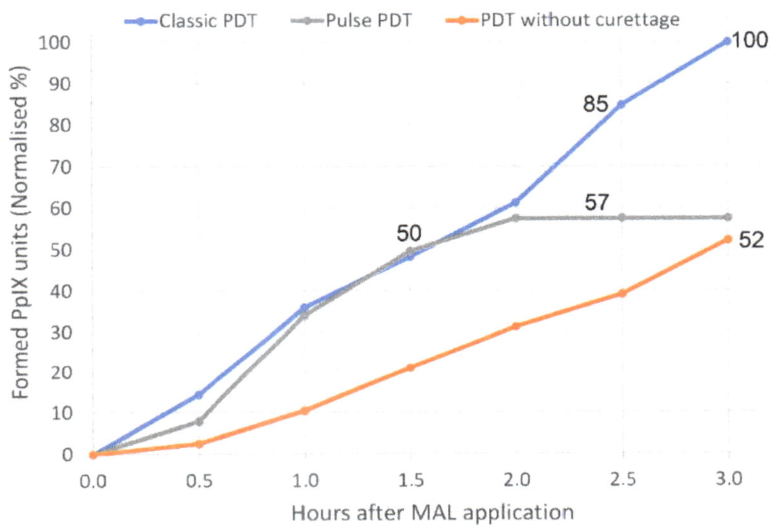

Figure 2. Time-dependent protoporphyrin IX (PpIX) formation without light exposure, when using different photodynamic therapy (PDT) treatment modalities. PpIX formation was measured by surface fluorescence. The PpIX units available for photoactivation are shown for classic PDT (blue line), daylight PDT (blue line up to 2.5 h), daylight pulse PDT (grey line), and when no curettage was performed before methyl aminolevulinate (MAL) application (orange line). PpIX formed 3 h after curettage and MAL application has been normalized to 100%. The activatable PpIX units in the used treatment modalities are shown at 2.5 h and 3 h after MAL application. The PpIX units formed after the suggested shorter treatment time of 1½ h is shown to be 50 PpIX units (50% of the PpIX formed in classic PDT).

2.2. *PpIX Photoactivation*

During photoactivation, PpIX is converted into a nonfluorescent product. In classic PDT, approximately 9 min of illumination with red diode light (Aktilite® 68 mW/cm^2) is performed after 3 h of MAL incubation. The PpIX level just before the start of illumination was set at 100%. All 100 PpIX units were photoactivated after 7.75 min, 80 PpIX units were photoactivated after 2.5 min, and 50 PpIX units were photoactivated after 1 min of illumination. All data on PpIX photoactivation are shown in Figure 3.

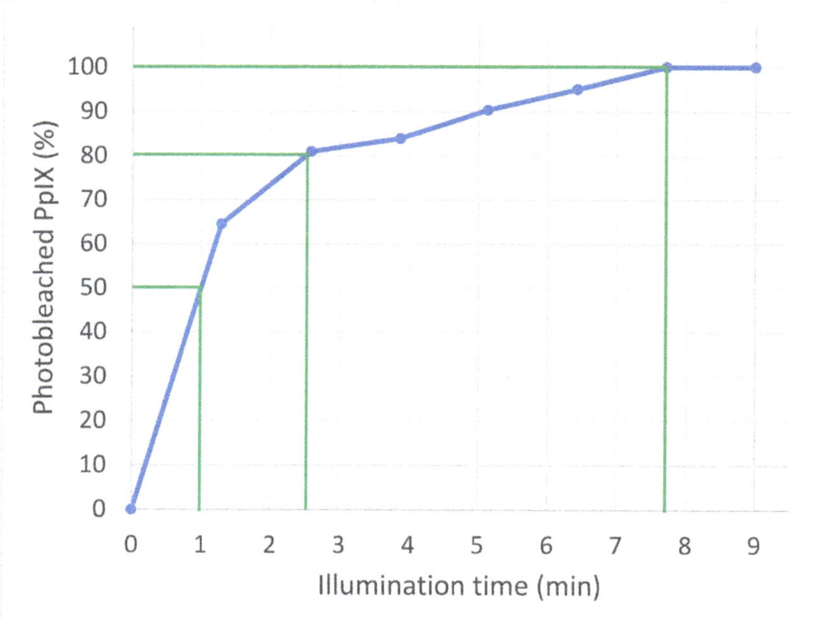

Figure 3. The curve shows the amount of photobleached protoporphyrin IX (PpIX) during 9 min of illumination by red diode light (Aktilite® 68 mW/cm^2) as used in classic photodynamic therapy. Illumination was performed after 3 h of methyl aminolevulinate incubation. Decrease in PpIX fluorescence (photobleaching) was interpreted as photoactivated PpIX.

3. Discussion

A certain amount of PpIX must be formed and activated to achieve optimal efficacy and the light intensity must be high enough to activate all formed PpXI [5,11]. Activation speed has been measured and illustrated in Figure 3. This allows us to estimate how many units of PpIX are activated per time unit. There is no indication, however, of activation speed being of importance as lasers, daylight, and classic red light with very different illumination times may all be used. This indicates that the activation curve is different when using lamps with different light intensities and different light spectra.

It has previously been shown that increasing the amount of PpIX to a higher level than that of classic PDT does not increase treatment efficacy on AK of the hands [10]. Further, the pain associated with PDT is considerably worse when more PpIX is activated during illumination, or if the daylight exposure time is prolonged, which also results in more PpIX activation [6]. Therefore, a sensible course could be reducing the illumination time with red light 3 h after MAL application to last just long enough to activate the necessary amount of PpIX. Fernández-Guarino et al. [12] halved the illumination time and still obtained the full treatment effect on AK. We have previously determined the minimal total PpIX weighted daylight dose for full efficacy to be 4–8 J/cm^2 [6,11]. A similar light dose is obtained after 1–2 min of illumination with 68 mW/cm^2, indicating that 2 min of illumination should be acceptable (Figure 3). This is less than one-fourth of the recommended time.

It is well known that the concentration of PpIX becomes higher when PDT is performed on the face and scalp, compared to other body locations, and illumination length may depend on the location of AK. Nevertheless, many studies have been performed with the purpose of increasing PpIX formation. Such attempts could be irrelevant, unless the distribution of PpIX in the skin is more optimal, or goes deeper into dermis, which might be of advantage in some cases [13].

The full obtainable effect on AK is achieved by the classic PDT approach. This is also the case when performing continuous activation with daylight PDT, even though only 85% of PpIX is activated compared to the amount activated in classic PDT. The full effect on AK of the face and scalp is also observed after activation of 57–52 PpIX units, as in pulse PDT and PDT without curettage [8,9]. Thus, continuous activation of 50 PpIX units specifically located intracellularly may be close to the minimum level of PpIX activation needed to obtain the full effect of the PDT treatment, but it is possible that the necessary PpIX activation may be even lower.

In pulse PDT, the amount of activatable PpIX is stable, between 2 and 3 h after MAL application (Figure 2), and red-light illumination 2 h after MAL or daylight up to 2 h after MAL application should be effective. As seen in Figure 2, the treatment may be shortened even more, as 50 PpIX units are already found 1.5 h after MAL application. This will further minimize the side effects from light exposure after the end of treatment. MAL in the skin is not removed or metabolized chemically by exposure to light and PpIX is formed while MAL is present, also after the end of treatment. Compared to the other modalities, the advantage of pulse PDT is the presence of very little residual MAL at the end of treatment. This lessens the risk of post-treatment inflammation due to random daylight exposure.

PDT without curettage still necessitates illumination up to 3 h after MAL application when 52 PpIX units are activated. The advantages of this modality are less pain and inflammation, and the avoidance of bleeding following pre-treatment curettage, which may present a problem in patients treated with anticoagulants. MAL provokes oozing and bleeding when applied to curettaged skin, which may necessitate reapplication of cream. When curettage is omitted, these problems are avoided.

Our calculations depend on the correctness of the PpIX formation curve in classic PDT as the other results are related to this. It is likely that this curve is credible as it is identical to the PpIX formation curve in pulse PDT for the first 2 h after MAL application (Figure 2), and to previous measurements (by use of a different technique) performed by Wiegell et al. [14,15].

The effectiveness of each PpIX unit depends on the high specificity of the drug to mainly target diseased cells. Most likely, the PpIX formed in the mitochondria must remain there for the PpIX units to have the greatest effect. This is best achieved by using continuous activation by light in order to activate every PpIX molecule as soon as it is formed in the mitochondria. Daylight PDT was developed for this purpose. Additionally, pulse PDT was developed to reduce PpIX formation after the end of treatment. When the PpIX formation time is prolonged before the start of illumination, PpIX leaks to other cell components, as well as extracellularly, resulting in excessive inflammation and possibly rendering each PpIX unit less effective. Treatment with MAL with a short incubation time represents a way to keep the PpIX in the mitochondria of the abnormal cells, thereby lessening the pain and inflammation while upholding efficacy. This may be obtained by use of the data presented here.

Earlier attempts to reduce treatment time were not based on a clear pharmacologic rationale. In 2008, Braathen et al. [16] used MAL cream (160 mg/g) and compared the cure rate of AK when illuminating 1 h after MAL application with red light at 75 J/cm^2 (Curelight®, PhotoCure, Norway) to the cure rate of AK with illumination 3 h after MAL application by the same light source (classic PDT). Compared to illumination after 1 h, a moderately better cure rate was obtained with the classic method, indicating that one hour of PpIX formation is too short. If Figure 2 is applied, only around 35 PpIX units are formed after 1 h. The authors also indicate that debridement before MAL improves efficacy in both modalities.

In an uncontrolled study by von Dobbeler et al. [17], a full effect was obtained by use of a nanoemulsion of 5-aminolevulinic acid (ALA) (Ameluz® 78 mg/g), after pre-treatment with curettage, degreasing, and ALA on the skin for 1 h, after which the skin was illuminated with white LED for 1 h (20 J/cm^2). If the result is comparable to the results obtained by MAL, we might expect an activation of around 60 PpIX units after two hours (Figure 2). As ALA typically forms PpIX faster than MAL, even more PpIX may have been activated [18]. However, after ALA, more PpIX may be found in normal skin structures than when using MAL and this extra PpIX may not result in improved efficacy. As the ALA treatment was effective in this uncontrolled study on AK in the face and scalp, it indicates that 60% activation of PpIX is a rather safe level when enough light is present to activate all formed PpIX within the time range [11].

On the basis that activation of 50 PpIX units is sufficient to achieve the full effect of the treatment, the modalities used in the face and scalp may be translated into one of the following procedures:

Classic PDT including curettage, MAL application, followed by illumination for 9 min with Aktilite® 68 mW/cm^2, 37 J/cm^2, 1.5 h after MAL application.

Classic PDT including curettage, MAL application, followed by illumination for 2 min with Aktilite® 68 mW/cm^2, in total 8 J/cm^2, 3 h after MAL.

Classic daylight PDT including curettage, MAL application and continuous activation for 1 h of daylight, starting 30 min after application.

Pulse PDT including curettage followed by MAL application, removal of cream residue after 30 min, and illumination for 9 min with Aktilite® 68 mW/cm^2, 37 J/cm^2, 1.5 h after MAL application.

Daylight pulse PDT including curettage, followed by MAL application, removal of cream residue after 30 min, and continuous illumination with daylight for 1 h, starting 30 min after MAL application.

It is unknown whether the shortened time of exposure to MAL and shortened time of illumination with red diode light can be combined. However, it is likely that illumination with red light for 9 min is still necessary when only 50 PpIX units are activated (after 1.5 h) if the activation speed is still logarithmic (Figure 3). Using a different measuring technique, Ericson et al. [19] have shown similar photobleaching as in Figure 3 and a relative independence of light intensity and light spectrum. This issue needs further investigation.

The efficacy of these radical changes in protocol needs testing by clinical trials before translation into clinical praxis.

4. Materials and Methods

4.1. PpIX Measurements

PpIX is an endogenous fluorophore emitting red light when excited by blue light [3]. PpIX was measured noninvasively by surface fluorescence using a handheld photometer (FluoDerm, Dia Medico ApS, Gentofte, Denmark) [20,21]. The photometer illuminates the skin with blue light (400–420 nm), matching the Soret band of PpIX at 407 nm. By measuring excitation wavelengths at 610–700 nm, the corresponding red PpIX fluorescence was detected. By this method, the PpIX concentration in the epidermis was roughly estimated [22].

4.2. PpIX Formation

Background fluorescence was measured before MAL application. Subsequently, fluorescence was measured during MAL incubation at different time points. Increase in measured fluorescence was presented as PpIX formation. Data on PpIX formation originate partly from a previous published study [23], supplemented with new data on the development of PpIX in 5 individuals older than 55 years of age during classic PDT of AK on the face and scalp.

In this study, we introduce the following definitions: −Cur3hourMAL represents PpIX formation during PDT without curettage [9], +Cur30MinMAL represents PpIX formation during pulse PDT [8], and +Cur3hourMAL represents PpIX formation during classic PDT.

During routine classic PDT treatment of AKs at the Department of Dermatology, Bispebjerg Hospital, we measured (i) PpIX formation every 30 min after curettage during 3 h of MAL incubation (+Cur3hourMAL) before illumination without interfering with the planned standard treatment. Further, we extracted data on PpIX formation every

30 min for 3 h: (ii) during MAL incubation without prior curettage (−Cur3hourMAL) and (iii) 30 min MAL incubation after curettage (+Cur30MinMAL) from a previous study [23]. All PpIX data were measured without light exposure and the measured PpIX represents the PpIX available for activation during PDT. All measurements in the different settings were performed by the same investigator (I.M.H.) and the highest amounts of PpIX measured were normalized to 100 PpIX units.

4.3. PpIX Photoactivation

All data on PpIX photoactivation were extracted from a previous study [24]. During photoactivation, PpIX is converted into a nonfluorescent photobleached product. PpIX fluorescence was measured after 3 h of MAL incubation just before illumination start and 7 times during illumination. As in classical PDT, the skin was illuminated by red diode light (Aktilite® 68 mW/cm^2) for 9 min. Decrease in PpIX fluorescence (photobleaching) was interpreted as photoactivated PpIX.

4.4. Statistics

All reported PpIX values were mean PpIX values.

5. Conclusions

In conclusion, most PDT treatment modalities can probably be performed after a considerably shorter MAL exposure time or a shorter illumination time without losing efficacy. When performing classic PDT, the illumination time can probably be shortened from around 9 min to a few minutes without losing efficacy. This indicates that cessation of illumination after a few minutes due to pain may not constitute a lack of efficacy in classic PDT.

Author Contributions: Conceptualization, H.C.W.; methodology, all authors; validation, all authors; formal analysis, all authors; investigation, H.C.W. and I.M.H.; writing—original draft preparation, H.C.W.; writing—review and editing, all authors. All authors have read and agreed to the published version of the manuscript.

Funding: This research received no external funding.

Institutional Review Board Statement: The study was conducted according to the guidelines of the Declaration of Helsinki. Investigations on off-label PDT modalities were approved by the Ethics Committee of Region Hovedstaden (protocol code H-16049191, approved on 17 February 2017). Data on classical PDT include only noninvasive measurements done without interfering with planned standard treatment.

Informed Consent Statement: Informed consent was obtained from all subjects involved in the study.

Data Availability Statement: The data presented in this study are available on request from the corresponding author. The data presented in this study are available in this article.

Acknowledgments: The authors wish to thank Louise Holbæk Kaihøj for secretarial assistance.

Conflicts of Interest: The authors declare no conflict of interest.

References

1. Grapengiesser, S.; Gudmundsson, F.; Larkö, O.; Ericson, M.; Rosen, A.; Wennberg, A.M. Pain caused by photodynamic therapy of skin cancer. *Clin. Exp. Dermatol.* **2002**, *27*, 493–497. [CrossRef]
2. Ang, J.M.; Riaz, I.B.; Kamal, M.U.; Paragh, G.; Zeitouni, N.C. Photodynamic therapy and pain: A systematic review. *Photodiagnosis Photodyn. Ther.* **2017**, *19*, 308–344. [CrossRef]
3. Peng, Q.; Warloe, T.; Berg, K.; Moan, J.; Kongshaug, M.; Giercksky, K.E.; Nesland, J.M. 5-Aminolevulinic acid-based photodynamic therapy: Clinical research and future challenges. *Cancer* **1997**, *79*, 2282–2308. [CrossRef]
4. Szeimiesa, R.M.; Karrera, S.; Radakovic-Fijanb, S.; Tanewb, A.; Calzavara-Pintonc, P.G.; Zanec, C.; Sidoroffd, A.; Hempele, M.; Ulrichf, J.; Proebstleg, T.; et al. Photodynamic therapy using topical methyl 5-aminolevulinate compared with cryotherapy for actinic keratosis: A prospective, randomized study. *J. Am. Acad. Dermatol.* **2002**, *47*, 258–262. [CrossRef]
5. Lerche, C.M.; Heerfordt, I.M.; Heydenreich, J.; Wulf, H.C. Alternatives to Outdoor Daylight Illumination for Photodynamic Therapy—Use of Greenhouses and Artificial Light Sources. *Int. J. Mol. Sci.* **2016**, *17*, 309. [CrossRef] [PubMed]

6. Wiegell, S.R.; Haedersdal, M.; Eriksen, P.; Wulf, H.C. Photodynamic therapy of actinic keratoses with 8% and 16% methyl aminolaevulinate and home-based daylight exposure: A double-blinded randomized clinical trial. *Br. J. Dermatol.* **2009**, *160*, 1308–1314. [CrossRef] [PubMed]
7. Wiegell, S.R.; Petersen, B.; Wulf, H.C. Pulse photodynamic therapy reduces inflammation without compromising efficacy in the treatment of multiple mild actinic keratoses of the face and scalp: A randomized clinical trial. *Br. J. Dermatol.* **2016**, *174*, 979–984. [CrossRef] [PubMed]
8. Wiegell, S.; Johansen, U.; Wulf, H. Pulse–Daylight–Photodynamic Therapy in Combination with Corticosteroid and Brimonidine Tartrate for Multiple Actinic Keratoses: A Randomized Clinical Trial. *Acta Derm. Venereol.* **2019**, *99*, 242–243. [CrossRef] [PubMed]
9. Heerfordt, I.M.; Wulf, H.C. Daylight photodynamic therapy of actinic keratosis without curettage is as effective as with curettage: A randomized clinical trial. *J. Eur. Acad. Dermatol. Venereol.* **2019**, *33*, 2058–2061. [CrossRef]
10. Nissen, C.V.; Heerfordt, I.M.; Wiegell, S.R.; Mikkelsen, C.S.; Wulf, H.C. Increased protoporphyrin IX accumulation does not improve the effect of photodynamic therapy for actinic keratosis: A randomized controlled trial. *Br. J. Dermatol.* **2017**, *176*, 1241–1246. [CrossRef]
11. Wiegell, S.R.; Heydenreich, J.; Fabricius, S.; Wulf, H.C. Continuous ultra-low-intensity artificial daylight is not as effective as red LED light in photodynamic therapy of multiple actinic keratoses. *Photodermatol. Photoimmunol. Photomed.* **2011**, *27*, 280–285. [CrossRef] [PubMed]
12. Fernández-Guarino, M.; Fonda Pascual, P.; Lizuain Gomez, P.; Harto Castaño, A.; Jaén Olasolo, P. Split-face study comparing conventional MAL photodynamic therapy in multiple actinic keratosis with complete time vs. half-time red light LED conventional illumination. *J. Eur. Acad. Dermatol. Venereol.* **2019**, *33*, 1529–1534. [CrossRef] [PubMed]
13. Togsverd-Bo, K.; Haak, C.; Thaysen-Petersen, D.; Wulf, H.; Anderson, R.; Haedesdal, M. Intensified photodynamic therapy of actinic keratoses with fractional CO_2 laser: A randomized clinical trial. *Br. J. Dermatol.* **2012**, *166*, 1262–1269. [CrossRef] [PubMed]
14. Wiegell, S.R.; Fabricius, S.; Stender, I.M.; Berne, B.; Kroon, S.; Andersen, B.L.; Mørk, C.; Sandberg, C.; Jemec, G.B.E.; Mogensen, M.; et al. A randomized, multicentre study of directed daylight exposure times of 1½ vs. 2½ h in daylight-mediated photodynamic therapy with methyl aminolaevulinate in patients with multiple thin actinic keratoses of the face and scalp. *Br. J. Dermatol.* **2011**, *164*, 1083–1090. [CrossRef]
15. Wiegell, S.R.; Haedersdal, M.; Philipsen, P.A.; Eriksen, P.; Enk, C.D.; Wulf, H.C. Continuous activation of PpIX by daylight is as effective as and less painful than conventional photodynamic therapy for actinic keratoses; a randomized, controlled, single-blinded study. *Br. J. Dermatol.* **2008**, *158*, 740–746. [CrossRef] [PubMed]
16. Braathen, L.R.; Paredes, B.E.; Saksela, O.; Fritsch, C.; Gardlo, K.; Morken, T.; Frølich, K.W.; Warloe, T.; Solér, A.M.; Ros, A.M. Short incubation with methyl aminolevulinate for photodynamic therapy of actinic keratoses. *J. Eur. Acad. Dermatol. Venereol.* **2009**, *23*, 550–555. [CrossRef]
17. von Dobbeler, C.; Schmitz, L.; Dicke, K.; Szeimies, R.M.; Dirschka, T. PDT with PPIX absorption peaks adjusted wavelengths: Safety and efficacy of a new irradiation procedure for actinic keratoses on the head. *Photodiagnosis Photodyn. Ther.* **2019**, *27*, 198–202. [CrossRef] [PubMed]
18. Nissen, C.V.; Philipsen, P.A.; Wulf, H.C. Protoporphyrin IX formation after topical application of methyl aminolaevulinate and BF-200 aminolaevulinic acid declines with age. *Br. J. Dermatol.* **2015**, *173*, 760–766. [CrossRef]
19. Ericson, M.B.; Sandberg, C.; Stenquist, B.; Gudmundson, F.; Karlsson, M.; Ros, A.M.; Rosén, A.; Larkö, O.; Wennberg, A.M.; Rosdahl, I. Photodynamic therapy of actinic keratosis at varying fluence rates: Assessment of photobleaching, pain and primary clinical outcome. *Br. J. Dermatol.* **2004**, *151*, 1204–1212. [CrossRef]
20. Bjerring, P.; Christiansen, K.; Troilius, A.; Bekhor, P.; de Leeuw, J. Skin fluorescence controlled photodynamic photorejuvenation (wrinkle reduction). *Lasers Surg. Med.* **2009**, *41*, 327–336. [CrossRef]
21. Haak, C.S.; Christiansen, K.; Erlendsson, A.M.; Taudorf, E.H.; Thaysen-Petersen, D.; Wulf, H.C.; Haedersdal, M. Ablative fractional laser enhances MAL-induced PpIX accumulation: Impact of laser channel density, incubation time and drug concentration. *J. Photochem. Photobiol. B Biol.* **2016**, *159*, 42–48. [CrossRef] [PubMed]
22. Togsverd-Bo, K.; Idorn, L.W.; Philipsen, P.A.; Wulf, H.C.; Hædersdal, M. Protoporphyrin IX formation and photobleaching in different layers of normal human skin: Methyl- and hexylaminolevulinate and different light sources. *Exp. Dermatol.* **2012**, *21*, 745–750. [CrossRef] [PubMed]
23. Heerfordt, I.M.; Bieliauskiene, G.; Wulf, H.C. Protoporphyrin IX formation after application of methyl aminolevulinate on the face and scalp with and without prior curettage. *Photodiagnosis Photodyn. Ther.* **2018**, *22*, 155–157. [CrossRef] [PubMed]
24. Wulf, H.C. Daylight PDT acts by continuous activation of PpIX. *Photodiagnosis Photodyn. Ther.* **2019**, *27*, A1–A2. [CrossRef] [PubMed]

pharmaceuticals

MDPI

Article

Photophysical Properties of Protoporphyrin IX, Pyropheophorbide-a, and Photofrin® in Different Conditions

Bauyrzhan Myrzakhmetov [1,2], Philippe Arnoux [1], Serge Mordon [3], Samir Acherar [4], Irina Tsoy [2] and Céline Frochot [1,*]

[1] LRGP UMR 7274, CNRS, University of Lorraine, 54000 Nancy, France; bauyrzhanmyrzakhmetov86@gmail.com (B.M.); philippe.arnoux@univ-lorraine.fr (P.A.)
[2] Department of Chemistry and Chemical Technology, M.Kh. Dulaty Taraz Regional University, Taraz 080012, Kazakhstan; tsoyirinagen@mail.ru
[3] ONCO-THAI U1189, INSERM, CHU Lille, University of Lille, 59000 Lille, France; serge.mordon@inserm.fr
[4] LCPM UMR 7375, CNRS, University of Lorraine, 54000 Nancy, France; samir.acherar@univ-lorraine.fr
* Correspondence: celine.frochot@univ-lorraine.fr; Tel.: +33-372743780

Abstract: Photodynamic therapy (PDT) is an innovative treatment of malignant or diseased tissues. The effectiveness of PDT depends on light dosimetry, oxygen availability, and properties of the photosensitizer (PS). Depending on the medium, photophysical properties of the PS can change leading to increase or decrease in fluorescence emission and formation of reactive oxygen species (ROS) especially singlet oxygen ($^1O_2$). In this study, the influence of solvent polarity, viscosity, concentration, temperature, and pH medium on the photophysical properties of protoporphyrin IX, pyropheophorbide-a, and Photofrin® were investigated by UV-visible absorption, fluorescence emission, singlet oxygen emission, and time-resolved fluorescence spectroscopies.

Keywords: photodynamic therapy; protoporphyrin IX; pyropheophorbide-a; Photofrin®; absorption; fluorescence; singlet oxygen

Citation: Myrzakhmetov, B.; Arnoux, P.; Mordon, S.; Acherar, S.; Tsoy, I.; Frochot, C. Photophysical Properties of Protoporphyrin IX, Pyropheophorbide-a, and Photofrin® in Different Conditions. *Pharmaceuticals* **2021**, *14*, 138. https://doi.org/10.3390/ph14020138

Academic Editor: Luís M. T. Frija

Received: 26 January 2021
Accepted: 7 February 2021
Published: 9 February 2021

1. Introduction

Photodynamic therapy (PDT) is a targeted technique for the treatment of malignant or diseased tissues that relies on three non-toxic elements—a light-activated drug (photosensitizer, PS), light, and molecular oxygen. Illumination of the PS induces the production of the triplet excited state $^3PS^*$ which is able to transfer protons, electrons, or energy, leading to the formation of reactive oxygen species (ROS). ROS cause apoptosis or necrosis of tumor cells by photochemical oxidation [1–3].

A PS should ideally possess some valuable properties including (i) absorption peak in the near-infrared (NIR) region (700–1000 nm) of the UV-visible spectrum that provides enough penetration of light into deep tissues and energy to excite molecular oxygen to its singlet state efficiently, (ii) minimal skin photosensitivity, (iii) no dark toxicity, (iv) selective uptake by cancer tissues, thereby enabling the decrease of side effects [4,5], and (v) fast elimination.

The self-assembly or aggregation of PS in aqueous environments can be caused by different reasons and is favored for amphiphilic PSs showing a negligible PDT activity due to the emission reduction of the $^3PS^*$ state in aggregated form [6,7]. The physico-chemical properties of aggregates differ from those of monomers. They exhibit a broadened Soret band and red-shifted Q bands in the UV-visible absorption spectra, low fluorescence intensity and lifetime [8–13], and low singlet oxygen ($^1O_2$) production.

The hematoporphyrin derivative (HPD) and its purified form Photofrin® (PF) were the first used PSs in PDT and PF was approved for the treatment of solid tumors [14–17]. PF was also indicated as a specific and selective radiosensitizing agent by several in vitro and in vivo studies [18–22]. PF is being used for the treatment of esophageal and non-small-cell lung and pancreatic cancers as well as a possible therapy against Karposi's sarcoma and brain, breast, skin, and bladder cancers [23].

The selective accumulation of protoporphyrin IX (PpIX) in the tumor cells following administration of 5-aminolevulinic acid (5-ALA) has made this PS precursor very popular for skin cancer PDT and fluorescent diagnostics of tumor tissues [24,25]. Topical, oral, or intravenous administration of 5-ALA prodrug in excess leading to the formation and accumulation of PpIX in vivo [26] is used by dermatologists to treat several malignant neoplasms of the skin, such as Bowen disease or actinic keratosis [27].

Pyropheophorbide-a (PPa) is a natural second-generation bacteriochlorin PS which presents a significant absorption in the far-red spectral region and high $^1O_2$ formation upon light illumination, suggesting it for PDT [28,29].

This study aimed to explore the different parameters (i.e., solvent polarity, concentration, temperature, and pH medium) that influence the photophysical properties (absorption, fluorescence emission, and $^1O_2$ formation) of the three PSs (PpIX, PPa, and PF).

2. Results and Discussion

Chemical structures of PpIX, PPa, and PF are shown in Figure 1. PpIX has two ionizable propionate groups and a hydrophobic ring core, which gives it amphiphilic properties leading to an aggregation through π–π stacking interaction and vesicle formation [30]. PPa has only one propionate group and a hydrophobic ring core, it can also aggregate in aqueous solutions [31]. PF is composed of monomers, dimers, and some very large oligomers [32].

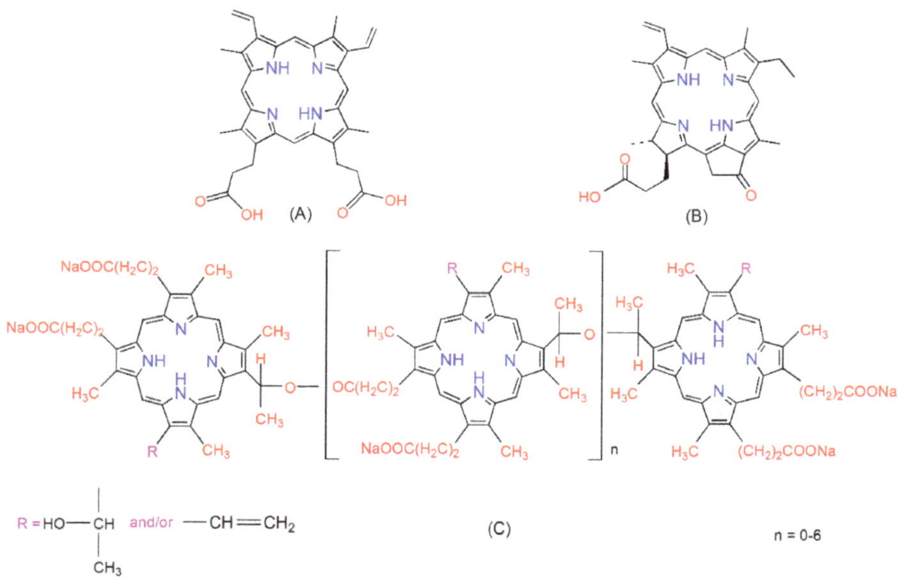

Figure 1. Chemical structures of protoporphyrin IX (PpIX) (**A**), pyropheophorbide-a (PPa) (**B**), and Photofrin®(PF) (**C**).

2.1. *Influence of the Solvent*

The structure of the molecule, ionic strength, pH, and temperature should play a main role in the photophysical properties [33]. UV-visible absorption spectra of PSs presented in Figure 2, were recorded in different solvents. ET(30) is a solvent polarity parameter that characterized the polarity of the different solvents. The bigger the ET(30) value, the more solvent polarity it is associated with.

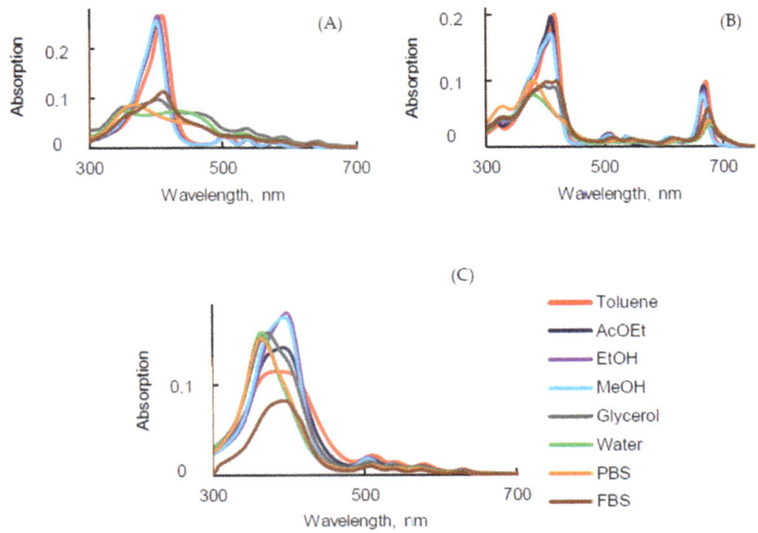

Figure 2. UV-visible absorption spectra of PpIX (**A**), PPa (**B**) and PF (**C**) in different solvents (c = 1.87 µM).

As expected, UV-visible absorption spectra of PpIX (Figure 2A) exhibited an intense Soret band centered at around 406 nm and four weaker Q bands in the visible range in toluene, ethyl acetate (AcOEt), ethanol (EtOH), and methanol (MeOH). These similar spectra are typical of monomeric PpIX. Nevertheless, in glycerol, water, phosphate-buffered solution (PBS), and fetal bovine serum (FBS), the Soret band was split into two bands [34]. This can be explained by the fact that PpIX is aggregated in aqueous solutions. In polar solvents the QI band was red-shifted compared to the QI band in less polar solvents (629 nm and 641 nm in EtOH and PBS, respectively) (Table 1), and the intensity decreased drastically. In the literature, it is often claimed that PpIX should be excited in vitro or in vivo at 630 nm. This wavelength of excitation is based on the absorption spectrum in EtOH. As it can be seen in Table 1 in water, PBS, and FBS, the QI band was located at 641, 641, and 640 nm, respectively.

Table 1. Soret and Q bands (nm) of PpIX, PPa, and PF in different solvents at room temperature (c = 1.87 µM).

Solvent	ET(30)	PpIX					PPa					PF				
		Soret	QIV	QIII	QII	QI	Soret	QIV	QIII	QII	QI	Soret	QIV	QIII	QII	QI
Toluene	33.9	409	506	540	577	632	415	510	539	612	671	388	509	539	578	628
AcOEt	38.1	402	503	536	575	630	410	506	536	608	667	394	503	536	574	625
EtOH	51.9	402	503	537	575	629	411	509	539	609	667	398	503	536	574	625
MeOH	55.4	401	502	537	574	628	409	507	538	608	665	394	503	536	574	625
Glycerol	57.0	404	536	561	590	642	415	512	544	616	671	371	507	536	574	625
Water	63.1	352	532	557	589	641	380	522	554	625	677	365	507	542	567	616
PBS	≈63.1	365	532	557	589	641	379	526	558	630	677	365	507	542	567	616
FBS	-	409	506	537	585	640	405	515	546	617	675	391	506	538	573	624

The UV-visible absorption spectra of PPa in various solvents exhibited a Soret band and four Q bands in the spectral range 300–700 nm (Figure 2B and Table 1). The Soret band in non-aqueous solution was located between 409 nm and 415 nm whereas it was blue-shifted to 380 nm in water and PBS and 405 nm in FBS. QIV, QIII, QII, and QI were red-shifted from, respectively, 510 nm to 526 nm, 539 nm to 558 nm, 612 nm to 630 nm, and 671 nm to 677 nm, from toluene to PBS. The UV-visible absorption spectra of PPa showed a broad Soret band in glycerol, water, PBS, and FBS due to the formation of aggregates. Interestingly, all shifts of PPa in glycerol and FBS showed close values.

UV-visible absorption spectra of PF in the different solvents were less impacted by the change of the polarity than PpIX and PPa. The shape of the Soret band was broader in toluene, AcOEt and it became intense when the polarity of solvent increased. The Soret band was blue-shifted by 23 nm from toluene to PBS. The positions of QI and QII bands in water and PBS were blue-shifted by 12 nm, respectively, compared to toluene (Figure 2C). The maxima of the absorption bands are presented in Table 1. In FBS, due to the presence of proteins (30–45 $g \cdot L^{-1}$), the behavior was different to than in water. The absorption spectra in FBS were similar to those in toluene.

On the basis of the UV-visible absorption spectra of PSs, molar extinction coefficients (ε) were calculated for all observed bands in all solvents. For PpIX and PPa, there was a single abrupt jump of ε (for the Soret band) when moving from toluene, AcOEt, EtOH, or MeOH into glycerol, water, or PBS. That was not observed for PF due to the fact that PF is a mixture of different compounds that do not all behave in the same way (Table 2). The high value of ε for the QI band of PPa is interesting for PDT applications. ε for the QI band of PPa was 3.5 times higher than the one of PpIX and 16.5 times higher than the one of PF [35].

Table 2. Molar extinction coefficient (ε, ($M^{-1} \cdot cm^{-1}$)) of PpIX, PPa, and PF in different solvents at room temperature.

Solvent	PpIX					PPa					PF				
	Soret	**QIV**	**QIII**	**QII**	**QI**	**Soret**	**QIV**	**QIII**	**QII**	**QI**	**Soret**	**QIV**	**QIII**	**QII**	**QI**
Toluene	143,590	13,746	10,448	6481	5204	107,765	10,718	8589	7108	52,500	45,454	8535	5544	4705	2381
AcOEt	140,975	13,148	10,432	6113	5302	106,642	10,691	8455	6202	49,416	55,780	7351	4655	3854	2056
EtOH	142,440	13,305	10,615	6476	5067	92,874	9303	8403	7851	45,435	71,905	7832	4779	3661	1891
MeOH	139,797	12,410	10,048	6111	4579	94,334	9555	8771	8202	43,430	70,829	7333	4495	3497	1722
Glycerol	38,582	14,342	9180	8029	4909	46,057	5215	4989	4603	20,415	61,111	5151	3891	3531	1392
Water	40,603	15,326	9447	8177	5005	41,993	3900	3489	4179	17,311	61,617	3456	2806	2751	1044
PBS	47,433	11,565	7964	6945	4152	39,431	3699	3467	3937	15,444	59,023	3970	2786	2524	931
FBS	61,034	13,332	12,969	7436	4346	54,518	7172	6478	6808	31,092	45,952	5447	3515	2956	1387

Fluorescence emission spectra presented in Figure 3 were recorded in different solvents at room temperature and at a concentration of 1.87 μM.

PpIX was excited at 400 nm. The highest was the polarity of the solvent and the blue-shift of the two fluorescence emission bands in agreement with the UV-visible absorption spectra (Figure 3A). All the maximum wavelengths are in the supporting information. Moreover, fluorescence intensity decreased with increase of the polarity of the solvent due to the lack of solubility of PpIX in aqueous media.

PPa was excited at 415 nm. The fluorescence emission spectra presented two bands and they were blue-shifted in polar solvents in accordance with the blue shift observed in the UV-visible absorption spectra. In water, PBS, and FBS at this concentration, the fluorescence intensity was very weak (Figure 3B), which could be explained by the aggregation [31].

PF was excited at 400 nm and a different behavior was observed in fluorescence emission spectra. Very weak maximum emission peaks at 633 nm and 696 nm were observed in toluene. The PF cores might have been in their highly quenched state, which did not generate fluorescence. The fluorescence intensity increased around 10 times in EtOH, and decreased again in polar solvents (Figure 3C). The emission bands of PF in water and PBS were blue-shifted for 15 ± 2 nm (Table S1) like the UV-visible absorption spectra indicating highly ordered aggregated structures [36]. In FBS, the fluorescence spectrum was similar to the one in non-polar solvent.

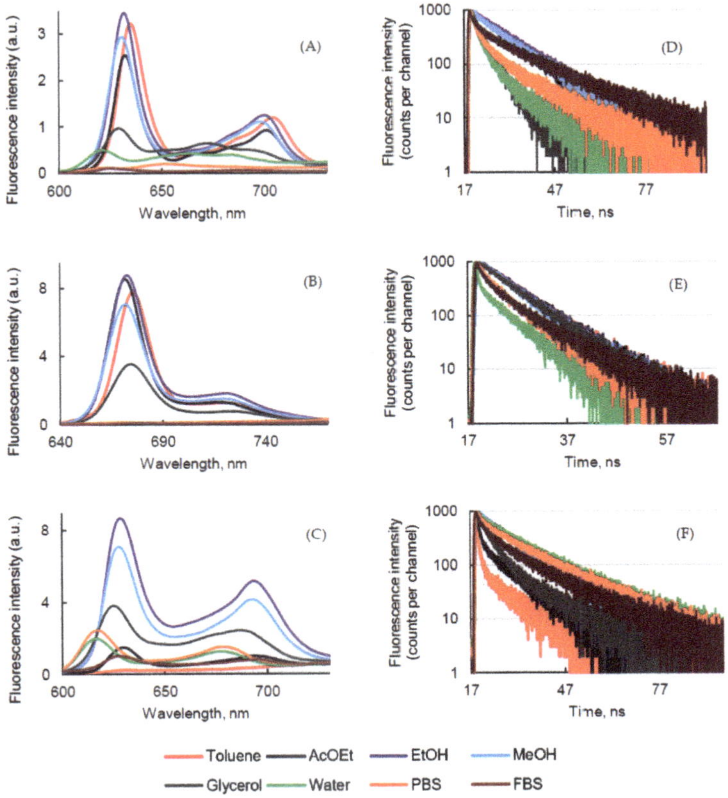

Figure 3. Fluorescence emission and decay (λ_{exc} = 408 nm) of PpIX (**A**,**D**), PPa (**B**,**E**), and PF (**C**,**F**) in different solvents (c = 1.87 μM).

The fluorescence quantum yield (Φ_f) of PpIX was evaluated to be higher in less polar solvents than in water, PBS, and FBS (Table 3). This is in good agreement with the fact that PpIX tends to aggregate in aqueous media [37]. Among the three PSs, PPa presented the best Φ_f which was 0.39 in toluene and EtOH. Φ_f of PF was low (below 0.1) and the highest was obtained in EtOH and MeOH, possibly due to a better solubilization (Table 3).

Table 3. Φ_f of PpIX, PPa, and PF in different solvents at room temperature (c = 1.87 μM).

Solvent	ET(30)	Φ_f (± 0.01)		
		PpI	**PPa**	**PF**
Toluene	33.9	0.09	0.39	<0.01
AcOEt	38.1	0.06	0.34	<0.01
EtOH	51.9	0.08	0.39	0.07
MeOH	55.4	0.07	0.31	0.05
Glycerol	57.0	0.04	0.20	0.02
Water	63.1	<0.01	<0.01	0.01
PBS	≈63.1	<0.02	<0.01	0.01
FBS	-	<0.01	<0.01	<0.01

The fluorescence lifetime (τ_f) of PSs was measured by time-resolved fluorescence after excitation at 408 nm. An exponential decay was fitted with an $R^2 \approx 1.000$. A bi-exponential decay of PpIX in polar solvents confirmed the presence of two populations—

monomers (long decay) and aggregates (short decay). PpIX exhibited mono-exponential decay (Figure 3D) in non-polar solvents, though the monomer–aggregate equilibrium was observed in glycerol, water, PBS, and FBS solutions: τ_f values of 10.3–15.9 ns and 2.5–3.0 ns for PpIX monomer and aggregates, respectively (Table 4).

Table 4. Fluorescence lifetimes (τ_f) of PpIX, PPa, and PF in different solvents at room temperature (c = 1.87 μM, λ_{exc} = 408 nm).

Solvent	τ_f **(ns)**			
	PpIX	**PPa**	**PF**	**References**
Toluene	11.2 ± 0.06	6.7 ± 0.02	8.7 ± 0.2	This work
AcOEt	10.3 ± 0.1	6.6 ± 0.01	2.4 ± 0.2 (11%) 9.3 ± 0.2 (89%)	This work
EtOH	11.6 ± 0.1	6.6 ± 0.01	10.8 ± 0.1	This work
MeOH	11.0 ± 0.02	6.1 ± 0.01	10.2 ± 0.1	10.0 ± 0.6 [38] [PF]
Glycerol	2.9 ± 0.02 (66%) 6.4 ± 0.1 (34%)	6.5 ± 0.01	3.0 ± 0.01 (64%) 12.0 ± 0.1 (36%)	This work
Water	2.5 ± 0.05 (54%) 9.6 ± 0.4 (46%)	0.3 ± 0.02 (1%) 5.5 ± 0.1 (99%)	3.4 ± 0.08 (13%) 14.5 ± 0.08 (87%)	This work
PBS	3.0 ± 0.2 (27%) 13.0 ± 0.3 (73%)	1.4 ± 0.06 (15%) 5.7 ± 0.07 (85%)	2.2 ± 0.08 (12%) 14.1 ± 0.1 (88%)	13.2 ± 2.0 [38] [PF] 14.7 [39] [PF]
FBS	2.9 ± 0.2 (10%) 15.9 ± 0.3 (90%)	2.1 ± 0.2 (18%) 7.5 ± 0.1 (82%)	3.2 ± 0.2 (21%) 15.0 ± 0.2 (79%)	This work

As it is known that aggregates reduce the inter-system crossing (ISC) transition from $^1PS^*$ to $^3PS^*$ and τ_f of aggregated PPa was shorter. PPa exhibited mono-exponential and bi-exponential decay in toluene, AcOEt, EtOH, MeOH, glycerol, and in water, PBS, and FBS, respectively (Figure 3E), confirming the presence of two forms—monomers (long decay) and aggregates (short decay). The τ_f value of PPa was between 6.1 to 7.5 ns for monomers and 0.3 to 2.1 ns for aggregates (Table 4).

The solution of PF in toluene, EtOH, and MeOH exhibited mono-exponential decay with τ_f values of 8.7, 10.8, and 10.2 ns, respectively, which was in good agreement with literature values [38,39], and two decays in the other solvents (Figure 3F). Once again, the short decay corresponded to the aggregated parts with lifetime of 2.4, 3.0, 3.4, 2.2, and 3.2 ns in AcOEt, glycerol, water, PBS, and FBS, respectively, since there was a longer decay for monomers (Table 4).

The $^1O_2$ production of PSs in different solvents was carried out and $^1O_2$ emission was detected at 1270 ± 5 nm after excitation at 400 nm for PpIX, PF, and 415 nm for PPa (Figure 4).

As expected, it was not possible to determine the Φ_Δ of PpIX in aqueous solutions due to possible aggregation of the PS, but it generated $^1O_2$ very efficiently in toluene, AcOEt, EtOH, and MeOH (Figure 4A). The same observation could be made with PPa. In our conditions we could not detect $^1O_2$ emission in glycerol, water, FBS, or PBS (Figure 4B). On the contrary, PF (Figure 4C) generated $^1O_2$ in EtOH, AcOEt, and MeOH (Table 5. The detection of $^1O_2$ was performed in D_2O, since the τ_Δ value is higher than in H_2O. Indeed, solvents with high vibrational frequencies are more able to quench $^1O_2$ [40]. However, no emission could be detected for PpIX and PPa whereas a Φ_Δ of 0.15 was obtained for PF. Additionally, $^1O_2$ generation from PpIX, PPa, and PF in D_2O was monitored by using the most common fluorescence probe Singlet Oxygen Sensor Green (SOSG), which is not sensitive to hydroxyl radicals or superoxide. It clearly appeared that fluorescence emission intensity of SOSG increased during the time due to the production of $^1O_2$ after excitation of PpIX, PPa, and PF. Sodium azide quenched $^1O_2$ very efficiently and fluorescence emission intensity of SOSG in the presence of quencher decreased (Figure 5).

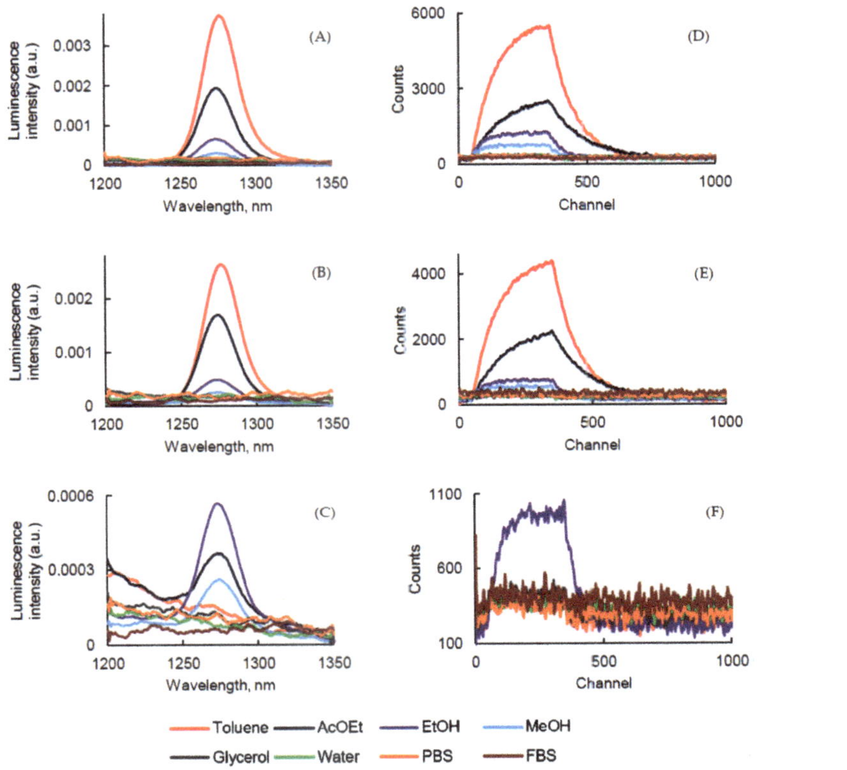

Figure 4. $^1O_2$ luminescence emission and decay (λ_{exc} = 400 nm for PpIX and PF and 415 nm for PPa) of PpIX (**A**,**D**), PPa (**B**,**E**), and PF (**C**,**F**) in different solvents (c = 1.87 μM).

Table 5. Φ_Δ of PpIX, PPa, and PF in different solvents at room temperature (c = 1.87 μM).

Solvent	Φ_Δ (± 0.10)		
	PpIX	PPa	PF
Toluene	0.68	0.49	0.01
EtOH	0.92	0.53	0.80
MeOH	0.92	0.42	0.61
D_2O	-	-	0.15

$^1O_2$ lifetime (τ_Δ) was determined. In solution, τ_Δ is governed by solvent deactivation through electronic-vibrational energy transfer [41]. If no reaction happens between $^1O_2$ and PS, τ_Δ value should be the same for the three PSs in each solvent. What we can observe is in good relation with the literature data (Table 6).

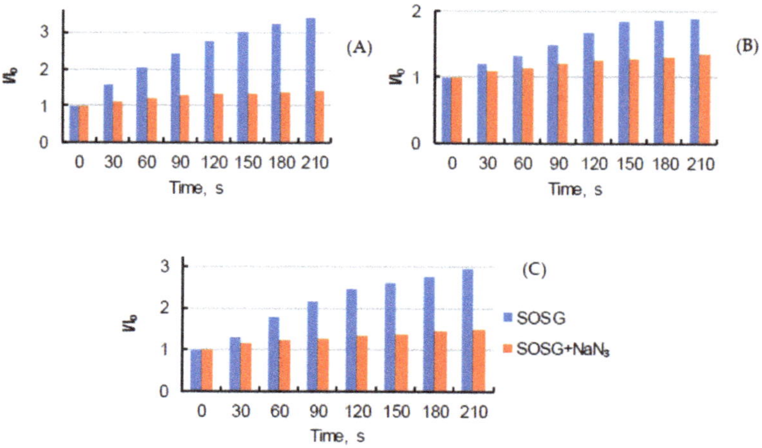

Figure 5. Comparison of fluorescence emission intensity of Singlet Oxygen Sensor Green (SOSG) and SOSG + NaN_3 in D_2O for $^1O_2$ detection after excitation of PpIX (**A**), PPa (**B**), and PF (**C**) (λ_{exc} = 400 nm for PS and 495 nm for SOSG) (c = 3.1 μM).

Table 6. $^1O_2$ lifetime of PpIX, PPa, and PF in different solvents at room temperature (c = 1.87 μM).

Solvent	τ_Δ (μs)			Literature Values
	PpIX	PPa	PF	
Toluene	30.4 ± 0.2	30.7 ± 0.2	-	30.5 ± 0.6 [42] [PS: 1H-phenalen-1-one-2- sulfonic acid (PNS)]
AcOEt	44.1 ± 0.6	43.2 ± 0.5	-	45 ± 1.5 [42] [PS: 1H-phenalen-1-one-2- sulfonic acid (PNS)]
EtOH	14.9 ± 0.6	14.6 ± 0.9	14.7 ± 0.8	15.3 ± 0.8, 12 [42,43] PS: hydrogen peroxide
MeOH	12.6 ± 0.9	8.9 ± 1.3	9.1 ± 1.0	9.9 ± 0.3, 7 [42,43] PS: ozone-triphenylphosphite

2.2. *Influence of the Medium Viscosity*

To evaluate the influence of the viscosity on the photophysical properties, a water/glycerol (W/G) mixture at various ratios was used. The higher the glycerol concentration, the higher the viscosity of the medium. The UV-visible absorption and fluorescence emission spectra of all PSs are shown in Figure 6.

For the three PSs, fluorescence emission decreased with addition of water. The highest was the viscosity and the lowest was the non-radiative decay.

The Soret band became larger and split in the solutions of PpIX with high concentrations of water, but the maximum wavelengths of four Q bands were not affected. A thin Soret band at 406 nm was only observed in 100% glycerol. This might be due to the fact that with the increase of the viscosity, the movement of the molecules was reduced and the formation of aggregates decreased or just due to the fact that aggregation occurred in water (Figure 6A).

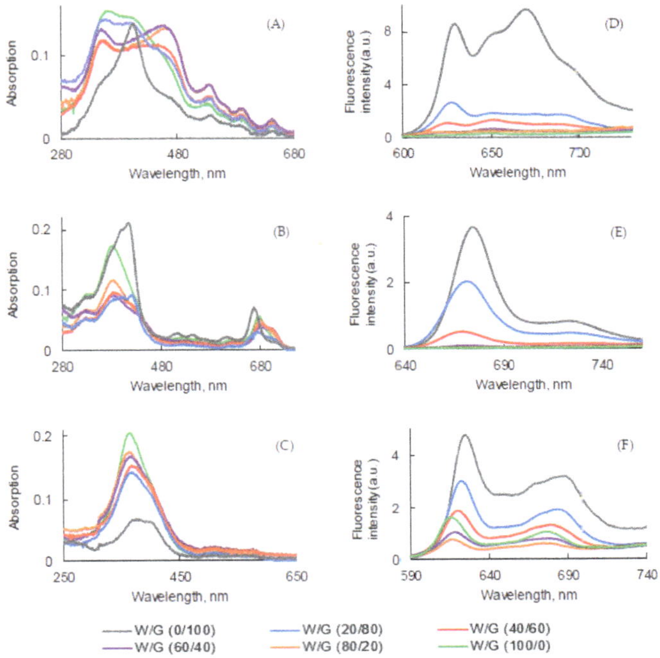

Figure 6. UV-visible absorption and fluorescence emission spectra ((λ_{exc} = 400 nm for PpIX, PF and 415 nm for PPa) of PpIX (**A**,**D**), PPa (**B**,**E**), and PF (**C**,**F**) in water/glycerol (W/G) mixtures (c = 3.1 μM) at room temperature.

UV-visible absorption spectra of PPa in water and in the mixture of water/glycerol showed a blue-shifted Soret band and weak, red-shifted Q bands. These band shifts might have been a result of the viscosity, which reduced the molecule's mobility for aggregate formation (Figure 6B).

A totally different behavior was observed for PF. The intensity of the Soret band of PF decreased by increasing the viscosity of the medium and the Soret band became wider in 100% glycerol with a red-shift of the maximum of absorption (Figure 6C).

The intensity of fluorescence emission of PpIX in the W/G mixture increased with the viscosity of the medium (Figure 6D). Φ_f value of PpIX in the W/G mixture increased in highly viscous media due to the fact that the formation of aggregates was less important. PPa in glycerol showed two emission bands located at 675 nm and 724 nm (Figure 6E). The fluorescence emission intensity increased with the viscosity of medium. The viscous medium might prevent non radiative deactivation. The fluorescence emission intensity of PF also increased with the viscosity of the medium (except in water) and was red-shifted (Figure 6F) for 10 nm. It is interesting to note that for PF when the viscosity increased, fluorescence emission increased but absorption decreased. The highest Φ_f value for all PSs was calculated for the solution in glycerol (Figure 7).

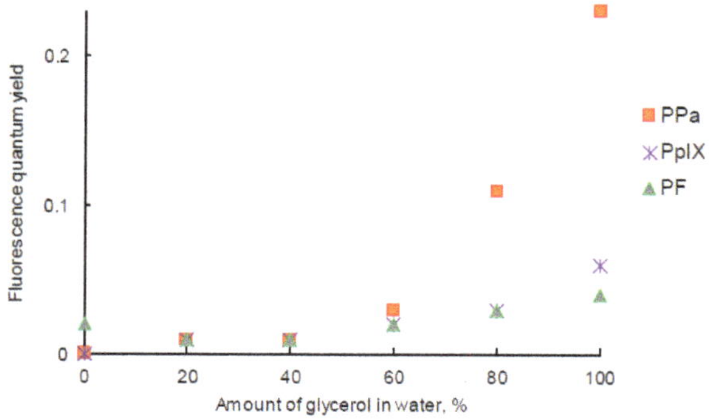

Figure 7. Fluorescence quantum yield of PpIX, PPa, and PF in the water/glycerol (W/G) mixture (c = 3.1 μM).

Fluorescence emission decays presented in Figure S1 were measured in the different media. τ_f were evaluated and are presented in Table 7. In all mixtures two lifetimes were detected, probably because of the presence of both monomers and aggregates. The τ_f value of PpIX increased with the viscosity (Figure S1A). The solution of PPa in W/G (100/0, 80/20, and 60/40) ratio exhibited two decays, but starting at a ratio of 40/60 showed mono-exponential decay and τ_f increased in line with the medium viscosity (Figure S1B). The solution of PF exhibited bi-exponential decay (Figure S1C) in all W/G mixtures.

Table 7. Fluorescence lifetimes of PpIX, PPa, and PF (λ_{exc} = 408 nm, c = 3.1 μM) at room temperature.

(W/G, *v*/*v*)	τ_f (ns)		
	PpIX	**PPa**	**PF**
W/G (100/0)	2.7 ± 0.02; 8.0 ± 0.1	0.3 ± 0.01; 5.5 ± 0.1	3.3 ± 0.1; 13.8 ± 0.2
W/G (80/20)	2.4 ± 0.04; 9.2 ± 0.3	0.1 ± 0.01; 5.6 ± 0.03	3.7 ± 0.02; 12.8 ± 0.04
W/G (60/40)	3.0 ± 0.01; 10.3 ± 0.1	0.1 ± 0.01; 5.7 ± 0.03	4.2 ± 0.02; 12.6 ± 0.04
W/G (40/60)	3.1 ± 0.02; 10.9 ± 0.2	5.5 ± 0.03	3.8 ± 0.04; 12.3 ± 0.1
W/G (20/80)	3.1 ± 0.05; 12.3 ± 0.4	5.9 ± 0.02	3.6 ± 0.02; 14.4 ± 0.05
W/G (0/100)	3.2 ± 0.02; 12.5 ± 0.2	6.5 ± 0.02	3.5 ± 0.01; 15.2 ± 0.05

Unfortunately, no correlation could be established between the fraction of monomers/aggregates and the viscosity of the medium. One reason might be that the polarity of the medium also changes when different amounts of glycerol and water are mixed. Therefore, the changes observed in W/G mixtures cannot only be attributed to the solution viscosity.

2.3. *Influence of the Concentration*

The influence of the concentration of PSs in PBS and FBS on photophysical properties was evaluated. As expected, the increase in concentration induced an increase in intensity, but no change in the absorption band maximum wavelength was observed in this concentration range for all PSs (Figure 8A–C).

The concentration increase led to a decrease of the fluorescence emission intensity for all PSs. Aggregation was higher in concentrated solutions (Figure 9D–F). The higher was the concentration, the lower was the fluorescence. Φ_f of all PSs in PBS at different concentrations were measured and were all less than 1%. However, the results obtained in FBS turned out to be the opposite in comparison with PBS. As the concentration of all PSs increased, fluorescence emission of all PSs increased. The aggregation process might be lower in FBS than in PBS due to the interaction with the proteins.

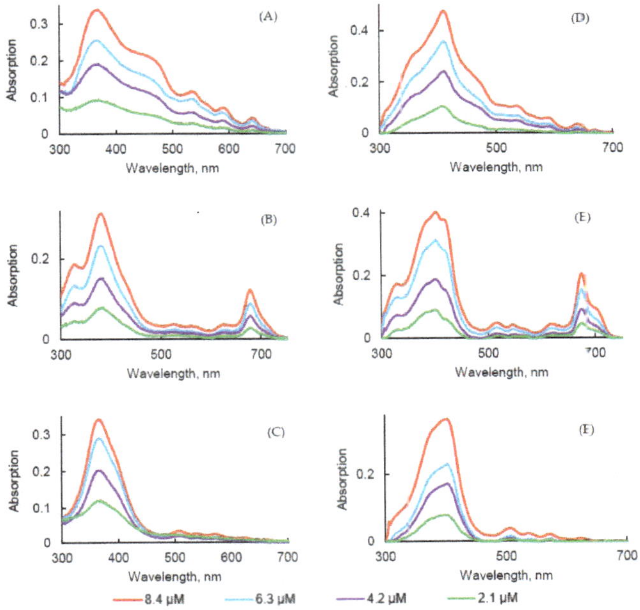

Figure 8. UV-visible absorption spectra of PpIX (**A**,**D**), PPa (**B**,**E**), and PF (**C**,**F**) in PBS (**A**–**C**) and FBS (**D**–**F**) at different concentrations.

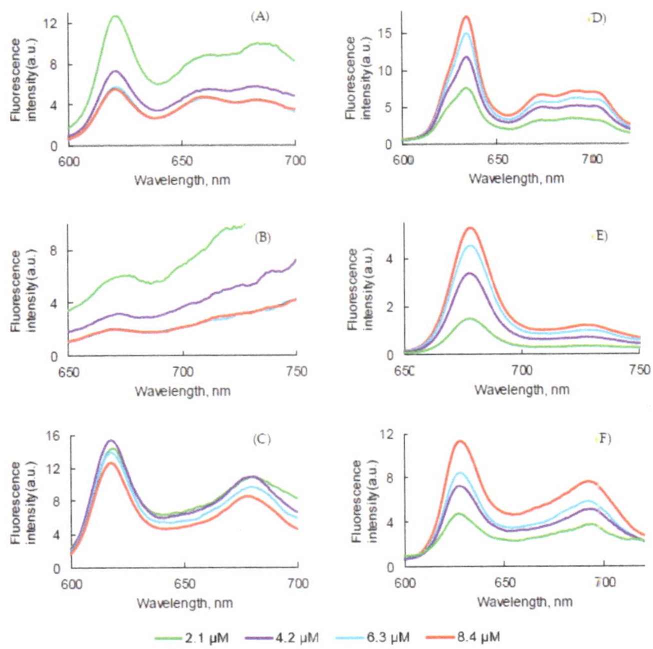

Figure 9. Fluorescence emission spectra (λ_{exc} = 400 nm for PpIX, PF and 415 nm for PPa) of PpIX (**A**,**D**), PPa (**B**,**E**), and PF (**C**,**F**) in PBS (**A**–**C**) and FBS (**D**–**F**) at different concentrations.

Fluorescence decays were recorded (Figure S2A) and τ_f were evaluated (Table 8). PpIX in PBS or FBS at different concentrations exhibited bi-exponential decay. The longest τ_f likely corresponded to the monomer decay time and the shorter lifetime was likely due to aggregates' decay time. PpIX in PBS at different concentrations exhibited bi-exponential decay. We could observe a slight increase of the ratio aggregate/monomers with the concentration increase in PBS but not in FBS. For PPa, only one population was observed in PBS between 5.6 and 6.8 ns. In FBS, a bi-exponential decay suggested the presence of both aggregates and monomers. For PF, no effect of the concentration could be observed. In PBS, 8% of aggregates and 92% of monomers can be evaluated whereas it was 14–17% of aggregates and 83–86% of monomers in FBS.

Table 8. Fluorescence lifetimes of PpIX, PPa, and PF in PBS and FBS (λ_{exc} = 408 nm) at room temperature.

Concentration (μM)	τ_f (ns)		
	PpIX	PPa	PF
	PBS		
2.1	3.8 ± 0.2 (21%) 14.5 ± 0.2 (79%)	6.8 ± 0.1	2.1 ± 0.3 (8%) 14.0 ± 0.2 (92%)
4.2	3.6 ± 0.2 (22%) 13.8 ± 0.2 (78%)	5.9 ± 0.1	2.7 ± 0.4 (8%) 14.8 ± 0.2 (92%)
6.3	3.5 ± 0.2 (25%) 12.8 ± 0.2 (75%)	5.8 ± 0.1	2.4 ± 0.3 (8%) 14.7 ± 0.2 (92%)
8.4	3.8 ± 0.2 (27%) 13.7 ± 0.2 (73%)	5.6 ± 0.1	2.6 ± 0.4 (8%) 14.8 ± 0.2 (92%)
	FBS		
2.1	3.7 ± 0.2 (13%) 17.0 ± 0.1 (87%)	3.5 ± 0.3 (13%) 8.2 ± 0.1 (87%)	3.3 ± 0.3 (17%) 15.3 ± 0.2 (83%)
4.2	3.6 ± 0.3 (12%) 17.1 ± 0.2 (88%)	2.5 ± 0.6 (5%) 8.2 ± 0.1 (95%)	3.6 ± 0.3 (15%) 15.5 ± 0.2 (85%)
6.3	3.9 ± 0.4 (11%) 17.8 ± 0.2 (89%)	2.6 ± 0.5 (5%) 8.2 ± 0.1 (95%)	3.4 ± 0.3 (14%) 15.6 ± 0.2 (86%)
8.4	3.9 ± 0.4 (11%) 17.8 ± 0.2 (89%)	3.2 ± 0.2 (4%) 8.5 ± 0.2 (96%)	3.4 ± 0.3 (14%) 15.7 ± 0.2 (86%)

2.4. Influence of the Temperature

The influence of temperature on UV-visible absorption, fluorescence emission, and lifetime of PpIX, PPa, and PF in aqueous media was evaluated. UV-visible absorption and fluorescence emission spectra of the PSs in PBS and FBS after heating from 10 °C to 40 °C are presented in Figure 10.

For PpIX, in PBS, the intensity of the Soret band decreased with the increase of temperature (Figure 10A), whereas the intensity of the Q bands increased, except QI. In FBS (Figure 10D), a different behavior could be observed with a decrease of all band intensities, except QI. For PPa in PBS (Figure 10B), the intensity of the Soret band decreased with the increase of temperature and the QI band shape changed and was red-shifted from 680 nm to 712 nm with an isobestic point at 685 nm. In FBS (Figure 10E), the intensity of the Soret band increased with the increase of temperature as well as the QI band, with a change of shape and an isobestic point at the same 685 nm. For PF, almost no change could be observed in PBS (Figure 10C) whereas in FBS, a blue shift of the Soret band and a decrease of intensity was observed by increasing the temperature, as well as an increase of QI intensity (Figure 10F).

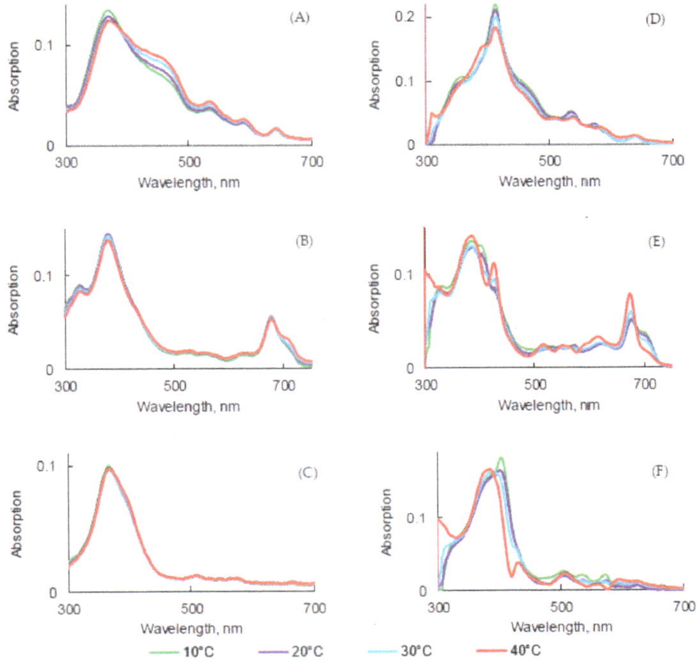

Figure 10. UV-visible absorption spectra (λ_{exc} = 400 nm for PpIX and PF and 415 nm for PPa) of PpIX (**A**,**D**), PPa (**B**,**E**), and PF (**C**,**F**) in PBS (**A–C**) and FBS (**D–E**) at different temperatures (c = 3.1 µM).

Fluorescence emission spectra were recorded in PBS and FBS at different temperatures (Figure 11). Whatever PS, the fluorescence emission intensity increased when the temperature rose from 10 to 40 °C. This might be have been because more monomers were in solution exhibiting fluorescence. For PpIX in PBS a slight red shift could be observed for the first band (Figure 11A) whereas in PBS, it was a slight blue-shift. No shift was detected for PPa (Figure 11B,E). Concerning PF, a red shift was observed both in PBS and FBS.

Fluorescence decays were recorded (Figure S3) and τ_f were evaluated (Table 9). PpIX in PBS of FBS at different temperatures exhibited bi-exponential decay. The longest τ_f likely corresponded to the monomer decay time and the shorter lifetime and was likely due to the aggregate decay time. A decrease of the shortest τ_f could be observed with the increase of the temperature in both solutions. Moreover, the ratio aggregate/monomer also seemed to decrease with the increase of the temperature. For PPa, only one τ_f was calculated in PBS. At low temperature (10 and 2 °C), both monomers and aggregates were present in FBS whereas aggregates disappeared at high temperature (30 and 40 °C). For PF, no effect of temperature was detected in PBS, whereas both short and long τ_f decreased with temperature increase in FBS.

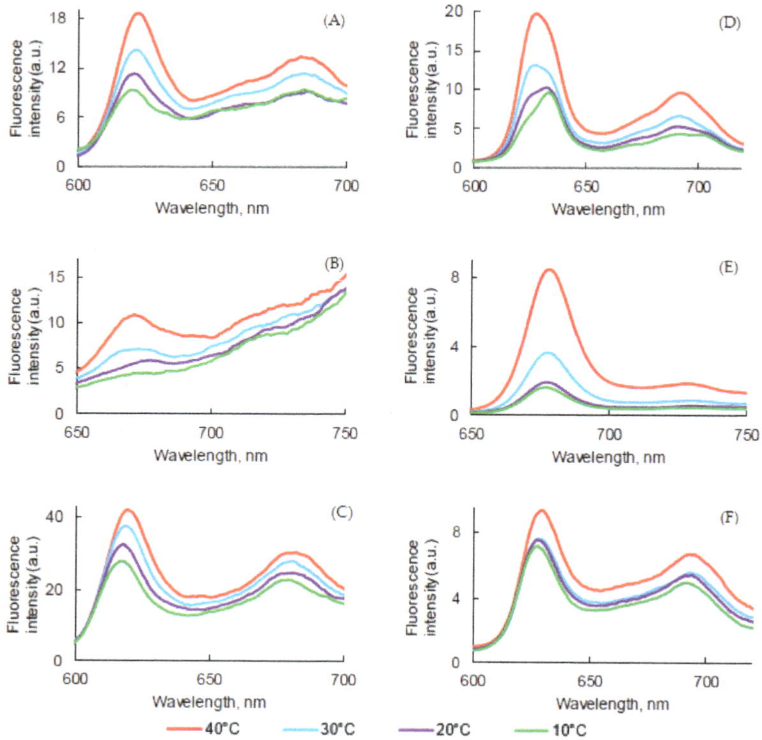

Figure 11. Fluorescence emission spectra (λ_{exc} = 400 nm for PpIX and PF and 415 nm for PPa) of PpIX (**A**,**D**), PPa (**B**,**E**), and PF (**C**,**F**) in PBS (**A**–**C**) and FBS (**D**–**F**) at different temperatures (c = 3.1 μM).

Table 9. Fluorescence lifetimes of PpIX, PPa, and PF (λ_{exc} = 408 nm, c = 3.1 μM).

Temperature, °C	Fluorescence Lifetime (ns)		
	PpIX	**PPa**	**PF**
		PBS	
10	5.7 ± 0.4 (19%) 16.4 ± 0.3 (81%)	5.8 ± 0.1	2.4 ± 0.3 (9%) 14.6 ± 0.2 (91%)
20	4.6 ± 0.3 (16%) 15.4 ± 0.2 (84%)	5.6 ± 0.1	2.9 ± 0.3 (9%) 14.8 ± 0.2 (91%)
30	4.6 ± 0.4 (13%) 13.9 ± 0.2 (87%)	5.5 ± 0.1	2.6 ± 0.2 (9%) 14.6 ± 0.2 (91%)
40	4.0 ± 0.4 (13%) 14.1 ± 0.2 (87%)	5.3 ± 0.1	2.6 ± 0.3 (9%) 14.6 ± 0.2 (91%)
		FBS	
10	3.9 ± 0.5 (9%) 18.0 ± 0.2 (91%)	2.8 ± 0.3 (10%) 7.8 ± 0.1 (90%)	3.0 ± 0.2 (16%) 15.0 ± 0.2 (84%)
20	3.4 ± 0.4 (8%) 18.0 ± 0.2 (92%)	2.0 ± 0.3 (7%) 7.7 ± 0.1 (93%)	2.8 ± 0.2 (16%) 14.7 ± 0.2 (84%)
30	3.3 ± 0.4 (7%) 17.9 ± 0.2 (93%)	7.3 ± 0.1	2.8 ± 0.2 (16%) 14.7 ± 0.2 (84%)
40	3.2 ± 0.4 (6%) 17.6 ± 0.2 (94%)	7.2 ± 0.1	2.6 ± 0.2 (16%) 13.8 ± 0.2 (84%)

2.5. Influence of pH Medium

pH could also have an influence on photophysical properties. The UV-visible absorption, fluorescence emission, and lifetime of all PSs in PBS with a concentration of 3.1 μM were measured under different pH conditions (pH 5.0–8.0). For PpIX, increasing pH from 5 to 8 led to a red-shifted Soret band from 354 nm to 375 nm whereas the Q bands were pH-independent (Figure 12A).

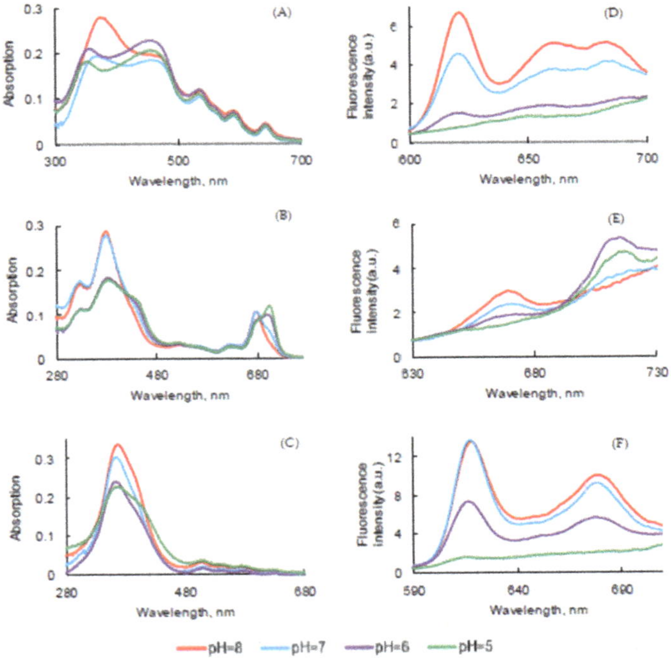

Figure 12. UV-visible absorption (**A–C**) and fluorescence emission (**D–F**) spectra (λ_{exc} = 400 nm for PpIX and PF and 415 nm for PPa) of PpIX (**A**,**D**), PPa (**B**,**E**), and PF (**C**,**F**) in PBS under different pH (c = 3.1 μM) at room temperature.

By increasing the pH, PPa showed an increase of the Soret and QI band intensities (Figure 12B). Two isobestic points could be observed at 415 nm and 685 nm, exactly the same as those observed by changing the temperature. This is in good agreement with the presence of two different species that could be monomers or aggregates. The Soret band of PF increased with pH and the maximum of absorption and Q bands were not affected (Figure 12C).

For all PSs (Figure 12D–F), fluorescence increased with the increase of pH in relation to the formation of monomers and disappearance of aggregates [44–46] but we could also observe a decrease of the band at 717 nm for PPa (Figure 12E). Φ_f of all PSs in PBS under different pH medium were below 0.01.

τ_f value of PpIX, PPa, and PF in PBS was also measured at different pH (Figure S4A). For PpIX, no aggregation could be observed at pH = 5 whereas aggregation occurred at pH 6–8 with the appearance of a short decay. At pH 5 and 6, the height of fast decay of PPa was higher than at pH 7 and 8, which was due to aggregation. At pH 5 and 6, the PPa was more aggregated with low τ_f value, so only τ_f of longer decay is given in the Table 10. For PF, τ_f value of fast decay was around 3.0 ns and long decay 14.5 ns (Table 10).

Table 10. Fluorescence lifetimes of PpIX, PPa, and PF in PBS (λ_{exc} = 408 nm, c = 3.1 μM) at room temperature.

pH	τ_f (ns)		
	PpIX	PPa	PF
5	5.0 ± 0.2	3.5 ± 0.4	2.2 ± 0.2 (21%) 11.2 ± 0.2 (79%)
6	5.7 ± 0.4 (42%) 13.1 ± 0.5 (58%)	5.1 ± 0.3	3.6 ± 0.5 (9%) 14.8 ± 0.2 (91%)
7	4.3 ± 0.06 (30%) 14.6 ± 0.1 (70%)	5.5 ± 0.1	2.7 ± 0.3 (9%) 15.0 ± 0.2 (91%)
8	3.7 ± 0.2 (23%) 14.6 ± 0.2 (77%)	5.6 ± 0.09	2.4 ± 0.2 (8%) 14.4 ± 0.2 (92%)

3. Materials and Methods

Protoporphyrin IX, Pyropheophorbide-a, and Porfimer sodium (Photofrin®) were purchased from Sigma (Saint-Louis, MO, USA), BOC Sciences (Shirley, NY, USA), and Oncothai (Lille, France), respectively, and used without further purification. The stock solution of PpIX and PPa was prepared in dimethylsulfoxide (DMSO), and PF in methanol (MeOH). FBS was purchased from Sigma (Saint-Louis, MO, USA). PBS was prepared by mixing the exact volume of 0.2 M sodium phosphate, dibasic dehydrate and 0.2 M sodium phosphate, monobasic, monohydrate, and pH was adjusted to 7.4. The stock solution of SOSG in methanol was prepared by dissolving 100 μg vial in 33.0 μL of methanol and sodium azide solution was prepared in water with concentration of 0.15 M.

3.1. Spectroscopic Measurements

UV-visible absorption spectra were recorded on a UV-3600 UV-visible double beam spectrophotometer (Shimadzu, Marne La Vallee, France). Fluorescence spectra were recorded on a Fluorolog FL3-222 spectrofluorimeter (Horiba JobinYvon, Longjumeau, France) equipped with 450 W Xenon lamp, a thermo-stated cell compartment (25 °C), a UV-visible photomultiplier R928 (Hamamatsu, Japan) and an InGaAs infrared detector (DSS-16A020L Electro-Optical System Inc, Phoenixville, PA, USA). The excitation beam was diffracted by a double ruled grating SPEX monochromator (1200 grooves/mm blazed at 330 nm). The emission beam was diffracted by a double-ruled grating SPEX monochromator (1200 grooves/mm blazed at 500 nm). The $^1O_2$ phosphorescence detection was measured with a HORIBA SpectraLED emitting at 415 nm, by a Multi-Channel Scaling (MCS) technique. The excitation pulse length was 102 μs and 600,000 pulses were averaged. $^1O_2$ emission was detected through a double-ruled grating SPEX monochromator (600 grooves/mm blazed at 1 μm) and a long-wave pass (780 nm). All spectra were measured in 4-face quartz cuvettes. All the emission spectra (fluorescence and $^1O_2$ luminescence) were displayed with the same absorbance (less than 0.2) with the lamp and photomultiplier correction.

Fluorescence quantum yield (Φ_f) was calculated with tetraphenylporphyrin (TPP) in toluene as reference (Φ_f = 0.11) [47], using the following Equation (1):

$$\Phi_f = \Phi_{f0} \times \frac{I_f}{I_{f0}} \times \frac{DO}{DO_0} \times \left(\frac{n}{n_0}\right)^2 \quad (1)$$

where Φ_f and Φ_{f0}, I_f and I_{f0}, DO and DO_0, and n and n_0 are the quantum yields, fluorescence emission intensities, optical densities, and refraction indices of the sample and reference, respectively.

$^1O_2$ quantum yield ($\Phi\Delta$) was measured with TPP in toluene ($\Phi\Delta$ = 0.68), rose Bengal in ethanol (EtOH) ($\Phi\Delta$ = 0.68) and MeOH ($\Phi\Delta$ = 0.76) as references [48,49] by Equation (2):

$$\Phi_\Delta = \Phi_{\Delta 0} \times \frac{I_f}{I_{f0}} \times \frac{DO}{DO_0} \quad (2)$$

where Φ_Δ and $\Phi_{\Delta 0}$, I and I_0, and DO and DO_0 are the luminescence quantum yields of singlet oxygen, the luminescence intensities, and the optical densities of the sample and references, respectively.

3.2. Fluorescence and Luminescence Decays

Time-resolved experiments were performed using, for excitation, a pulsed laser diode emitting at 408 nm (LDH-P-C-400M, FWHM < 70 ps, 1 MHz) coupled with a driver PDL 800-D (both PicoQuant GmbH, Berlin, Germany) and for detection, an avalanche photodiode SPCM-AQR-15 (EG&G, Vaudreuil, QC, Canada) coupled with a 550 nm long-wave pass filter as detection system. The acquisition was performed by a PicoHarp 300 module with a 4-channel router PHR-800 (both PicoQuant GmbH, Berlin, Germany). The fluorescence decays were recorded using the single photon counting method. Data were collected up to 1000 counts accumulated in the maximum channel and analyzed using Time Correlated Single Photon Counting (TCSPC) software Fluofit (PicoQuant GmbH, Berlin, Germany) based on iterative deconvolution using a Levensberg–Marquandt algorithm. $^1O_2$ lifetime (τ_Δ) measurements were performed on a TEMPRO-01 spectrophotometer (Horiba Jobin Yvon, Palaiseau, France). The apparatus was composed of a pulsed diode excitation source SpectralLED-415 emitting at 415 nm, a cuvette compartment, a Seya–Namioka type emission monochromator (between 600 and 2000 nm) and a H10330-45 near-infrared photomultiplier tube with a thermoelectric cooler (Hamamatsu, Massy, France) for the detection. The system was monitored by a single-photon counting controller FluoroHub-B and the software DataStation and DAS6 (Horiba Jobin Yvon, Palaiseau, France).

4. Conclusions

This study focalized on three PSs that are used clinically (PpIX and PF) or for in vivo experiments (PPa). Our team proposed PPa coupled to folic acid to treat ovarian metastases by PDT (Patent WO/2019/016397).

By analyzing the photophysical properties of these three PSs in different conditions, we highlighted the fact that each PS is unique and reacts very differently depending on its chemical structure and concentration.

If the change of the medium polarity does not greatly affect the UV-visible absorption spectrum of PF, there is a drastic change for PpIX and PPa. In the literature, it is often claimed that PpIX should be excited at 630 nm in vitro or in vivo. This excitation wavelength is based on the absorption spectrum in ethanol. In FBS and PBS, which are aqueous media more similar to physiological media, the QI band is located at 641 nm.

Depending on the localization of the PS in the cells, the local viscosity can be very different. We could also observe that modifying the solvent viscosity did not greatly affect the maximal wavelengths of absorption of QI in PpIX and PF but it was blue-shifted for PPa for 10 nm (from 678 nm to 668 nm).

Temperature change slightly affected the UV-visible absorption spectra of PpIX and PF but drastically modified the UV-visible absorption of PPa in the range of 10 to 40 °C.

Finally, modifying pH also induced a shift of QI band for PPa of 25 nm (from 704 nm to 679 nm).

Perhaps the most interesting results are the Φ_Δ obtained in different solvents. Depending on the solvent, the values were totally different. In toluene, we could not detect any $^1O_2$ whereas the Φ_Δ were quite good for PpIX and PPa 0.68 and 0.49, respectively. In EtOH, the Φ_Δ was 0.92, 0.53, and 0.80 for PpIX, PPa, and PF, respectively. If we switched to D_2O, we could not detect any $^1O_2$ of PpIX or PPa and the Φ_Δ was 0.15 for PF. Moreover, in real-life applications, the PS is ideally in a cellular context. The presence of protein, lipid, and other biomolecules molecules will also affect the photophysics of the PS. This raised the question of what type of experiments and which solvent should be used in the solution when performing in vitro studies.

Supplementary Materials: The following are available online at https://www.mdpi.com/1424-8247/14/2/138/s1, Figure S1: Fluorescence decay of PpIX (A), PPa (B), and PF (C) in water/glycerol mixture (λ_{exc} = 408 nm, c = 3.1 μM). Figure S2: Fluorescence decay of PpIX (A), PPa (B), and PF (C) in PBS at different concentrations (λ_{exc} = 408 nm). Figure S3: Fluorescence decay of PpIX (A), PPa (B), and PF (C) in PBS at different temperatures (λ_{exc} = 408 nm, c = 3.1 μM). Figure S4: Fluorescence decay of PpIX (A), PPa (B), and PF (C) in PBS under different pH (λ_{exc} = 408 nm, c = 3.1 μM). Table S1: Fluorescence emission bands (nm) of PpIX, PPa, and PF in different solvents (c =1.87 μM).

Author Contributions: Conceptualization, B.M., S.M., C.F., and P.A.; methodology, C.F.; software, P.A.; validation, P.A., C.F., and S.A.; formal analysis, B.M.; investigation, B.M.; resources, P.A. and C.F.; data curation, B.M.; writing—original draft preparation, B.M.; writing—review and editing, P.A., S.M., S.A., and C.F.; visualization, C.F.; supervision, C.F. and I.T.; project administration, C.F. and I.T.; funding acquisition, C.F. and P.A. All authors have read and agreed to the published version of the manuscript.

Funding: This research was funded by ERASMUS+.

Institutional Review Board Statement: Not applicable.

Informed Consent Statement: Not applicable.

Data Availability Statement: The data presented in this study are available on request from the corresponding author.

Acknowledgments: We would like to thank the program ERASMUS + (mobility for study) for the financial support.

Conflicts of Interest: The authors declare no conflict of interest.

References

1. Price, M.; Heilbrun, L.; Kessel, D. Effects of the oxygenation level on formation of different reactive oxygen species during photodynamic therapy. *Photochem. Photobiol.* **2013**, *89*, 683–686. [CrossRef]
2. Castano, A.P.; Mroz, P.; Hamblin, M.R. Photodynamic therapy and anti-tumor immunity. *Nat. Rev. Cancer* **2006**, *6*, 535–545. [CrossRef] [PubMed]
3. Sansaloni-Pastor, S.; Bouilloux, J.; Lange, N. The Dark Side: Photosensitizer Prodrugs. *Pharmaceuticals* **2019**, *12*, 148. [CrossRef] [PubMed]
4. Abrahamse, H.; Hamblin, M.R. New Photosensitizers for Photodynamic Therapy. *Biochem. J.* **2016**, *473*, 347–364. [CrossRef]
5. Stamati, I.; Kuimova, M.K.; Lion, M.; Yahioglu, G.; Phillips, D.; Deonarain, M.P. Novel Photosensitisers Derived from Pyropheophorbide-a: Uptake by Cells and Photodynamic Efficiency in Vitro. *Photochem. Photobiol. Sci.* **2010**, *9*, 1033–1041. [CrossRef]
6. Bonneau, C.; Vever-Bizet, C. Tetrapyrrole photosensitisers, determinants of subcellular localisation and mechanisms of photodynamic processes in therapeutic approaches. *Exp. Opin. Therap. Pat.* **2008**, *18*, 1011–1025. [CrossRef]
7. Mojzisova, H.; Bonneau, S.; Brault, D. Structural and physico-chemical determinants of the interactions of macrocyclic photosensitizers with cells. *Eur. Biophys. J.* **2007**, *36*, 943–953. [CrossRef]
8. Kelbauskas, L.; Dietel, W. Internalization of Aggregated Photosensitizers by Tumor Cells: Subcellular Time-resolved Fluorescence Spectroscopy on Derivatives of Pyropheophorbide-a Ethers and Chlorin e6 under Femtosecond One- and Two-photon Excitation. *Photochem. Photobiol.* **2002**, *76*, 686–694. [CrossRef]
9. Shiah, J.G.; Konák, C.; Spikes, J.D.; Kopecek, J. Influence of pH on aggregation and photoproperties of n-(2-hydroxypropyl) methacrylamide copolymer-meso-chlorin e6 conjugates. *Drug Deliv.* **1998**, *5*, 119–126. [CrossRef]
10. Roeder, B.; Wabnitz, H. Time-resolved fluorescence spectroscopy of hematoporphyrin, mesoporphyrin, pheophorbide *a* and chlorin e_6 in ethanol and aqueous solution. *J. Photochem. Photobiol. B* **1987**, *1*, 103–113. [CrossRef]
11. Margalit, R.; Cohen, S. Studies of hematoporphyrin and hematoporphyrin derivative equilibria in heterogeneous systems. Porphyrin-liposome binding and porphyrin aqueous dimerization. *Biochim. Biophys. Acta* **1983**, *736*, 163–170. [CrossRef]
12. Margalit, R.; Rotenberg, M. Thermodynamics of porphyrin dimerization in aqueous solutions. *Biochem. J.* **1984**, *219*, 445–450. [CrossRef] [PubMed]
13. Eichwurzel, I.; Stiel, H.; Röder, B. Photophysical studies of the pheophorbide a dimer. *J. Photochem. Photobiol. B* **2000**, *54*, 194–200. [CrossRef]
14. Macdonald, J.I.; Dougherty, J.T. Basic principles of photodynamic therapy. *J. Porphyr. Phthalocyanines* **2001**, *5*, 105–129. [CrossRef]
15. Niedre, M.J.; Yu, C.S.; Patterson, M.; Wilson, B.C. Singlet oxygen luminescence as an in vivo photodynamic therapy dose metric: Validation in normal mouse skin with topical amino-levulinic acid. *Br. J. Cancer* **2005**, *92*, 298–304. [CrossRef] [PubMed]
16. Kohoutova, D.; Haidry, R.; Banks, M.; Butt, M.A.; Dunn, J.; Thorpe, S.; Lovat, L. Long-term outcomes of the randomized controlled trial comparing 5-aminolaevulinic acid and Photofrin photodynamic therapy for Barrett's esophagus related neoplasia. *Scand. J. Gastroenterol.* **2018**, *53*, 527–532. [CrossRef]

17. Schaffer, M.; Schaffer, P.M.; Vogesser, M.; Ertl-Wagner, B.; Rauch, J.; Oberneder, R.; Jori, G.; Hofstetter, A.; Dühmke, E. Application of Photofrin II as a specific radiosensitising agent in patients with bladder cancer—A report of two cases. *Photochem. Photobiol. Sci.* **2002**, *1*, 686–689. [CrossRef]
18. Schaffer, M.; Ertl-Wagner, B.; Schaffer, P.M.; Kulka, U.; Jori, G.; Dühmke, E.; Hofstetter, A. The Application of Photofrin II® as a Sensitizing Agent for Ionizing Radiation—A New Approach in Tumor Therapy. *Curr. Med. Chem.* **2005**, *12*, 1209–1215. [CrossRef]
19. Schaffer, M.; Ertl-Wagner, B.; Schaffer, P.M.; Kulka, U.; Jori, G.; Wilkowski, R.; Hofstetter, A.; Dühmke, E. Feasibility of Photofrin II as a Radiosensitizing Agent in Solid Tumors—Preliminary Results. *Onkologie* **2006**, *29*, 514–519. [CrossRef]
20. Kou, J.; Dou, D.; Yang, L. Porphyrin photosensitizers in photodynamic therapy and its applications. *Oncotarget* **2017**, *8*, 81591–81603. [CrossRef]
21. Schaffer, P.; Batash, R.; Ertl-Wagner, B.; Hofstetter, A.; Asna, N.; Schaffer, M. Treatment of cervix carcinoma FIGO IIIb with Photofrin II as a radiosensitizer: A case report. *Photochem. Photobiol. Sci.* **2019**, *18*, 1275–1279. [CrossRef]
22. Schaffer, M.; Schaffer, P.M.; Corti, L.; Gardiman, M.; Sotti, G.; Hofstetter, A.; Dühmke, E. Photofrin as a specific radiosensitizing agent for tumors: Studies in comparison to other porphyrins, in an experimental in vivo model. *J. Photochem. Photobiol. B* **2002**, *66*, 157–164. [CrossRef]
23. Chang, C.J.; Cheng, S.M.; Nelson, J.S. Microvascular effects of Photofrin®-induced photodynamic therapy. *Photodiagn. Photodyn. Ther.* **2007**, *4*, 95–99. [CrossRef] [PubMed]
24. Berger, Y.; Greppi, A.; Siri, O.; Neier, R.; Juillerat-Jeanneret, L. Ethylene glycol and amino acid derivatives of 5-aminolevulinic acid as new photosensitizing precursors of protoporphyrin IX in cells. *J. Med. Chem.* **2000**, *43*, 4738–4746. [CrossRef] [PubMed]
25. Collaud, S.; Juzeniene, A.; Moan, J.; Lange, N. On the selectivity of 5-aminolevulinic acid-induced protoporphyrin IX formation. *Curr. Med. Chem. Anticancer Agents* **2004**, *4*, 301–316. [CrossRef]
26. Moan, J.; Bech, Ø.; Gaullier, J.M.; Stokke, T.; Steen, H.B.; Ma, L.; Berg, K. Protoporphyrin IX accumulation in cells treated with 5-aminolevulinic acid: Dependence on cell density, cell size and cell cycle. *Int. J. Cancer* **1998**, *75*, 134–139. [CrossRef]
27. Choudhary, S.; Nouri, K.; Elsaie, M.L. Photodynamic therapy in dermatology: A review. *Lasers Med. Sci.* **2009**, *24*, 971–980. [CrossRef]
28. Guelluy, P.H.; Marie-Pierre, F.A.; Grammenos, A.; Lecart, S.; Piette, J.; Hoebeke, M. Optimizing photodynamic therapy by liposomal formulation of the photosensitizer pyropheophorbide-a methyl ester: In Vitro and ex vivo comparative biophysical investigations in a colon carcinoma cell line. *Photochem. Photobiol. Sci.* **2010**, *9*, 1252–1260. [CrossRef]
29. Siwawannapong, K.; Zhang, R.; Lei, H.; Jin, Q.; Tang, W.; Dong, Z.; Lai, R.Y.; Liu, Z.; Kamkaew, A.; Cheng, L. Ultra-small Pyropheophorbide-a Nanodots for Near-infrared Fluorescence/Photoacoustic Imaging-guided Photodynamic Therapy. *Theranostics* **2020**, *10*, 62–73. [CrossRef] [PubMed]
30. Scolaro, L.M.; Castriciano, M.; Romeo, A.; Patanè, S.; Cefalì, A.; Allegrini, M.G. Aggregation Behavior of Protoporphyrin IX in Aqueous Solutions: Clear Evidence of Vesicle Formation. *J. Phys. Chem. B* **2002**, *106*, 2453–2459. [CrossRef]
31. Delanaye, L.; Bahri, M.; Tfibel, F.; Fontaine-Aupart, M.; Mouithys-Mickalad, A.; Heine, B.; Piette, J.; Hoebeke, M. Physical and chemical properties of pyropheophorbide-a methyl ester in ethanol, phosphate buffer and aqueous dispersion of small unilamellar dimyristoyl-L-α-phosphatidylcholine vesicles. *Photochem. Photobiol. Sci.* **2006**, *5*, 317–325. [CrossRef]
32. Gomes, A.T.P.C.; Neves, M.G.P.M.S.; Cavaleiro, J.A.S. Cancer, Photodynamic Therapy and Porphyrin-Type Derivatives. *An. Acad. Bras. Cienc.* **2018**, *90*, 993–1026. [CrossRef]
33. Kubát, P.; Lang, K.; Procházková, K.; Anzenbacher, P. Self-Aggregates of Cationic *meso*-Tetratolylporphyrins in Aqueous Solutions. *Langmuir* **2003**, *19*, 422–428. [CrossRef]
34. Ding, H.; Sumer, B.D.; Kessinger, C.W.; Dong, Y.; Huang, G.; Boothman, D.A.; Gao, J. Nanoscopic micelle delivery improves the photophysical properties and efficacy of photodynamic therapy of protoporphyrin IX. *J. Control Release* **2011**, *151*, 271–277. [CrossRef]
35. Al-Omari, S.; Ali, A. Photodynamic activity of pyropheophorbide methyl ester and pyropheophorbide-a in dimethylformamide solution. *Gen. Physiol. Biophys.* **2009**, *28*, 70–77. [CrossRef]
36. Teng, K.W.; Lee, S.H. Characterization of Protoporphyrin IX Species In Vitro Using Fluorescence Spectroscopy and Polar Plot Analysis. *J. Phys.Chem. B* **2019**, *123*, 5832–5840. [CrossRef] [PubMed]
37. Zampini, G.; Planas, O.; Marmottini, F.; Gulías, O.; Agut, M.; Nonell, S.; Latterini, L. Morphology effects on singlet oxygen production and bacterial photoinactivation efficiency by different silica-protoporphyrin IX nanocomposites. *RSC Adv.* **2017**, *7*, 14422–14429. [CrossRef]
38. Russell, J.A.; Diamond, K.R.; Collins, T.J.; Tiedje, H.F.; Hayward, J.; Farrell, T.; Patterson, M.S.; Fang, Q. Characterization of Fluorescence Lifetime of Photofrin and Delta-Aminolevulinic Acid Induced Protoporphyrin IX in Living Cells Using Single- and Two-Photon Excitation. *IEEE J. Sel. Top. Quant.* **2008**, *14*, 158–166. [CrossRef]
39. Cubeddu, R.; Ramponi, R.; Bottiroli, G. Time-resolved fluorescencespectroscopy of hematoporphyrin derivatives in micelles. *Chem. Phys. Lett.* **1986**, *128*, 439–442. [CrossRef]
40. Foote, C.S.; Krasnovsky, A.A.; Fu, Y.; Selke, M.; Karney, W.L. Singlet oxygen dimol-sensitized luminescence and reactions of singlet oxygen with organometallics. In *The Activation of Dioxygen and Homogeneous Catalytic Oxidation*; Barton, D.H.R., Marttell, A.E., Sawyer, D.T., Eds.; Plenum Press: New York, NY, USA; London, UK, 1993; pp. 411–422.
41. Thorning, F.; Jensen, F.; Ogilby, P.R. Modeling the Effect of Solvents on Nonradiative Singlet Oxygen Deactivation: Going beyond Weak Coupling in Intermolecular Electronic-to-Vibrational Energy Transfer. *J. Phys. Chem. B* **2020**, *124*, 2245–2254. [CrossRef]

42. Bregnhøj, M.; Westberg, M.; Minaev, B.F.; Ogilby, P.R. Singlet Oxygen Photophysics in Liquid Solvents: Converging on a Unified Picture. *Acc. Chem. Res.* **2017**, *50*, 1920–1927. [CrossRef]
43. Malik, S.; Khan, S.A.; Ahuja, P.; Arya, S.K.; Sahu, S.; Sahu, K. Singlet oxygen-mediated synthesis of malarial chemotherapeutic agents. *Med. Chem. Res.* **2013**, *22*, 5633–5653. [CrossRef]
44. Sobczynski, J.; Tønnesen, H.H.; Kristensen, S. Influence of aqueous media properties on aggregation and solubility of four structurally related *meso*-porphyrin photosensitizers evaluated by spectrophotometric measurements. *Die Pharmazie.* **2013**, *68*, 100–109. [PubMed]
45. Siggel, U.; Bindig, U.; Endisch, C.; Komatsu, T.; Tsuchida, E.; Voigt, J.; Fuhrhop, J.H. Photophysical and Photochemical Properties of Porphyrin Aggregates. *Ber. Bunsenges. Phys. Chem.* **1996**, *100*, 2070–2075. [CrossRef]
46. Inamura, I.; Uchida, K. Association Behavior of Protoporphyrin IX in Water and Aqueous Poly(N-vinylpyrrolidone) Solutions. Interaction between Protoporphyrin IX and Poly(N-vinylpyrrolidone). *Bull. Chem. Soc. Jpn.* **1991**, *64*, 2005–2007. [CrossRef]
47. Giovanetti, R.; Alibabaei, L.; Petetta, L. Aggregation behaviour of a tetracarboxylic porphyrin in aqueous solution. *J. Photoch. Photobio. A.* **2010**, *211*, 108–114. [CrossRef]
48. Figueiredo, T.L.C.; Johnstone, R.A.W.; Sørensen, A.M.P.S.; Burget, D.; Jacques, P. Determination of Fluorescence Yields, Singlet Lifetimes and Singlet Oxygen Yields of Water-Insoluble Porphyrins and Metalloporphyrins in Organic Solvents and in Aqueous Media. *Photochem. Photobiol.* **1999**, *69*, 517–528. [CrossRef]
49. Redmond, R.W.; Gamlin, J.N. A compilation of singlet oxygen yields from biologically relevant molecules. *Photochem. Photobiol.* **1999**, *70*, 391–475. [CrossRef]

pharmaceuticals

Review

Current Prospects for Treatment of Solid Tumors via Photodynamic, Photothermal, or Ionizing Radiation Therapies Combined with Immune Checkpoint Inhibition (A Review)

Sanjay Anand [1,2,3,*], Timothy A. Chan [3,4,5], Tayyaba Hasan [6,7] and Edward V. Maytin [1,2,3,*]

1 Department of Biomedical Engineering, Lerner Research Institute, Cleveland Clinic, Cleveland, OH 44195, USA
2 Dermatology and Plastic Surgery Institute, Cleveland Clinic, Cleveland, OH 44195, USA
3 Cleveland Clinic Lerner College of Medicine, Case Western Reserve University, Cleveland, OH 44106, USA; chant2@ccf.org
4 Center for Immunotherapy and Precision Immuno-Oncology, Cleveland Clinic, Cleveland, OH 44195, USA
5 Department of Radiation Oncology, Cleveland Clinic, Cleveland, OH 44195, USA
6 Wellman Center for Photomedicine, Massachusetts General Hospital and Harvard Medical School, Boston, MA 02115, USA; thasan@mgh.harvard.edu
7 Division of Health Sciences and Technology, Harvard University and Massachusetts Institute of Technology (Harvard–MIT), Cambridge, MA 02139, USA
* Correspondence: anands@ccf.org (S.A.); maytine@ccf.org (E.V.M.); Tel.: +1-(216)-444-7015 (S.A.); +1-(216)-445-6676 (E.V.M.)

Abstract: Photodynamic therapy (PDT) causes selective damage to tumor cells and vasculature and also triggers an anti-tumor immune response. The latter fact has prompted the exploration of PDT as an immune-stimulatory adjuvant. PDT is not the only cancer treatment that relies on electromagnetic energy to destroy cancer tissue. Ionizing radiation therapy (RT) and photothermal therapy (PTT) are two other treatment modalities that employ photons (with wavelengths either shorter or longer than PDT, respectively) and also cause tissue damage and immunomodulation. Research on the three modalities has occurred in different "silos", with minimal interaction between the three topics. This is happening at a time when immune checkpoint inhibition (ICI), another focus of intense research and clinical development, has opened exciting possibilities for combining PDT, PTT, or RT with ICI to achieve improved therapeutic benefits. In this review, we surveyed the literature for studies that describe changes in anti-tumor immunity following the administration of PDT, PTT, and RT, including efforts to combine each modality with ICI. This information, collected all in one place, may make it easier to recognize similarities and differences and help to identify new mechanistic hypotheses toward the goal of achieving optimized combinations and tumor cures.

Keywords: photodynamic therapy; photothermal therapy; radiation therapy; immunotherapy; immune checkpoint inhibition; murine models; clinical trials

Citation: Anand, S.; Chan, T.A.; Hasan, T.; Maytin, E.V. Current Prospects for Treatment of Solid Tumors via Photodynamic, Photothermal, or Ionizing Radiation Therapies Combined with Immune Checkpoint Inhibition (A Review). *Pharmaceuticals* **2021**, *14*, 447. https://doi.org/10.3390/ph14050447

Academic Editor: Serge Mordon

Received: 14 April 2021
Accepted: 3 May 2021
Published: 10 May 2021

1. Introduction

Cancer, one of the most serious public health problems, has been precisely described as "The Emperor of All Maladies" [1]. The incidence of cancer is increasing worldwide at an alarming rate, with approximately 1.9 million cases diagnosed and 608,570 cases of death expected in the United States alone, according to American Cancer Society estimates for 2021 [2]. Numerous modalities for cancer treatment are currently in use, including chemotherapy, hormonal therapy, and immunotherapy. Several treatments that employ various wavelengths of radiation, from short wavelengths (radiation therapy, RT), visible wavelengths (photodynamic therapy, PDT), or infrared/heat (photothermal therapy, PTT), are also available and undergoing rapid research and development in an attempt to better manage cancer progression and mortality. Despite best efforts, metastatic spread is often undetected until the disease is very advanced, resulting in cancer treatment failure and accounting for nearly 90% of cancer-related mortality. When treatment fails, each of the individual treatment modalities mentioned above can be used for palliation in patients with advanced metastases. However, the extension of survival is often modest, pointing to a need for additional approaches in order to cure cancer. In principle, we need therapeutic strategies that offer high tumor-specificity, minimize off-target normal tissue damage, and achieve long-term cure. Toward the latter goal, research over the past few decades has led to new immunotherapeutic approaches that have been creating much excitement because they exploit the body's natural defense systems in order to target tumor cells [3–5]. Some immunotherapy approaches under investigation include vaccine therapy, cytokine therapy, and most recently, immune checkpoint blockade (ICB) therapy, also known as immune checkpoint inhibition (ICI), which targets cell membrane receptors (such as programmed cell death protein 1, PD-1, programmed cell death protein 1 ligand 1, PD-L1, and cytotoxic T lymphocyte antigen 4, CTLA4) expressed on the surface of tumor cells and tumor-infiltrating immune cells, and whose interactions regulate anti-tumor immune responses [6–10]. While ICI is able to bring about complete cures in some cancer patients, the actual proportion of patients who respond to ICI is unfortunately rather small. This has led to efforts to further stimulate therapeutic responses by combining ICI with more traditional therapies such as chemotherapy, or with radiation-based modalities such as the three mentioned above (PDT, PTT, and RT) [11–16]. Research combining ICI with the radiation-based strategies (light, heat, or ionizing radiation) is currently at a very early stage, and the findings are being published in widely disparate specialty journals. However, there could be great value in considering these modalities side by side, i.e., comparing the ability of each treatment to stimulate anti-tumor immunity, and asking whether those changes are leveraged by ICI administered at the appropriate time, resulting in improved therapeutic outcomes. A recent study by our group, in addition to a few studies by others, demonstrated that anti-tumor immunity generated by PDT may play a relatively larger role in the therapeutic outcomes, as compared to direct PDT-induced cell death within the primary tumor, than was previously thought [17–21]. This has major implications because the development of long-term anti-tumor immunity is the desired outcome and ultimate goal for generating durable cancer cures. In this review, we have collected the existing literature pertinent to PDT, PTT, and RT, and described what is known about how each treatment contributes to the development of anti-tumor immunity. We have also described preclinical and clinical studies in which PDT, PTT, or RT were combined with ICI, and the outcomes of those studies. ICI combination with currently available cancer treatment options is a rapidly evolving area. While our review is by no means exhaustive, we hope that by providing information about ICI and the three different radiation-based modalities all in one place, that commonalities and differences may become apparent, possibly leading to insights about how each tissue-damaging approach might best be combined with ICI in order to improve cancer treatment outcomes.

2. Immune Checkpoint Inhibition Therapy

Tumors that are resistant to mainline or monotherapies such as chemotherapy and RT often carry a treatment challenge by the upregulation of inhibitory genes and pathways which favor tumor growth in an immunosuppressive tumor microenvironment. Another challenge for the success of immunotherapy is the uncertain relationship between the tumor and its host immune microenvironment, a hot area in the contemporary cancer immunotherapy research field. The majority of tumors thrive in their host environment by neutralizing anti-tumor immunoregulatory signals, such as PD1/PDL1 and CTLA4, that block the cytotoxicity of immune cells and result in immunotherapy failure. In the past two decades, a therapeutic approach called immune checkpoint inhibition (ICI), also known as immune checkpoint blockade (ICB), has been developed to reduce or overcome these inhibitory factors and has been successfully translated to the clinic in combination with mainline treatment therapies for various cancers [22–26]. Immune checkpoint molecules are a subset of inhibitory receptors on the surface of both tumors and T cells which antagonize T cell-mediated killing, thereby evading immune recognition and favoring tumor growth. Some of the most common immune checkpoint receptors are PD1, PDL1, CTLA4, indoleamine-pyrrole 2,3-dioxygenase (IDO) and lymphocyte-activation gene 3 (LAG3) [9,27,28]. PD1, also known as PDCD1 and CD279, is a transmembrane receptor expressed mainly on activated T cells and B cells, and in some cases on macrophages, natural killer cells, and cells of myeloid lineage. PD1 is expressed during T cell activation to counter positive signals that occur through T cell receptor (TCR) and CD28. PD1 engages a specific set of ligands, either PD-L1 (also known as CD274 and B7-DC) or PD-L2 (also known as CD273 and B7-DC), which are expressed on a variety of cell types including cancer cells [8,29,30]. Ligand-bound PD1 receptors function as "brakes" or "immune checkpoints" for T cell-mediated adaptive immune responses, a signal that effector T cells must overcome to exert their cytotoxic activities [30]. CTLA4, also known as CD152, is another negative regulator (and the first one identified) which is induced in Tregs and also in some other T cell types. CTLA4 directly competes with the T cell co-stimulatory activator CD28 for the ligands CD80 (B7-1) and CD86 (B7-2) [8,31]. Blockade of PD1 or CTLA4 by immune checkpoint inhibitory agents (anti-PD1 or anti-CTLA4 antibodies) has been shown to restore tumoricidal activities of T lymphocytes and enhance the therapeutic effects of mainline monotherapies (e.g., RT or chemotherapy) when given as a combination therapy [9,25]. This new approach has revolutionized cancer therapy over the past decade. To date, immune checkpoint inhibitors targeting PD1/PDL1 and CTLA4 have transformed the care of patients with advanced-stage cancers, most effectively for melanoma, renal, head and neck, bladder, and Hodgkin lymphoma [9,27,28].

3. Photodynamic Therapy (PDT)

Photodynamic therapy (PDT) is a treatment modality that has been successfully utilized to treat cancer and non-cancerous conditions in the clinic [20,32–34]. PDT is a multi-step procedure that involves systemic or topical administration of a light-sensitive photosensitizer (PS), its selective uptake by the tumor, followed by excitation of the PS within the tumor tissue by illumination with visible light. The light source can be either a broadband source or a laser of the appropriate wavelength, i.e., corresponding to a major peak within the absorption spectrum of the PS. The energy generated from the excited state of PS in the presence of oxygen results in the production of cytotoxic singlet oxygen ($^1O_2$) and reactive oxygen species (ROS), triggering a cascade of events leading to tumor cell death and the destruction of tumor vasculature [32,33,35–38]. The therapeutic effects of PDT depend upon the cumulative response of three well-characterized and mechanistically linked events that occur in sequence. First, PDT directly kills cancer cells by triggering signaling cascades that lead to cell death via apoptosis, necroptosis, autophagy and/or pyroptosis [32,36,38]. Then, PDT-mediated damage of tumor-associated vasculature limits the blood supply and induces hypoxia, resulting in tumor destruction by starvation [39–42]. These two events are directly responsible for destruction of the primary tumor, activation/release of damage-associated molecular patterns (DAMPs), and the production of cellular debris which primes a third event involving the immune system. This third event, the triggering of inflammation and activation of the immune system, can last from days to weeks and can eventually exert a systemic (abscopal) effect; the latter constitutes the physiological basis for the concept of PDT-generated vaccines for cancer therapy [12,43–47].

3.1. PDT-Induced Immunogenic Cell Death (ICD) and Activation of Damage-Associated Molecular Patterns (DAMPs)

Over the past decade, it has been realized that certain chemotherapeutic drugs and cancer therapies, such as PDT, PTT, and RT, induce a form of cell death which triggers an immune response, which is hence referred to as immunogenic cell death (ICD) [36,47–50]. ICD activates innate and adaptive arms of the immune system, resulting in efficient elimination of tumors by generating long-term immunological memory [51,52]. The immune response generated by ICD is determined by the antigenicity and adjuvanticity of target cancer cells. Antigenicity of the tumor cells is determined by the tumor-specific antigens (TSA), which are essentially PDT-generated tumor debris. The ICD is usually accompanied by the release of adjuvant-like DAMPs which reside in the cells as a part of their normal functions, but once released, act as danger signals [53,54]. DAMPs, either secreted or exposed extracellularly on the surface of dying cells, are recognized by innate pattern recognition receptors (PRRs) such as Toll-like receptors (TLRs) expressed on immune cells, thereby promoting the recruitment of antigen-presenting cells (APCs) [50,52]. Dendritic cells (DCs), being the predominant APCs in most scenarios, take up and process the TSAs and present them to naïve T cells, thereby activating long-term adaptive immunity [36]. The list of DAMPs is continuously growing and includes calreticulin (CRT), high-mobility group box 1 (HMGB1), heat shock proteins (HSPs) 70 and 90, and ATP as some of the common members activated following PDT [21,53,54].

3.2. *PDT-Induced Inflammation and Activation of Innate Immunity*

PDT-induced oxidative stress and cell death trigger an acute inflammatory response that is often seen as edema at the treatment site. Localized inflammatory responses following PDT involve the upregulation and release of inflammatory cytokines such as interleukin 1 beta (IL-1β), IL-6, tumor necrosis factor alpha (TNFα), macrophage inflammatory protein 2 (MIP2) or chemokine (C-X-C motif) ligand 2 (CXCL2), and activation of complement protein C3 [20,36,47,55]. A rather complex balance between the levels of pro-inflammatory and anti-inflammatory cytokines has been linked to the anti-tumoral immune response following PDT. For example, blockade of pro-inflammatory cytokine IL-1β levels diminished the therapeutic effects of PDT [56], but neutralization of transforming growth factor beta (TGFβ) or IL-10 by antibody depletion significantly enhanced the therapeutic effects of PDT [40]. Release of inflammatory cytokines results in the rapid infiltration of immune cells at the site of damage; these cells attack and remove dying tumor cells. Neutrophils are the first population of cells of the innate immune system to enter the damage site, followed by macrophages, DCs, natural killer (NK) cells and lymphocytes. This resets a "cold" tumor microenvironment (non-immunogenic, immunosuppressive) into a "hot" (immunogenic) tumor microenvironment [12,47,57–59].

Neutrophils, being the most predominant leukocytes early on, have been reported to accumulate in high numbers within less than 5 min following PDT, and remain present at the site until 24 h post-PDT [21,57]. In addition to infiltrating treated tumors, neutrophils have also been reported to accumulate in tumor-draining lymph nodes (DLN). The induction of IL-17 levels by T helper cells (Th17) following PDT has been linked to this migration of neutrophils [60,61]. By secreting alarmins and TNFα, neutrophils have also been reported to help with the maturation and activation of DCs, which trigger adaptive immunity by the stimulating of CD8+ cells [60,62].

Dendritic cells (DCs) are the major APCs in the PDT-treated tumor microenvironment (TME), where they phagocytize and process the tumor cell debris, differentiate into APCs, and present TSAs to naïve T cells, resulting in the activation and proliferation of T cells involved in long-term adaptive immunity. Accumulation of CD11c+ and CD1a+ DCs in the treated tumor sites have been reported around 24 h post Photofrin-PDT and Aminolevulinic acid-mediated PDT (ALA-PDT), respectively [63,64]. A recent study by Lamberti et al. showed the critical involvement of the interferon 1 (IFN-1) pathway in regulation of the functions of DCs. PDT-treated B16-OVA murine melanoma cells induced IFN-1-dependent maturation of DCs by enhancing co-stimulatory signals (CD80 and MHC-II) and tumor-directed chemotaxis [65].

Another class of immune cell in the TME is macrophages, also referred to as tumor-associated macrophages (TAMs), which differentiate from monocytes and acquire an ability to activate immune effector functions following PDT [21,66–68]. In an unperturbed tumor microenvironment, the majority of macrophages belong to an anti-inflammatory M2 phenotype that promotes immunosuppression, growth, angiogenesis and metastasis.

Following PDT, the majority of M2 macrophages are removed from the TME and replaced by a fresh population of M1 macrophages, derived from monocytes present in the tumor and surrounding vasculature. These M1 macrophages provide an immunostimulatory environment by secreting pro-inflammatory cytokines (IL-1, IL-6, IL-12 and TNFα) that promote tumor regression [68–71].

Natural killer (NK) cells are another type of cytotoxic lymphocyte, part of the innate immunity system that responds to local inflammation following PDT. Studies by Belicha-Villanueva et al. using human and murine colon carcinoma cells showed an increased expression of MHC class I-like molecules (MICA) and natural killer group 2D (NKG2D) ligands following PDT, which corresponded to enhanced NK cell-mediated killing [72]. These two molecules serve as ligands for activation receptors on NK cells that support their roles in anti-tumor immunity [47]. Additionally, Kabingu et al. reported that the reduction in distant tumors by CD8+ T cells, following PDT of a primary tumor (abscopal effect), was improved in the presence of NK cells, supporting the role of these cells in PDT-mediated anti-tumor immunity [18].

3.3. Activation of the Adaptive Immune System by PDT

PDT-induced differentiation, activation, and transformation of DCs into APCs is the step which connects the innate and adaptive arms of the immune system; adaptive long-term immunity involves CD4+ (helper), CD8+ (cytotoxic) and regulatory (Tregs) T lymphocytes [12,20,47]. The observation that anti-tumor effects exerted by ICD-inducing therapies involve DC recruitment and activation has resulted in the combination of DC-based immunotherapies with PDT to enhance the treatment outcome. A study by Ji et al., using a murine model of cutaneous SCC, showed the enhancement of anti-tumor activity of DC-based vaccines in mice by immunogenic apoptotic cells induced by ALA-PDT. Higher levels of IFNγ and IL-12 and the suppression of immunosuppressive IL-10 were reported to be associated with functional DC maturation and T cell proliferation [73,74]. A role for adaptive immunity in generating anti-tumor responses after PDT was established by observing a diminished or absent long-term tumor treatment response in immunocompromised mice, and an increase in PDT efficacy after replenishment of these mice with bone marrow or T cells from immunocompetent mice [67,75]. PDT can activate both B cell- (humoral) and T cell-mediated anti-tumor immune responses; however, while both types of responses have been investigated, the major research focus over the past three decades has been upon immune responses involving T lymphocytes [75–77]. For T cells, when mature dendritic cells are activated following PDT, the APCs interact with T cells through a complex mechanism of co-stimulation that involves major histocompatibility antigens (MHC class I or II) on the APC, and the T cell receptor (TCR) and co-receptors CD8 or CD4 on the T cell; this process ensures that the proper type of antigen is presented to the appropriate class of T cell, thereby avoiding the development of autoimmunity. Regarding MHC, there are two separate antigen-loading pathways: (1) MHC-I displays peptides that are endogenously derived (e.g., capsid proteins in virally infected cells) within almost any kind of cell, including cancer cells; (2) MHC-II displays peptides that are produced within lysosomes of immune system cells after they ingest foreign proteins ("exogenous" peptides). MHC-I antigen complexes bind to the TCR only when also bound to CD8, thereby activating CD8+ T cells that play a major role in anti-tumor immunity by their tumoricidal/cytotoxic properties. MHC-II complexes will bind TCR only on T cells that express CD4; activation of these CD4+ T helper T cells plays a supportive role in amplifying the adaptive immune response [77–81].

Interactions between APCs and naïve T cells can lead to the development of several different subclasses of T cell subsets. Anti-tumor responses after PDT are generally thought to involve three subsets of T cells, namely: (1) CD8+ cytotoxic T lymphocytes (CTL); (2) CD4+ T helper (Th) cells; and (3) regulatory T cells (Tregs). A role for CD8+ CTLs in PDT-induced anti-tumor immunity was first demonstrated by Korbelik et al., who showed that the depletion of CD8+ T cells in EMT6 mammary carcinoma model resulted in a 50%

reduction in tumor clearance compared to controls [17]. Similarly, the adoptive transfer of CD8+ T cells from PDT-cured animals protected naïve recipients from cancer cells of the same origin [82]. PDT-induced elevations in the number of CD8+ T cells and increases in their antigen-specific cytotoxic activities have been reported in several preclinical studies; for example, Abdel-Hady et al. reported a direct correlation between treatment response and increased levels of CD8+ cells in lesions following PDT [19,21,60,83,84].

The second class of T cell involved in PDT anti-tumor responses, the CD4+ T cells, facilitate the activation of B cells and CD8+ T cells. Involvement of three subtypes of T helper cell populations, i.e., Th1, Th2 and Th17, have been reported in PDT-induced anti-tumor immunity [85,86]. Th1 cells, characterized by the production of IFNγ, can activate CTLs and mediate direct cell killing by the release of cytokines and activation of death receptors on tumor cells [87,88]. Th2 cells secrete cytokines such as IL-4, IL-5, IL-9, and IL-13, which regulate humoral immunity and coordinate immune responses to extracellular pathogens by B cell isotype switching [89]. Th17 cells, defined by their secretion of IL-17 cytokine, are interesting due a dichotomy related to their origin. In an inflammatory TME, levels of TGFβ regulate the differentiation of T cells into either Tregs or Th17 cells. While low levels of TGFβ promote differentiation into Th17 cells, high levels favor their differentiation into Tregs [90]. Experimental immune-depletion of CD4+ cells in mice has shown mixed results. When Korbelik et al. used antibodies against CD4, CD25 and a combination of both to deplete T helper cells, a reduction in treatment response was seen [17,67]. However, a study by Kabingu et al. showed no effect of CD4+ T cell depletion on therapeutic response and systemic anti-tumor immunity [18].

The third subtype of T cells involved in anti-tumor responses after PDT are a unique subpopulation in the CD4+ category which are regulatory or suppressive in nature; they are also referred to as suppressor T cells, or Tregs. The most common type of Tregs is CD25+ CD4+ FoxP3+ T cells [91]. At a molecular level, these cells constitutively express high levels of the transmembrane protein CD25, CTLA4, and forkhead box P3 (FoxP3) transcription factor, also known as scurfin [92,93]. By suppressing the differentiation of effector T cells (Teff), Tregs maintain the Teffs in an intermediate stage by favoring IL-2 production. Tregs, by withholding IL-2 and producing TGFβ, prevent full T-effector differentiation during the acute phase of the CD8+ T cell response, blocking differentiation into tumor-specific cytotoxic CD8+ T cells [77,94]. Tregs were shown to be involved in anti-tumor immunity induced by PDT, in two studies showing that suppression of Tregs using a cyclophosphamide-PDT combination led to improved therapeutic efficacy (enhanced tumor regression and long-term survival) in murine models of reticulum cell carcinoma and colon carcinoma [95,96]. Oh et al. showed that intra-tumoral depletion of Tregs using CD25-targeted photodynamic therapy in a mouse melanoma model induced antitumor immune responses, possibly due to increased infiltration of CD8+ effector T cells and the expression of interferon gamma (IFNγ) and CD107a, a marker of cytotoxicity [97].

A brief summary of the important properties of PDT, PTT and RT, and of subsequent events that result in induction of anti-tumor immunity (based on studies in murine tumor models) is provided in Figure 1.

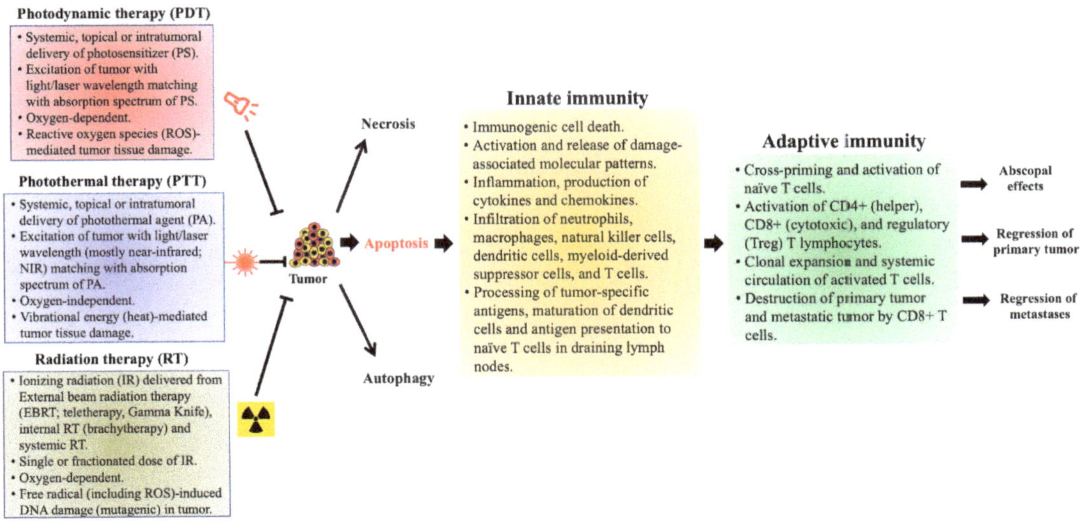

Figure 1. Photodynamic therapy- (PDT), photothermal therapy- (PTT), and radiation therapy (RT)-induced effects on anti-tumor immunity, based on preclinical studies in murine tumor models. Treatment of tumors with PDT, PTT or RT leads to cell death within the primary tumor by apoptotic, necrotic and autophagic mechanisms. Apoptosis induced by these therapies also generates an immune response, referred to as immunogenic cell death (ICD), within the tumor microenvironment (TME). Damage-associated molecular patterns (DAMPs), expressed on the surface of dying cells and released into the TME, promote the recruitment and maturation of antigen-presenting cells (APCs), primarily dendritic cells (DCs). Various cytokines and chemokines (IL-1β, IL-6 and TNFα), released by the photodamaged cells, induce local inflammation and recruit cells associated with innate immunity (neutrophils, macrophages, natural killer cells, and mast cells). The DCs engulf and process tumor-specific antigens (TSAs), then migrate to draining lymph nodes (DLNs) to present the processed TSAs to naïve T cells, thereby triggering the adaptive arm of anti-tumor immunity. Activated T cell subsets (CD4+, CD8+ and FoxP3+) undergo clonal expansion and differentiation within the TME, mediating tumor regression via cytotoxic activities of CD8+ cytotoxic T lymphocytes (CTLs). Optimally, these activated T cells may enter the systemic circulation and travel to distant metastases, mediating a more widespread (abscopal) effect. The immunological events discussed here were observed in PDT-treated murine tumor models, but similar mechanisms of anti-tumor immunity have also been reported in pre-clinical studies using PTT and RT (see text for details).

3.4. Combination of Immune Checkpoint Inhibition with PDT

Tumors that are resistant to PDT as a monotherapy pose a therapeutic challenge by the upregulation of inhibitory genes and pathways which favor tumor growth in an immunosuppressive tumor microenvironment. Combinations of ICI with PDT have been explored mainly in preclinical studies in the past decade for their potential to overcome the inhibitory effects of immune checkpoints, with the ultimate goal of future translation into the clinic. In this section, we describe a few preclinical studies investigating the combination of ICI with PDT to improve the therapeutic efficacy. A combination of PDT and ICI using antibodies against PD1/PDL1, CTLA4 and IDO have been investigated in preclinical studies using murine cancer models of breast, colon, renal, lung and skin, to show significant improvements in therapeutic efficacies [98–102]. For example, Zhang et al. used chlorin 6-mediated PDT combined with either an inhibitor for PD1/PDL1 interaction called Bristol Mayers Squibb 202 (BMS-202), or an anti-PDL1 antibody treated 4T1 murine breast cancer model and showed that tumor regression was associated with the inhibition of lung metastasis. The therapeutic effects by combination regimens were possibly achieved by enhanced maturation of DCs and infiltration of CD8+ T cells, along with increased levels of IFNγ, IL-6, and TNFα cytokines [103]. In another study using a mouse model for renal carcinoma which developed lung metastasis, O'Shaughnessy et al. showed that the combination approach had synergistic effect over tumor regression and metastasis to lungs, compared to the outcome with either treatments given alone. Furthermore, CD8+:Tregs and CD4+:Tregs ratios were increased in both primary tumors and lung metastasis in the combination treatment mice [99]. Details of a few recent studies using immune checkpoint inhibition combination with PDT in murine tumor models showing the elimination of primary and distant (abscopal effect) tumors, reduction in metastases, and the involvement of immune cells in the observed outcome have been listed in Table 1. A timeline of immunological events that contribute to anti-tumor immunity after PDT, along with the time frame for effective ICI, is shown in Figure 2.

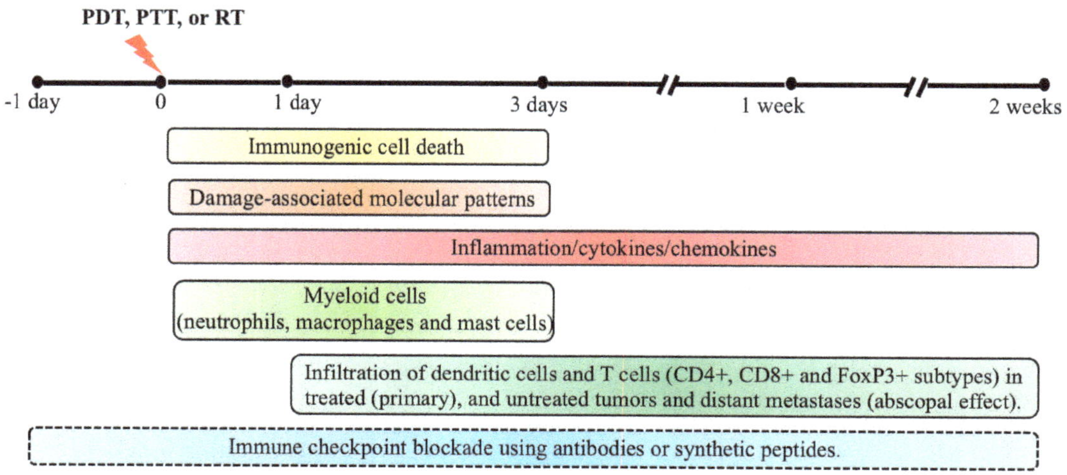

Figure 2. Timeline of immunological events contributing to anti-tumor immunity by photodynamic therapy (PDT), based on preclinical studies in murine tumor models. Following PDT, tumor cells undergo cell death mainly by apoptosis, necrosis, and autophagy as PDT's primary therapeutic mechanism. In parallel, immunogenic cell death (ICD) triggers anti-tumor immunity by inducing inflammation and activation and release of damage-associated molecular patterns (DAMPs). While effects of ICD and DAMPs can last from 1 day to 3 days post-PDT, involvement of cytokines and chemokines can last from the time of light exposure until 2 weeks post-treatment in some cases. A robust neutrophil infiltration occurs within minutes following PDT, followed by infiltration of macrophages and mast cells, most prominently in the first 3 days post-PDT. One day post-PDT, dendritic cells (DCs) along with lymphocytes start infiltrating the treated tumor site. Maturation of DCs by exposure to tumor-specific antigens, presentation of processed antigens to naïve T cells, and elevated levels of TNFα, IFNγ and IL-6 in the tumor microenvironment (TME) trigger the adaptive immune response during the two weeks post-PDT. Activated T cells (CD4+, CD8+ and FoxP3+) undergo clonal expansion and reach the primary tumor and metastatic sites through the systemic circulation to induce the regression of primary and metastatic tumors (abscopal effect). The combination of immune checkpoint inhibition (ICI) with PDT, PTT or RT has been explored in pre-clinical models by injecting antibodies against PD1/PDL1, CTLA4 and IDO, at a variety of times (from 1 day prior up to 2 weeks post-therapy) with the reinjection of antibodies every 2–3 days until the endpoint. The optimal sequence and timing of these combinations is still under exploration. Although the timeline and immunological events discussed above were observed in PDT-treated mouse models, similar mechanisms of anti-tumor immunity have also been reported in pre-clinical studies using PTT and RT, as described in the text.

Table 1. A list of selected pre-clinical studies using combinations of immune checkpoint inhibition and photodynamic therapy.

Checkpoint Inhibitor Target	Photosensitizer/Construct	Murine Tumor Model	Immune Effector Cells	Cytokines	Therapeutic/Immune Response	Ref.
PD-1	$\alpha v\beta 6$ integrin-specific phthalocyanine dye labeled probe	4T1 breast tumor	DC, CD8+ T cells	IL-1β, IL-12	Reduced primary tumor growth and lung metastasis. Abscopal effect.	[98]
PD-1	Pheophorbide A, given together with a tumor-specific peptide vaccine adjuvanted with TLR5 antagonist	BF16-F10 murine melanoma model	DC, CD8+ T cells	IFNγ	Reduced primary tumor growth and lung metastasis	[101]
PD-L1	IRD700, conjugated to Fab fragment of anti-αCD276 antibody	4T1 breast tumor	CD8+ T cells	Not analyzed	Reduced primary tumor growth and lung metastasis	[104]
PD-L1	EGFR-targeted porphyrin-containing nanoliposomes conjugated with IRDye800CW and DOTA-Gd	Subcutaneous CT26 colon cancer	Not analyzed	Not analyzed	Tumor regression	[100]
PD-L1	Verteporfin	4T1 breast tumor	DC, CD8+ T cells	Not analyzed	Regression of primary tumors by destruction of tumor-associated lymphatic vessels	[105]
PD-L1 and BMS202 PD1/PDL1 inhibitor	Chlorin 6 NPs	4T1 breast tumor	DC, CD8+ T cells	IFNγ, IL-6, TNFα	Regression of primary tumors, reduced lung metastases	[103]
PD1 + PD-L1	WST11	Renal cell carcinoma line that develops lung metastases	CD8+, CD4+FoxP3-T cells	Not analyzed	Regression of primary tumors, reduced lung metastases	[99]
CTLA4	Bremachlorin	Subcutaneous MC38 and CT26 colon cancer double tumor model	CD8+ T cells	Not analyzed	Significant improvement of therapeutic efficacy and survival, abscopal effect	[106]
CTLA4	Nanoparticles simultaneously loaded with chlorin e6 (photosensitizer) and imiquimod (Toll-like receptor-7 agonist)	Subcutaneous CT26 colon cancer	DCs, CD8+, CD4+FoxP3+ T cells	IFNγ, IL-12, TNFα	Therapeutic efficacy with abscopal effect. Prevented tumor recurrence, via immune memory effects	[107]
CTLA4	OR141	Ab1 and Ab12 mesothelioma murine model	CD4+ and CD8+ T cells, DCs	Not analyzed	Inhibition of mesothelioma cell growth	[102]
IDO	Chlorin-based nanoscale metal–organic framework (nMOF)	Subcutaneous B16F10 melanoma and CT26 colon cancer double tumor model.	CD4+ and CD45+ T cells, neutrophils, and B cells	Not analyzed	Local and distant tumor rejection and T cell infiltration of TME. Compensatory roles of neutrophils and B cells in presenting TAAs to T cells	[108]
IDO	Verteporfin	4T1 breast tumor E0771 breast tumor	Myeloid cells	IL-6	Tumor regression	[109]

Finally, when thinking about how to optimize PDT and ICI for cancer treatment, there is some literature suggesting a role for vascular endothelial growth factor (VEGF). One of the most important clinical advances in the last few years has been the combination of anti-PD1 agents (e.g., pembrolizumab) with VEGF-targeting tyrosine kinase inhibitors such as axitinib [110–112] and lenvatinib [113]. In animal tumor models, PDT has been shown to transiently increase VEGF expression [114,115], providing a potential rationale for why anti-VEGF blockades could be helpful when designing effective combination therapies. However, any new approach must be approached with caution due to competing effects of PDF, VEGF, and the immune system. For example, because tumor-derived VEGF encourages the growth of lymphatic vessels, and the latter potentially increases the risk of metastasis, one might think that PDT-induced damage of tumor-draining lymphatics would be helpful. However, a recent study in a murine breast cancer model showed that verteporfin PDT does indeed destroy lymphatic vessels—a combination treatment with anti-VEGF blockade or with lenalidomide (a lymphangiogenesis inhibitor) actually reduced tumor responsiveness to PDT and abrogated the potentiation of therapy by anti-PD1 monoclonal antibodies (mAbs); these effects were largely due to the reduced migration of DCs from the tumor to the DLNs [105].

Clinical studies that use a combination of ICI with PDT are very few at the current time, consisting of one published case report and one clinical trial in ClinicalTrials.gov (accessed on 24 April 2021), as described in more detail in Section 5.4).

4. Photothermal Therapy (PTT)

Photothermal therapy (PTT), a non-invasive, local treatment modality for cancer, utilizes a combination of light-absorbing photothermal agents (PTAs) and their wavelength-matched light or laser source to generate heat, which results in the thermal ablation of tumors, causing cell death [116,117]. Similar to photosensitizers (PS), during PDT, PTAs absorb energy from incoming photons and undergo a transformation from an electronic ground state to an excited state. Upon returning to its ground state by vibrational relaxation, the excited photothermal agent emits kinetic energy, heating the surrounding tissue, and causing thermal damage to the tumor microenvironment. Unlike PDT, in which PS is excited with a specific wavelength light to generate ROS in the presence of oxygen, PTT does not require oxygen in order to interact with target cells or tissues [14,116,118]. Recently developed PTAs use longer wavelengths of light, which not only penetrate deeper into the tissue, but are also less energetic and therefore less harmful to surrounding cells and tissues. The list of novel PTAs, comprising engineered nanomaterials with unique activation mechanisms to provide tumor-specific targeting and reduce adverse off-target effects, is continuously evolving [13]. Due to their promise of limited side effects and relatively low drug resistance, PTT agents have evolved through four generations, including precious metal nanoparticles such as Au, Ag, Pt, carbon nanorods and graphene, metal and non-metal compounds such as CuS and ZnS, and organic and inorganic nanomaterials such as Prussian blue, Indoline green, and organic semiconducting pro-nano-stimulants (OSPSs), which are still in an exploratory phase for research and clinical applications [119].

Similar to PDT, nanoparticle-based PTT for the treatment of cancer offers the following unique and advantageous features: (i) combination of a near-infrared (NIR) laser with nanoparticle-based photothermal agents allows target-focused and deeper penetration of the activating signal [119,120]; (ii) tumor-specific target molecules (peptides or nucleic acids) included in the nano-formulation can offer tumor-specific targeting and avoid off-target systemic effects [118,121]; (iii) by combining PTT with interventional technologies for the delivery of light, PTT is not limited to superficial tumors, but can treat internal malignancies such as prostate and pancreatic cancer [13,122–124]; (iv) PTT agents can be used for image-guided therapy using theranostic nanoparticles [13,125–127]. In preclinical studies, PTT has been successful as a monotherapy or in combination with other therapies for the treatment of several malignancies including breast cancer, melanoma, and liver cancer in murine models [116,118,128]. Based upon such preclinical studies, nanoparticle-based PTT has been successfully translated into the clinic to treat low- or intermediate-risk localized prostate cancer [123,129]. Several clinical trials using AuroLase, a type of PTT that combines silica–gold nanoshells (AuNS) with NIR light to treat tumors of the head and neck, lung and prostate have been reported as described in reference [130]. Similar to PDT, PTT causes destruction of the peritumoral extracellular matrix and vasculature, induces inflammation, and releases tumor antigens which trigger an anti-tumoral immune response by the recruitment of endogenous immune cells. Evidence that PTT elicits ICD and activation of inflammatory response, followed by innate and adaptive immune responses, is discussed in the following sections.

4.1. PTT-Induced Immunogenic Cell Death (ICD), Activation of Damage-Associated Molecular Patterns (DAMPs) and Activation of Anti-Tumor Immunity

Depending on the combination of photothermal agent and activating light wavelength utilized, PTT raises the temperature of the tumor microenvironment to 41–48 °C, which causes tissue damage, including damage to cellular architecture, degradation of proteins and nucleic acids (DNA/RNA), and eventually results in apoptosis [14,131,132]. Although both apoptotic and necrotic damage has been reported following PTT, it appears that one or other pathway may be favored, depending upon the hyperthermic temperature achieved during therapy [14,116]. When the temperature of the tumor microenvironment is raised to 41 °C, a heat shock response is initiated, which mitigates the effects of thermal damage on cellular machinery [133]. A rise in temperature to between 42 and 46 °C results in irreversible tissue damage and promotes apoptosis. Sustained hyperthermia (42–46 °C) that lasts beyond 10 min results in tissue damage by necrosis. A temperature above 60 °C often triggers instantaneous cell death by the denaturation of cellular components [116,134].

As discussed in the PDT section, cell death induced by PTT also triggers the induction of ICD that involves the release of TSAs and DAMPs from dying tumor cells, followed by the maturation of DCs that can activate T cells and trigger anti-tumor immunity. However, unlike PDT, PTT may only induce ICD within a specific thermal window. For example, a study by Sweeney et al. using Prussian blue nanoparticles (PBNP) for PTT in neuroblastoma cells showed that the induction of ICD was specific to an optimal thermal dose, at which the release of DAMPs (calreticulin, HMGB1 and ATP) was observed and a potent anti-tumor response was achieved. If the thermal dose was too low or too high, tumors cells were eliminated by cell death after PPBNP-PTT, but the dying cells did not trigger any ICD-mediated anti-tumor response [135].

In addition to inducing ICD, PTT can also activate innate immunity through macrophage reprogramming. Using a uniformly conjugated polymer nanoparticle for PTT, Wang et al. showed activation of a pro-inflammatory immune response (M1 macrophages) and inhibition of tumor growth in a murine tumor model [136]. Another study by Yu et al. used magnetic Fe_3O_4 photothermal nanoparticles (MNPs) coated with myeloid-derived suppressor cell membranes (MNP@MDSC) for PTT, demonstrating the enhancement of ICD (upregulation of HMGB1 and calreticulin) and reprogramming of infiltrating macrophages that involved increased CD86+ M1 macrophages and decreased CD206+ M2 macrophages in a B16/F10 murine melanoma model [137]. However, some recent studies utilizing conventional and nanomaterial-based PTAs have shown that immune-stimulation induced by PTT alone was not sufficient to effectively activate long-term anti-tumor immunity. Therefore, new efforts to combine nanoparticle-based PTT with immunoadjuvants or other immune response-promoting drugs are underway (see the next section).

4.2. Nanoparticle-Based Photothermal Immunotherapy

The term "photothermal immunotherapy" has been coined to encompass nanomaterial-based PTT that can not only eliminate primary tumors, but also reduce metastasis through sustained anti-tumor immune effects [13,14]. However, in several recent studies, it was shown that immunomodulation by PTT alone was not sufficient to activate long-term anti-tumor response. To overcome this limitation, immune response-promoting drugs have been added as immunoadjuvants to nanoparticle-based photothermal agents [13,14].

Photothermal nanoparticles loaded with immunoadjuvants have been shown to activate both innate and specific immune responses by significantly inducing the infiltration and maturation of NK cells and DCs in the tumor microenvironment, increasing the levels of immune-related cytokines in peripheral blood, and resulting in the inhibition of primary tumor growth and reductions in metastases [13,138,139]. A study by Guo et al., using a hollow copper sulfide nanomaterial for PTT in a murine model of breast cancer, showed no significant activation of the immune system; however, combining their formulation with CpG oligodeoxynucleotides, which activate Toll-like receptor 9 signaling in DCs, significantly increased the infiltration of NK cells and DCs in tumors and DLNs, with an increase in IFNγ and IL-2 secreted by CD8+ T cells in tumors and spleen [140]. A gold nanorod (GNR)-based hybrid nanomaterial (mPEG-GNRs@BSA/R837), involving the functionalization of BSA-bioinspired GNRs with imiquimod (R837, an immunoadjuvant recognized by Toll-like receptor 7), was used by Zhou et al. to treat melanoma in a murine model [141]. PTT with the mPEG-GNRs@BSA/R837 formulation enhanced the levels of cytokines (TNFα, IL-6 and IL-12), mature DCs, and CD8$^+$ T cell infiltration. In long-term analyses, prevention of lung metastasis and immunological memory with tumor re-challenge were also reported [141]. Another study by Zhou et al., using single-walled carbon nanotubes (SWCNTs) with an immunoadjuvant glycated chitosan (GC, an immunoadjuvant for the improvement of transport between the epithelium and promotion of phagocytosis) for PTT with a 980 nm laser in a murine model showed regression of the primary tumor, inhibition of metastasis, and a long-term anti-tumor immune response [142]. A detailed review of different types of immunoadjuvants combined with PTT and their therapeutic effects mediated by the immune response has been offered by others [13,14].

4.3. *Combination of Immune Checkpoint Inhibition (ICI) with Photothermal Therapy (PTT)*

Immune responses under normal physiological conditions are regulated by checkpoint receptors expressed on the surface of immune cells to maintain immune homeostasis and prevent autoimmunity. Photothermal therapy for cancer as a monotherapy is often insufficient to completely inhibit primary tumor growth, or to prevent distant metastasis; therefore, ICI together with PTT has been explored as a combination immunomodulatory approach for the treatment of tumors refractory to either PTT or ICI alone, in several preclinical studies [13,14,117,143]. A combination of PTT and ICI using antibodies against PD1/PDL1 and CTLA4 has been used in preclinical studies and shows significant improvement in therapeutic efficacies [14,117,144]. For example, Liu et al., using gold nanostar (GNS)-mediated PTT combined with an anti-PDL1 antibody, treated MB49 murine bladder cancer model and showed the complete clearance of primary tumors, along with distant untreated tumors (an abscopal effect) [145]. Treated mice showed long-term immunity (60 days) in re-challenge experiments with MB49 cells [145]. Wang et al. used PTT with SWNTs and demonstrated an increase in CD4+ Tregs with immunosuppressive characteristics [146]; a further combination with anti-CTLA4 antibody could reduce the Tregs and enhance the cytotoxic effects of T cells, thereby reducing the generation of primary tumors and distant metastasis [146]. Similarly, PTT using the organic nanocomposite PLGA-ICG-R837 combined with anti-CTLA4 exerted significant suppressive effects on primary and distant tumors, followed by the generation of memory T cells and inhibition of tumor recurrence and long-term tumor-free survival [147]. Table 2 provides a selected list which includes some other preclinical studies on photothermal immunotherapy that employ photothermal nanomaterials combined with ICI, and immunoadjuvants in some cases, along with study outcomes.

Table 2. A list of selected pre-clinical studies using a combination of immune checkpoint inhibition and photothermal therapy.

Checkpoint Inhibitor Target	Photothermal Agent/Construct	Murine Tumor Model	Immune Effector Cells	Cytokines	Therapeutic/Immune Response	Ref.
PD1	Hollow gold nanoshell (HAuNS)	4T1 breast tumor Colon cancer CT26	CD4+ and CD8+ T cells B cells	IFNγ, IL-2, TNFα	Reduced primary tumor growth and distant metastasis.	[148]
PD1	Black phosphorus quantum dots (BPQDs)	BF16-F10 murine melanoma 4T1 breast tumor	DCs, CD4+ and CD8+ T cells	IFNγ, TNFα	Reduced primary tumor growth and inhibition of lung metastasis.	[149]
PD1	A triple-layer nano-system AuNC@mSiO2@ copolymer∩vemurafenib (ASP∩V)	SMM103 melanoma tumors	CD3+, CD4+ and CD8+ T cells	Not analyzed	Primary tumor regression and distant tumor regression by abscopal effect.	[150]
PD1 CD47	ZIF-PQ-PDA-AUN	4T1 breast tumor	CD4+ and CD8+ T cells TAMs polarization from M2 to M1	Not analyzed	Primary tumor regression.	[151]
PDL1	Gold nanostar	Murine bladder cancer MB49	CD4+ and CD8+ T cells B cells	Not analyzed	Reduced primary tumor growth and distant metastasis. Long-term immunity in re-challenge experiments.	[145]
PDL1	Au@Pt nanoparticles	4T1 breast tumor	CD4+ and CD8+ T cells	IFNγ, IL-6, IL-12, TNFα	Regression of primary and distal tumors, inhibition of metastasis.	[152]
PDL1 and IDO	Reduced graphene oxide-based nanosheets	CT26 murine colon cancer	DCs, NK cells, CD45+ leukocytes, CD4+ and CD8+ T cells	IFNγ	Primary tumor regression and distant tumor regression by abscopal effect.	[153]
PDL1 and R837	Fe3O4-R837 spherical superparticles	4T1 breast tumor	DCs, NK cells, B cells, CD4+ and CD8+ T cells	IFNγ, IL-6, TNFα	Primary tumor regression and distant tumor regression by abscopal effect.	[154]
CTLA4	Single-walled nanotubes (SWNTs)	BF16-F10 murine melanoma 4T1 breast tumor	DCs, CD4+, CD8+, CD20+ T cells	IL-6, IL-12, IL-1β, TNFα	Reduced primary tumor growth and distant metastasis.	[146]
CTLA4	Prussian blue nanoparticles (PBNP)	Murine neuroblastoma cell Neuro2a	CD4+ and CD8+ T cells	Not analyzed	Lower tumor burden, synergistic effect on enhanced survival, development of immune memory in re-challenge experiments.	[155]
CTLA4 and R837	Indocyanine green and R837 co-encapsulated by poly (lactic-co-glycolic) acid (PLGA)	4T1 breast tumor Colon cancer CT26	DCs, CD4+, CD8+ T cells, memory T cells	IL-6, IL-12, IL-1β TNFα, IFNγ	Primary tumor regression and distant tumor regression by abscopal effect; inhibition of metastasis.	[147]

Regarding clinical trials, nanoparticle-based PTT as a monotherapy has been successfully translated into the clinic to treat low- or intermediate-risk localized prostate cancer [123,129] and several clinical trials using AuroLase, a type of PTT that combines AuNS with NIR light to treat tumors of the head and neck, lung, and prostate have been reported [130]. However, regarding combinations of ICI and PTT, no published reports nor any ongoing clinical trials listed on ClinicalTrials.gov (accessed on 24 April 2021) are currently available.

5. Radiation Therapy (RT)

Radiation therapy (RT) using ionizing radiation is a curative treatment for localized cancer and secondary metastasis. On average, 50–60% of patients with early- to mid- stage cancers of breast, prostate, cervical, endometrial, head and neck, lymphoid, etc., receive radiation therapy, either alone or in combination with surgery or chemotherapy [156]. Radiation therapy can be delivered from outside (external-beam radiation therapy; EBRT), by implanting radioactive sources inside the body (brachytherapy), or through systemic administration of radiopharmaceutical agents [157]. At the atomic level, the predominant interaction of tumor tissue with photons released during therapy is the so-called "Compton effect". After collision of a photon with an orbital electron, both are scattered, and while the photon continues on for additional interactions, the electron begins to ionize due to energy imparted by the photon, thereby allowing chemical reactions and destruction of tissue to occur [158].

5.1. Radiation Therapy-Induced Cell Death, Immunogenic Cell Death (ICD), and Activation of Anti-Tumor Responses

Ionizing radiation (IR), delivered in the form of X-rays, γ-rays, electrons, or protons, produces ROS and other types of ionizing free radicals upon interaction with tumor tissue, and results in DNA damage. Cells with damaged DNA undergo cell cycle arrest and lead eventually to either cellular senescence or cell death by activating apoptosis, necrosis, and autophagy, depending on the dose and schedule of the radiation therapy [16,159–161]. Ionizing radiation triggers cell death by apoptosis via intrinsic and extrinsic apoptotic pathways; the former involves activation of proapoptotic proteins/caspases, whereas the latter involves death receptors followed by downstream caspases, respectively. The DNA damage induced by IR results in cell cycle arrest and induction of senescence, both mediated by the activation of p53, leading to upregulation of p21. Ataxia-telangiectasia mutated (ATM) kinases activated by IR induce autophagy involving active p53 and damage-regulated autophagy modulator (DRAM) and is reviewed in [162]. Similar to PDT and PTT, RT is also known to exert anti-tumor immune responses, resulting in regression of the primary tumor, of distant untreated tumors (abscopal effect), as well as exerting anti-metastatic effects [163–165]. The involvement of the host immune system in anti-tumor effects of RT was first demonstrated by Stone et al. in a murine model of fibrosarcoma, showing that the dose of radiation required for tumor control was much higher in immunocompromised mice relative to immuno-sufficient mice [166]. Several studies afterwards elucidated the interplay between RT and anti-tumor immunity, both at the local and systemic levels, the latter being defined as "abscopal effect" [167]. Overall, the cascade of events involving IR-induced ICD are similar to those reported in PDT and PTT sections, and involve the activation of both innate and adaptive immunity. Briefly, the cell death induced by IR results in the release of DNA and RNA into the cytoplasm, triggering the activation of TLRs and transcription of type I interferon (IFN) gene. Type I interferon is essential for the activation of DCs, and for recruitment and regulating the effector function of CD8+ T cells [168,169]. The cytokines and chemokines produced in the tumor microenvironment following IR trigger the infiltration of immune cells such as DCs, macrophages, myeloid-derived suppressor cells (MDSCs), and regulatory and cytotoxic T cells. Activation and release of DAMPs following ICD trigger the activation of APCs, initiating adaptive immune response, reviewed elsewhere [15,16,161,162]. The anti-tumor effects of ICD are determined by the antigenicity and adjuvanticity of target cancer cells. While the antigenicity of tumor cells is determined by TSAs and tumor neoantigens (TNA), the adjuvant-like effects of ICD are mediated by the release of DAMPs [52,170,171]. A recent report by Lhuillier et al., using a 4T1 triple-negative murine breast tumor model, demonstrated radiotherapy-induced upregulation of the expression of genes containing immunogenic mutations in a poorly immunogenic model. Vaccination with neoepitopes encoded by these genes triggered a CD8+ and CD4+ mediated immune response, thereby improving the therapeutic efficacy of RT. The cytotoxic activity was mediated by the upregulation of MHC II molecules and death receptors FAS/CD95 and DR5 on the surface of tumor cells [172]. Dendritic cells, being professional APCs, serve as the link between innate and adaptive immune responses by their ability to stimulate unprimed naïve T cells and perform antigen cross-presentation [74]. A role for irradiated tumor-primed DCs in the prevention of local tumor growth involving CD4+ and CD8+ T cells was shown much before the introduction of the concept of ICD [173,174]. A recent study showed that X-ray-irradiated tumor cell lysates may work as effective antigen/adjuvant sources in DC vaccination studies. DCs, when incubated with X-ray-irradiated tumor cell lysates along with granulocyte-macrophage colony-stimulating factor (GMCF) and lipopolysaccharide (LPS)-containing media, led to reduced infiltration of Tregs, TAMs and MDSCs, along with enrichment of CD3+ T cells and strong infiltration of Th1 cells and CTLs [175].

Unlike PDT and PTT, which rely on systemic, local, or topical administration of PS or PTA, respectively, followed by irradiation with light, radiation delivered by an external or implanted source interacts directly with the tumor tissue without being limited by bioavailability, vascular permeability, and retention/efflux issues. Nevertheless, tumor resistance to RT represents an ongoing challenge for radiation oncologists. Possible reasons for this include hypoxia, and the presence of a significant proportion of growth-arrested/slow-dividing tumor cells that escape the therapeutic effects of RT. Normal tissue injury is an inherent consequence of radiation therapy, and hence a key consideration in the treatment design when using ionizing radiation. The effects of radiation on the tumor microenvironment can be regulated by the IR dose and methods of delivery; methods for improving anti-tumor efficacy include accelerated and hyper-fractionation of the radiation dose, in order to improve the tumor-killing effects while avoiding normal tissue damage [157]. In preclinical studies, the use of a high hyper-fractionated dose compared to a high single dose showed an advantage in terms of immunogenic effects of radiation therapy [176]. In addition to immunogenic effects, immunosuppressive effects of IR have been described that can counteract its anti-tumor immune effects. Thus, IR can switch the phenotype of infiltrating macrophages and alter the balance of Tregs and cytokines such as TGFβ, suppressing anti-tumor immunity [177–180]. In this scenario, radiation therapy alone may not be effective in generating robust immune response. Therefore, a number of combination approaches, including dose fractionation, immunotherapy, and different types of tumor and host factors, are currently being explored in preclinical murine tumor models [157]. One such combination of ICI with radiation therapy has been successfully trialed in preclinical murine models and is currently being utilized in the clinic to improve the therapeutic outcome of radiation therapy for different types of cancers, as described below.

5.2. Combination of Immune Checkpoint Inhibition (ICI) with Radiation Therapy (RT)

In the past decade, anticancer immunotherapy, specifically by ICI, has revolutionized the management of cancer, even in individuals with advanced-stage disease [22–26]. Both ICI and RT involve innate and adaptive immune systems; therefore, the effects of ICI may synergize with those of RT to improve the anti-tumor responses typically observed with either modality alone [15,16,181]. In this section, we discuss both preclinical and clinical studies in which ICI has been used in combination with RT.

5.3. Preclinical Scenario

Both the treatment regimens, ICI and RT, often fail to give a significant treatment response, when given individually. Therefore, finding combination treatment strategies to improve the clinical outcome would be highly desirable. There have been several preclinical studies using ICI plus RT in murine models for different types of cancer that have shown promising results. In a conditional Kras-driven genetically engineered mouse model (GEMM) of non-small-cell lung carcinoma (NSCLC), treatment with radiotherapy and an anti-PD1 antibody resulted in significant volume reduction (up to 70%) of the target lesion, and durable tumor regression (up to 12 weeks), along with an increase in inhibitory T cell markers [182]. In another C57BL/6 tumor xenograft mouse model of lung cancer, the combination of anti-PDL1 and IR treatment resulted in tumor regression as compared to either monotherapy alone [183]. In that study, increased infiltration of CD8+ T cells and reduced presence of MDSCs and inducible Tregs were reported only in the combination treatment group [183]. Combination of CTLA4 with RT in a dual murine model of mesothelioma enhanced the tumor regression, relative to either single treatment. Although RT alone increased both Tregs and CTL infiltration in primary tumors, the addition of CTLA4 reversed the proportion of Tregs to effector T cells, with increased CD8+ T cell activation [184]. Many of these preclinical studies used different doses of RT, with different schedules and delivery methodologies. Therefore, a better optimization of the radiotherapy component of the study is still needed for translation into the clinic. The sequence of combination may also be very important, because in a study using the CT26 colorectal murine model, an efficient systemic response to the combination was observed when anti-CTLA antibody was given prior to RT [185]; in contrast, another study using the 4T1 breast tumor model showed efficient tumor regression when anti-CTLA antibody was given after RT [186]. A list of selected preclinical studies using a combination of ICI with RT, and their outcomes, is presented in Table 3.

Table 3. A list of selected pre-clinical studies using combinations of immune checkpoint inhibition and radiation therapy.

Checkpoint Inhibitor Target	Radiation Therapy Dose (Fractions)	Murine Tumor Model	Immune Effector Cells	Cytokines	Therapeutic/Immune response	Ref.
PD1	8 Gy (4 fractions)	Metastatic melanoma in the brain	CD8+ T cells	Not analyzed	Reduced tumor growth and systemic immunity by abscopal effect	[187]
PD1	24 Gy (3 fractions)	Non-small-cell lung carcinoma	Neutrophils, CD4+ and CD8+ T cells	IL-5, IFNγ, TNFα	Higher lung injury score, increased inflammatory response	[188]
PD1	16 Gy (2 fractions)	B16-F10 melanoma TS/A mammary adeno-carcinoma	DCs, monocytes, macrophages and CD8+ T cells	IFNβ upregulated in abscopal tumors	Reduced tumor growth and systemic immunity by abscopal effect	[189]
PDL1	12 Gy	Pancreatic cancer	CD4+ and CD8+ T cells, myeloid-derived suppressor cells, tumor-associated macrophages	Not analyzed	Reduced primary tumor growth and systemic immunity by abscopal effect	[190]
PDL1	10 Gy	Head and neck squamous cell carcinoma	CD4+ and CD8+ T cells	Not analyzed	Enhanced tumor control and improved survival	[191]
PDL1	10 Gy	Hepatocellular carcinoma	CD8+ T cells	Not analyzed	Significant suppression of tumor growth and improved survival	[192]
CTLA4 along with immature dendritic cells (iDCs)	10 Gy	Colon cancer CT26	IFNγ-secreting T cells, CD8+ CTLs	IFNγ	Suppression of tumor growth and improved survival of tumor-bearing mice	[193]
CTLA4	10 Gy	Orthotopic glioma	CD4+ and CD8+ T cells	Not analyzed	Improved survival of treated mice	[194]
PD1 + CTLA4	20 Gy (either single dose or in fractions)	4T1 mammary carcinoma	APCs, CD4+ and CD8+ cells	IFNγ	Primary tumor regression, abscopal effect in fractionated dose	[195]
PD1 + CTLA4	10 Gy	LM8 osteosarcoma	CD8+ T cells	Not analyzed	Reduced primary tumor growth and lung metastasis, systemic immunity by abscopal effect	[196]

5.4. Clinical Trials with Combination of Immune Checkpoint Inhibition with Radiation, Photodynamic, or Photothermal Therapy

Amongst the three treatment modalities in this review, radiotherapy using ionizing radiation is the only one that has a long track record of widespread clinical use in oncology. Therefore, it is not surprising that physicians and clinical researchers have begun exploring different combinations of RT and various ICI agents, for a variety of cancers. To date, at least two dozen phase I and II trials to evaluate safety have shown that combinations of radiation plus an ICI is generally safe, with the possible exception of increased brain swelling and necrosis in patients with brain metastases treated with combined RT and ICI [16,161]. An excellent current listing of published results from these trials is provided in the supplementary table of the review by McLaughlin et al. [161]. In terms of unpublished results (from ongoing trials that are still recruiting or awaiting analysis), ~20 trials can be found on the ClinicalTrials.gov website (accessed 24 April 2021), as listed in Table 4. These trials involve a variety of internal malignancies, and they feature different brand-name ICI drugs (targeting PD1, PD-L1, or CTLA4, either alone or in combination). They also vary as to whether additional chemotherapeutic drugs are administered, and whether the checkpoint inhibitor is given before or after radiation (Table 4). Especially notable here is the AstraZeneca PACIFIC trial, which used adjuvant durvalumab following chemoradiotherapy for stage III NSCLC. This was phase III data showing that combination chemo/RT and anti-PD-L1 should be the new standard of care [197]. The results of other studies, once available, will provide further insight into the benefits of RT + ICI combination treatments, relative to RT or ICI alone, for the amelioration of human cancer.

Unlike RT and chemotherapy, PDT and PTT are still considered in the broader oncology arena to be palliative or investigational modalities at this time; therefore, the clinical role of combining ICI with PDT or PTT agents has not been widely tested. For PDT, there is an interesting clinical case report in which a patient's advanced head-and-neck cancer was cured via the administration of PDT (redaporfin/red light) followed by anti-PD1 antibody [198]. However, there are no published studies involving multiple patients. One ongoing PDT clinical trial involving PDT + ICI is listed in ClinicalTrials (identifier NCT04400539); in that trial, lung cancer patients (mesothelioma) will be treated with intrapleural PDT followed by injections of Nivolumab (anti-PD1). For PTT, no current publications nor any listings on ClinicalTrials that describe human trials with PTT + ICI combinations can be found.

Table 4. A list of ongoing clinical studies using a combination of immune checkpoint inhibition with radiation therapy [16,161].

Checkpoint Molecule Targeted for ICI	ICI Agent Used	Disease	Radiation Therapy Dose (Fractions)	Additional Drugs Used	Estimated Patient Accrual (n)	Timing of Radiotherapy	ClinicalTrials.gov for ICI Identifier *
PD1	Nivolumab	Glioblastoma	2 Gy × 30	Temozolomide	693	n/s	NCT02667587
PD1	Nivolumab	Glioblastoma	not specified	Temozolomide	550	n/s	NCT02617589
PD1	Pembrolizumab	HNSCC, locally advanced	2 Gy × 35	Cisplatin	780	ICI then RT (RT at cycle 2 of ICI)	NCT03040999
PD1	Nivolumab	HNSCC, locally advanced	n/s	Cisplatin, Cetuximab	1046	n/s	NCT03349710
PD1	Pembrolizumab	Breast cancer, triple negative	n/s	chemotherapy	1000	RT then ICI	NCT02954874
PD1	Nivolumab	NSCLC, Stage IV	4 Gy × 5	none	130	ICI then RT	NCT03044626
PD1	Pembrolizumab	Breast cancer, localized	8 Gy × 3 (alternate days)	± Flt3 ligand (CDX-301)	100	n/s	NCT03804944
PD1	Nivolumab	Pancreatic cancer (PDAC)	6.6 Gy × 5	± CCR2/CCR5 dual antagonist; ± GVAX	30	RT then ICI	NCT03767582
PD-L1	Durvalumab	Glioblastoma, recurrent	8 Gy × 3 once daily	none	62	RT then ICI (ICI starts on last day of RT)	NCT02866747
PD-L1	Durvalumab	Breast cancer, luminal B	SBRT 8 Gy × 2 fractions preoperatively	chemotherapy, ± anti-CD73 (oleclumab)	147	RT then ICI	NCT03875573
PD-L1	Avelumab	Hepatobiliary malignancy(advanced)	Hypofractionated in 5 fractions	DNA-PK inhibitor	92	RT then ICI	NCT04068194
PD-L1	Avelumab	Various advanced solid tumors	30 Gy in 10 fractions over 2 weeks	DNA-PK inhibitor	54	RT and ICI together (1st dose), then ICI continues	NCT03724890
CTLA4	Ipilimumab	Prostate cancer (metastatic)	n/s	none	988	RT then ICI	NCT00861614
PD1, and PD-L1	Nivolumab, and atezolizumab	RCC Stage IV, or UC Stage IV	3 Gy × 10	none	112	RT begins ±24 h of ICI start	NCT03115801
PD1, and CTLA4	Nivolumab, and Ipilimumab	NSCLC, Stage IV	n/s	none	270	ICI then RT	NCT03391869
PD-L1, and CTLA4	Durvalumab, and tremelimumab	NSCLC and colon cancer	High dose: 1 daily fraction × 3 days; Low dose: 2 fx daily on weeks 2, 6, 10, and 14	none	180	ICI then RT	NCT02888743
PD-L1, and CTLA4	Durvalumab, and tremelimumab	SCLC, relapsed	SBRT or hypofractionated RT over 3–5 days	none	20	RT then ICI	NCT02701400
PD-L1, and CTLA4	Durvalumab, and tremelimumab	SCLC, advanced stage	30 Gy in 10 fractions over 2 weeks	PARP inhibitor (olaparib)	54	RT then ICI	NCT03923270
PD-L1, and CTLA4	Durvalumab, and tremelimumab	Esophageal cancer, Stage III–IV	n/s	chemotherapy	75	ICI then RT	NCT02735239
Any ICI target	Any approved agent	Any metastatic cancer, with a lesion treatable with SBRT	SBRT 9.5 Gy × 3	none	146	ICI then RT	NCT02843165

* Verified on ClinicalTrials.gov (accessed on 24 April 2021). n/s, not specified.

6. Conclusions

In this review, we have surveyed the available literature on three treatment modalities that employ the electromagnetic spectrum, from very short (RT) to longer wavelengths (PDT and PTT), thereby causing tissue damage and stimulating a number of immune modulatory effects. Ongoing attempts to harness these effects by using immune checkpoint inhibitors were also reviewed. Many similarities and differences between the modalities can be identified. For example, although ICD responses and immune stimulation occur after both PDT and PTT, dose delivery may be a relatively more important factor in PTT, because the response appears to require reaching an optimal temperature range in tissue. Radiation therapy, although generally causing more limited damage targeted to the nucleus (as compared to cellular membrane damage caused with PDT), appears quite capable of inducing strong anti-tumor immune effects.

While our own particular research interest lies in PDT (still an investigational therapy in most human cancers), it is evident that much can be learned by comparison with the other modalities. For example, because RT is widely established and used in clinical practice, considerations of similarities and differences between RT and PDT or PTT could be instructive as ongoing clinical trial results are published and we learn which factors are critical for improving the effects of checkpoint inhibition. Clearly, with such a wide variety of cancers, each with different pathological features and different tumor microenvironments, one can anticipate an important role for each of these photon-involving modalities in combination with ICI in the future.

Author Contributions: Conceptualization, S.A. and E.V.M.; Writing—Original Draft Preparation, S.A. and E.V.M.; Writing—Review and Editing, S.A., T.A.C., T.H. and E.V.M.; Funding Acquisition, T.H. and E.V.M. All authors reviewed and approved the final version of the manuscript.

Funding: Financial support from the National Cancer Institute (NCI), National Institutes of Health (NIH) through: (i) a program project grant P01CA084203, co-directed by Tayyaba Hasan (Massachusetts General Hospital) and Brian Pogue (Dartmouth Medical Center); and (ii) R01CA204158 to Edward Maytin is gratefully acknowledged.

Conflicts of Interest: The authors declare no conflict of interest.

Abbreviations

Aminolevulinic acid-photodynamic therapy	ALA-PDT
Antigen-presenting cells	APC
Ataxia-telangiectasia mutated	ATM
Adenosine triphosphate	ATP
B7 protein family dendritic cell molecule	B7-DC
Bristol Mayers Squibb 202	BMS-202
Bursa of Fabricius cells	B cell
Calreticulin	CRT
Chemokine (C-X-C motif) ligand 2	CXCL2
Cluster of differentiation 8	CD8
Cytotoxic T lymphocyte	CTL
Cytotoxic T lymphocyte antigen 4	CTLA4
Damage-associated molecular patterns	DAMP
Damage-regulated autophagy modulator	DRAM
Dendritic cells	DC
Draining lymph nodes	DLN
External beam radiation therapy	EBRT
Forkhead box P3	FoxP3
Genetically engineered mouse model	GEMM
Glysated chitosan	GC
Gold nanorod	GNR
Gold nanostar	GNS
Granulocyte-macrophage colony-stimulating factor (GM-CSF)-transfected tumor cell vaccine	GVAX

Head and neck squamous cell carcinoma	HNSCC
Heat shock proteins	HSP
High mobility group box 1	HMGB1
Immune checkpoint blockade	ICB
Immune checkpoint inhibition	ICI
Immunogenic cell death	ICD
Indoleamine-pyrrole 2,3-dioxygenase	IDO
Interferon	IFN
Interferon gamma	IFNγ
Interleukin 1 beta	IL-1β
Interleukin 6	IL-6
Interleukin 12	IL-12
Interleukin 17	IL-17
Ionizing radiation	IR
Lymphocyte-activation gene 3	LAG3
Lipopolysaccharide	LPS
Monoclonal antibodies	mAbs
Macrophage inflammatory protein 2	MIP2
Magnetic Fe_3O_4 photothermal nanoparticle	MNP
Major histocompatibility complex I and II	MHC I and II
MHC class I polypeptide-related sequence A	MICA
Myeloid-derived suppressor cells	MDSC
Natural killer	NK
Natural killer group 2D	NKG2D
Near-infrared radiation	NIR
Non-small-cell lung carcinoma	NSCLC
Organic semiconducting pro-nano stimulant	OSPS
Pancreatic ductal adenocarcinoma	PDAC
Pattern recognition receptors	PRR
Photodynamic therapy	PDT
Photosensitizer	PS
Photothermal agents	PTA
Photothermal therapy	PTT
Poly(lactic-co-glycolic) acid-indocyanine green-R837	PLGA-ICG-R837
Programmed cell death protein 1	PD1/PDCD1
Programmed cell death protein 1 ligand 1	PDL1
Prussian blue nanoparticle	PBNP
Radiation therapy	RT
Reactive oxygen species	ROS
Regulatory T cells	Treg
Renal cell carcinoma	RCC
Stereotactic body radiation therapy	SBRT
Small-cell lung cancer	SCLC
Single-walled carbon nanotubes	SWCNT
Silica–gold nanoshell	AuNS
T cell receptor	TCR
T helper 17	Th17
T lymphocytes	T cell
Toll-like receptors	TLR
Transforming growth factor beta	TGFβ
Tumor-associated macrophages	TAM
Tumor microenvironment	TME
Tumor necrosis factor alpha	TNFα
Tumor neoantigens	TNA
Tumor-specific antigens	TSA
Urothelial carcinoma	UC
Vascular endothelial growth factor	VEGF

References

1. Mukerjee, S. *The Emperor of All Maladies: A Biography of Cancer*; Scribner: New York, NY, USA, 2010.
2. Siegel, R.L.; Miller, K.D.; Fuchs, H.E.; Jemal, A. Cancer Statistics, 2021. *CA Cancer J. Clin.* **2021**, *71*, 7–33. [CrossRef]
3. Rosenberg, S.A. Progress in the development of immunotherapy for the treatment of patients with cancer. *J. Intern. Med.* **2001**, *250*, 462–475. [CrossRef]
4. Rosenberg, S.A. Progress in human tumour immunology and immunotherapy. *Nature* **2001**, *411*, 380–384. [CrossRef]
5. Mellman, I.; Coukos, G.; Dranoff, G. Cancer immunotherapy comes of age. *Nature* **2011**, *480*, 480–489. [CrossRef] [PubMed]
6. Sheng, W.Y.; Huang, L. Cancer immunotherapy and nanomedicine. *Pharm. Res.* **2011**, *28*, 200–214. [CrossRef] [PubMed]
7. Matsuo, Y.; Takeyama, H.; Guha, S. Cytokine network: New targeted therapy for pancreatic cancer. *Curr. Pharm. Des.* **2012**, *18*, 2416–2419. [CrossRef]
8. Pardoll, D.M. The blockade of immune checkpoints in cancer immunotherapy. *Nat. Rev. Cancer* **2012**, *12*, 252–264. [CrossRef] [PubMed]
9. Sharma, P.; Allison, J.P. Immune checkpoint targeting in cancer therapy: Toward combination strategies with curative potential. *Cell* **2015**, *161*, 205–214. [CrossRef]
10. Havel, J.J.; Chowell, D.; Chan, T.A. The evolving landscape of biomarkers for checkpoint inhibitor immunotherapy. *Nat. Rev. Cancer* **2019**, *19*, 133–150. [CrossRef]
11. Cramer, G.M.; Moon, E.K.; Cengel, K.A.; Busch, T.M. Photodynamic Therapy and Immune Checkpoint Blockade. *Photochem. Photobiol.* **2020**, *96*, 954–961. [CrossRef]
12. Nath, S.; Obaid, G.; Hasan, T. The Course of Immune Stimulation by Photodynamic Therapy: Bridging Fundamentals of Photochem. ically Induced Immunogenic Cell Death to the Enrichment of T-Cell Repertoire. *Photochem. Photobiol.* **2019**, *95*, 1288–1305. [CrossRef]
13. Hou, X.; Tao, Y.; Pang, Y.; Li, X.; Jiang, G.; Liu, Y. Nanoparticle-based photothermal and photodynamic immunotherapy for tumor treatment. *Int. J. Cancer* **2018**, *143*, 3050–3060. [CrossRef] [PubMed]
14. Xu, P.; Liang, F. Nanomaterial-Based Tumor Photothermal Immunotherapy. *Int. J. Nanomed.* **2020**, *15*, 9159–9180. [CrossRef]
15. Weichselbaum, R.R.; Liang, H.; Deng, L.; Fu, Y.X. Radiotherapy and immunotherapy: A beneficial liaison? *Nat. Rev. Clin. Oncol.* **2017**, *14*, 365–379. [CrossRef]
16. Hwang, W.L.; Pike, L.R.G.; Royce, T.J.; Mahal, B.A.; Loeffler, J.S. Safety of combining radiotherapy with immune-checkpoint inhibition. *Nat. Rev. Clin. Oncol.* **2018**, *15*, 477–494. [CrossRef] [PubMed]
17. Korbelik, M.; Dougherty, G.J. Photodynamic therapy-mediated immune response against subcutaneous mouse tumors. *Cancer Res.* **1999**, *59*, 1941–1946.
18. Kabingu, E.; Vaughan, L.; Owczarczak, B.; Ramsey, K.D.; Gollnick, S.O. CD8+ T cell-mediated control of distant tumours following local photodynamic therapy is independent of CD4+ T cells and dependent on natural killer cells. *Br. J. Cancer* **2007**, *96*, 1839–1848. [CrossRef] [PubMed]
19. Shams, M.; Owczarczak, B.; Manderscheid-Kern, P.; Bellnier, D.A.; Gollnick, S.O. Development of photodynamic therapy regimens that control primary tumor growth and inhibit secondary disease. *Cancer Immunol. Immunother.* **2015**, *64*, 287–297. [CrossRef] [PubMed]
20. Beltran Hernandez, I.; Yu, Y.; Ossendorp, F.; Korbelik, M.; Oliveira, S. Preclinical and Clinical Evidence of Immune Responses Triggered in Oncologic Photodynamic Therapy: Clinical Recommendations. *J. Clin. Med.* **2020**, *9*, 333. [CrossRef]
21. Anand, S.; Govande, M.; Yasinchak, A.; Heusinkveld, L.; Shakya, S.; Fairchild, R.L.; Maytin, E.V. Painless Photodynamic Therapy Triggers Innate and Adaptive Immune Responses in a Murine Model of UV-induced Squamous Skin Pre-cancer. *Photochem. Photobiol.* **2020**. [CrossRef]
22. Ishida, Y.; Agata, Y.; Shibahara, K.; Honjo, T. Induced expression of PD-1, a novel member of the immunoglobulin gene superfamily, upon programmed cell death. *EMBO J.* **1992**, *11*, 3887–3895. [CrossRef]
23. Krummel, M.F.; Allison, J.P. CTLA-4 engagement inhibits IL-2 accumulation and cell cycle progression upon activation of resting T cells. *J. Exp. Med.* **1996**, *183*, 2533–2540. [CrossRef]
24. Freeman, G.J.; Long, A.J.; Iwai, Y.; Bourque, K.; Chernova, T.; Nishimura, H.; Fitz, L.J.; Malenkovich, N.; Okazaki, T.; Byrne, M.C.; et al. Engagement of the PD-1 immunoinhibitory receptor by a novel B7 family member leads to negative regulation of lymphocyte activation. *J. Exp. Med.* **2000**, *192*, 1027–1034. [CrossRef]
25. Sharma, P.; Allison, J.P. Dissecting the mechanisms of immune checkpoint therapy. *Nat. Rev. Immunol.* **2020**, *20*, 75–76. [CrossRef] [PubMed]
26. Kalbasi, A.; Ribas, A. Tumour-intrinsic resistance to immune checkpoint blockade. *Nat. Rev. Immunol.* **2020**, *20*, 25–39. [CrossRef] [PubMed]
27. Sharma, P.; Allison, J.P. The future of immune checkpoint therapy. *Science* **2015**, *348*, 56–61. [CrossRef] [PubMed]
28. Topalian, S.L.; Drake, C.G.; Pardoll, D.M. Immune checkpoint blockade: A common denominator approach to cancer therapy. *Cancer Cell* **2015**, *27*, 450–461. [CrossRef]
29. Buchbinder, E.I.; Desai, A. CTLA-4 and PD-1 Pathways: Similarities, Differences, and Implications of Their Inhibition. *Am. J. Clin. Oncol.* **2016**, *39*, 98–106. [CrossRef]
30. Sharpe, A.H.; Pauken, K.E. The diverse functions of the PD1 inhibitory pathway. *Nat. Rev. Immunol.* **2018**, *18*, 153–167. [CrossRef]
31. Greenwald, R.J.; Freeman, G.J.; Sharpe, A.H. The B7 family revisited. *Annu. Rev. Immunol.* **2005**, *23*, 515–548. [CrossRef]

32. Agostinis, P.; Berg, K.; Cengel, K.A.; Foster, T.H.; Girotti, A.W.; Gollnick, S.O.; Hahn, S.M.; Hamblin, M.R.; Juzeniene, A.; Kessel, D.; et al. Photodynamic therapy of cancer: An update. *CA Cancer J. Clin.* **2011**, *61*, 250–281. [CrossRef]
33. Anand, S.; Ortel, B.J.; Pereira, S.P.; Hasan, T.; Maytin, E.V. Biomodulatory approaches to photodynamic therapy for solid tumors. *Cancer Lett.* **2012**, *326*, 8–16. [CrossRef] [PubMed]
34. Yoo, S.W.; Oh, G.; Ahn, J.C.; Chung, E. Non-Oncologic Applications of Nanomedicine-Based Phototherapy. *Biomedicines* **2021**, *9*, 113. [CrossRef] [PubMed]
35. Celli, J.P.; Spring, B.Q.; Rizvi, I.; Evans, C.L.; Samkoe, K.S.; Verma, S.; Pogue, B.W.; Hasan, T. Imaging and photodynamic therapy: Mechanisms, monitoring, and optimization. *Chem. Rev.* **2010**, *110*, 2795–2838. [CrossRef]
36. Garg, A.D.; Nowis, D.; Golab, J.; Agostinis, P. Photodynamic therapy: Illuminating the road from cell death towards anti-tumour immunity. *Apoptosis* **2010**, *15*, 1050–1071. [CrossRef]
37. Anand, S.; Wilson, C.; Hasan, T.; Maytin, E.V. Vitamin D3 enhances the apoptotic response of epithelial tumors to aminolevulinate-based photodynamic therapy. *Cancer Res.* **2011**, *71*, 6040–6050. [CrossRef] [PubMed]
38. Kessel, D. Apoptosis, Paraptosis and Autophagy: Death and Survival Pathways Associated with Photodynamic Therapy. *Photochem. Photobiol.* **2019**, *95*, 119–125. [CrossRef]
39. Pogue, B.W.; O'Hara, J.A.; Demidenko, E.; Wilmot, C.M.; Goodwin, I.A.; Chen, B.; Swartz, H.M.; Hasan, T. Photodynamic therapy with verteporfin in the radiation-induced fibrosarcoma-1 tumor causes enhanced radiation sensitivity. *Cancer Res.* **2003**, *63*, 1025–1033.
40. Korbelik, M. PDT-associated host response and its role in the therapy outcome. *Lasers Surg. Med.* **2006**, *38*, 500–508. [CrossRef]
41. Preise, D.; Scherz, A.; Salomon, Y. Antitumor immunity promoted by vascular occluding therapy: Lessons from vascular-targeted photodynamic therapy (VTP). *Photochem. Photobiol. Sci.* **2011**, *10*, 681–688. [CrossRef]
42. Maas, A.L.; Carter, S.L.; Wileyto, E.P.; Miller, J.; Yuan, M.; Yu, G.; Durham, A.C.; Busch, T.M. Tumor vascular microenvironment determines responsiveness to photodynamic therapy. *Cancer Res.* **2012**, *72*, 2079–2088. [CrossRef]
43. Gollnick, S.O.; Vaughan, L.; Henderson, B.W. Generation of effective antitumor vaccines using photodynamic therapy. *Cancer Res.* **2002**, *62*, 1604–1608.
44. Korbelik, M.; Sun, J. Photodynamic therapy-generated vaccine for cancer therapy. *Cancer Immunol. Immunother.* **2006**, *55*, 900–909. [CrossRef]
45. Castano, A.P.; Mroz, P.; Hamblin, M.R. Photodynamic therapy and anti-tumour immunity. *Nat. Rev. Cancer* **2006**, *6*, 535–545. [CrossRef]
46. Korbelik, M. Cancer vaccines generated by photodynamic therapy. *Photochem. Photobiol. Sci.* **2011**, *10*, 664–669. [CrossRef]
47. Falk-Mahapatra, R.; Gollnick, S.O. Photodynamic Therapy and Immunity: An Update. *Photochem. Photobiol.* **2020**, *96*, 550–559. [CrossRef] [PubMed]
48. Kroemer, G.; Galluzzi, L.; Kepp, O.; Zitvogel, L. Immunogenic cell death in cancer therapy. *Annu. Rev. Immunol.* **2013**, *31*, 51–72. [CrossRef] [PubMed]
49. Legrand, A.J.; Konstantinou, M.; Goode, E.F.; Meier, P. The Diversification of Cell Death and Immunity: Memento Mori. *Mol. Cell* **2019**, *76*, 232–242. [CrossRef]
50. Ahmed, A.; Tait, S.W.G. Targeting immunogenic cell death in cancer. *Mol. Oncol.* **2020**, *14*, 2994–3006. [CrossRef] [PubMed]
51. Yatim, N.; Cullen, S.; Albert, M.L. Dying cells actively regulate adaptive immune responses. *Nat. Rev. Immunol.* **2017**, *17*, 262–275. [CrossRef]
52. Alzeibak, R.; Mishchenko, T.A.; Shilyagina, N.Y.; Balalaeva, I.V.; Vedunova, M.V.; Krysko, D.V. Targeting immunogenic cancer cell death by photodynamic therapy: Past, present and future. *J. Immunother. Cancer* **2021**, *9*. [CrossRef] [PubMed]
53. Garg, A.D.; Krysko, D.V.; Vandenabeele, P.; Agostinis, P. DAMPs and PDT-mediated photo-oxidative stress: Exploring the unknown. *Photochem. Photobiol. Sci.* **2011**, *10*, 670–680. [CrossRef] [PubMed]
54. Krysko, D.V.; Agostinis, P.; Krysko, O.; Garg, A.D.; Bachert, C.; Lambrecht, B.N.; Vandenabeele, P. Emerging role of damage-associated molecular patterns derived from mitochondria in inflammation. *Trends Immunol.* **2011**, *32*, 157–164. [CrossRef] [PubMed]
55. Yang, Y.; Hu, Y.; Wang, H. Targeting Antitumor Immune Response for Enhancing the Efficacy of Photodynamic Therapy of Cancer: Recent Advances and Future Perspectives. *Oxid Med. Cell Longev.* **2016**, *2016*, 5274084. [CrossRef] [PubMed]
56. Sun, J.; Cecic, I.; Parkins, C.S.; Korbelik, M. Neutrophils as inflammatory and immune effectors in photodynamic therapy-treated mouse SCCVII tumours. *Photochem. Photobiol. Sci.* **2002**, *1*, 690–695. [CrossRef]
57. Krosl, G.; Korbelik, M.; Dougherty, G.J. Induction of immune cell infiltration into murine SCCVII tumour by photofrin-based photodynamic therapy. *Br. J. Cancer* **1995**, *71*, 549–555. [CrossRef] [PubMed]
58. Nowis, D.; Stoklosa, T.; Legat, M.; Issat, T.; Jakobisiak, M.; Golab, J. The influence of photodynamic therapy on the immune response. *Photodiagnosis Photodyn. Ther.* **2005**, *2*, 283–298. [CrossRef]
59. Haanen, J. Converting Cold into Hot Tumors by Combining Immunotherapies. *Cell* **2017**, *170*, 1055–1056. [CrossRef] [PubMed]
60. Kousis, P.C.; Henderson, B.W.; Maier, P.G.; Gollnick, S.O. Photodynamic therapy enhancement of antitumor immunity is regulated by neutrophils. *Cancer Res.* **2007**, *67*, 10501–10510. [CrossRef]
61. Brackett, C.M.; Muhitch, J.B.; Evans, S.S.; Gollnick, S.O. IL-17 promotes neutrophil entry into tumor-draining lymph nodes following induction of sterile inflammation. *J. Immunol.* **2013**, *191*, 4348–4357. [CrossRef]

62. Yang, D.; de la Rosa, G.; Tewary, P.; Oppenheim, J.J. Alarmins link neutrophils and dendritic cells. *Trends Immunol.* **2009**, *30*, 531–537. [CrossRef]
63. Jalili, A.; Makowski, M.; Switaj, T.; Nowis, D.; Wilczynski, G.M.; Wilczek, E.; Chorazy-Massalska, M.; Radzikowska, A.; Maslinski, W.; Bialy, L.; et al. Effective photoimmunotherapy of murine colon carcinoma induced by the combination of photodynamic therapy and dendritic cells. *Clin. Cancer Res.* **2004**, *10*, 4498–4508. [CrossRef]
64. Wang, H.; Li, J.; Lv, T.; Tu, Q.; Huang, Z.; Wang, X. Therapeutic and immune effects of 5-aminolevulinic acid photodynamic therapy on UVB-induced squamous cell carcinomas in hairless mice. *Exp. Dermatol.* **2013**, *22*, 362–363. [CrossRef]
65. Lamberti, M.J.; Mentucci, F.M.; Roselli, E.; Araya, P.; Rivarola, V.A.; Rumie Vittar, N.B.; Maccioni, M. Photodynamic Modulation of Type 1 Interferon Pathway on Melanoma Cells Promotes Dendritic Cell Activation. *Front. Immunol.* **2019**, *10*, 2614. [CrossRef] [PubMed]
66. Korbelik, M.; Krosl, G. Enhanced macrophage cytotoxicity against tumor cells treated with photodynamic therapy. *Photochem. Photobiol.* **1994**, *60*, 497–502. [CrossRef] [PubMed]
67. Korbelik, M.; Cecic, I. Contribution of myeloid and lymphoid host cells to the curative outcome of mouse sarcoma treatment by photodynamic therapy. *Cancer Lett.* **1999**, *137*, 91–98. [CrossRef]
68. Korbelik, M.; Hamblin, M.R. The impact of macrophage-cancer cell interaction on the efficacy of photodynamic therapy. *Photochem. Photobiol. Sci.* **2015**, *14*, 1403–1409. [CrossRef] [PubMed]
69. Mosser, D.M.; Edwards, J.P. Exploring the full spectrum of macrophage activation. *Nat. Rev. Immunol.* **2008**, *8*, 958–969. [CrossRef]
70. Murray, P.J.; Wynn, T.A. Protective and pathogenic functions of macrophage subsets. *Nat. Rev. Immunol.* **2011**, *11*, 723–737. [CrossRef]
71. Hao, N.B.; Lu, M.H.; Fan, Y.H.; Cao, Y.L.; Zhang, Z.R.; Yang, S.M. Macrophages in tumor microenvironments and the progression of tumors. *Clin. Dev. Immunol.* **2012**, *2012*, 948098. [CrossRef] [PubMed]
72. Belicha-Villanueva, A.; Riddell, J.; Bangia, N.; Gollnick, S.O. The effect of photodynamic therapy on tumor cell expression of major histocompatibility complex (MHC) class I and MHC class I-related molecules. *Lasers Surg. Med.* **2012**, *44*, 60–68. [CrossRef]
73. Ji, J.; Fan, Z.; Zhou, F.; Wang, X.; Shi, L.; Zhang, H.; Wang, P.; Yang, D.; Zhang, L.; Chen, W.R.; et al. Improvement of DC vaccine with ALA-PDT induced immunogenic apoptotic cells for skin squamous cell carcinoma. *Oncotarget* **2015**, *6*, 17135–17146. [CrossRef]
74. Lamberti, M.J.; Nigro, A.; Mentucci, F.M.; Rumie Vittar, N.B.; Casolaro, V.; Dal Col, J. Dendritic Cells and Immunogenic Cancer Cell Death: A Combination for Improving Antitumor Immunity. *Pharmaceutics* **2020**, *12*, 256. [CrossRef]
75. Korbelik, M.; Krosl, G.; Krosl, J.; Dougherty, G.J. The role of host lymphoid populations in the response of mouse EMT6 tumor to photodynamic therapy. *Cancer Res.* **1996**, *56*, 5647–5652. [PubMed]
76. Preise, D.; Oren, R.; Glinert, I.; Kalchenko, V.; Jung, S.; Scherz, A.; Salomon, Y. Systemic antitumor protection by vascular-targeted photodynamic therapy involves cellular and humoral immunity. *Cancer Immunol. Immunother.* **2009**, *58*, 71–84. [CrossRef]
77. Maeding, N.; Verwanger, T.; Krammer, B. Boosting Tumor-Specific Immunity Using PDT. *Cancers* **2016**, *8*, 91. [CrossRef] [PubMed]
78. Sompayrac, L. *How the Immune System Works*, 6th ed.; John Wiley & Sons: Oxford, UK, 2019; p. 156.
79. Huang, Y.Y.; Tanaka, M.; Vecchio, D.; Garcia-Diaz, M.; Chang, J.; Morimoto, Y.; Hamblin, M.R. Photodynamic therapy induces an immune response against a bacterial pathogen. *Expert Rev. Clin. Immunol.* **2012**, *8*, 479–494. [CrossRef] [PubMed]
80. Reginato, E.; Wolf, P.; Hamblin, M.R. Immune response after photodynamic therapy increases anti-cancer and anti-bacterial effects. *World J. Immunol.* **2014**, *4*, 1–11. [CrossRef]
81. Anzengruber, F.; Avci, P.; de Freitas, L.F.; Hamblin, M.R. T-cell mediated anti-tumor immunity after photodynamic therapy: Why does it not always work and how can we improve it? *Photochem. Photobiol. Sci.* **2015**, *14*, 1492–1509. [CrossRef]
82. Wachowska, M.; Gabrysiak, M.; Muchowicz, A.; Bednarek, W.; Barankiewicz, J.; Rygiel, T.; Boon, L.; Mroz, P.; Hamblin, M.R.; Golab, J. 5-Aza-2′-deoxycytidine potentiates antitumour immune response induced by photodynamic therapy. *Eur. J. Cancer* **2014**, *50*, 1370–1381. [CrossRef]
83. Abdel-Hady, E.S.; Martin-Hirsch, P.; Duggan-Keen, M.; Stern, P.L.; Moore, J.V.; Corbitt, G.; Kitchener, H.C.; Hampson, I.N. Immunological and viral factors associated with the response of vulval intraepithelial neoplasia to photodynamic therapy. *Cancer Res.* **2001**, *61*, 192–196.
84. Mroz, P.; Hashmi, J.T.; Huang, Y.Y.; Lange, N.; Hamblin, M.R. Stimulation of anti-tumor immunity by photodynamic therapy. *Expert Rev. Clin. Immunol.* **2011**, *7*, 75–91. [CrossRef] [PubMed]
85. Pellegrini, C.; Orlandi, A.; Costanza, G.; Di Stefani, A.; Piccioni, A.; Di Cesare, A.; Chiricozzi, A.; Ferlosio, A.; Peris, K.; Fargnoli, M.C. Expression of IL-23/Th17-related cytokines in basal cell carcinoma and in the response to medical treatments. *PLoS ONE* **2017**, *12*, e0183415. [CrossRef]
86. Ding, B.; Shao, S.; Yu, C.; Teng, B.; Wang, M.; Cheng, Z.; Wong, K.L.; Ma, P.; Lin, J. Large-Pore Mesoporous-Silica-Coated Upconversion Nanoparticles as Multifunctional Immunoadjuvants with Ultrahigh Photosensitizer and Antigen Loading Efficiency for Improved Cancer Photodynamic Immunotherapy. *Adv. Mater.* **2018**, *30*, e1802479. [CrossRef]
87. Knutson, K.L.; Disis, M.L. Augmenting T helper cell immunity in cancer. *Curr. Drug Targets Immune Endocr. Metabol. Disord.* **2005**, *5*, 365–371. [CrossRef]
88. Knutson, K.L.; Disis, M.L. Tumor antigen-specific T helper cells in cancer immunity and immunotherapy. *Cancer Immunol. Immunother.* **2005**, *54*, 721–728. [CrossRef]

89. Medzhitov, R.; Janeway, C.A., Jr. Innate immunity: Impact on the adaptive immune response. *Curr. Opin. Immunol.* **1997**, *9*, 4–9. [CrossRef]
90. Zhou, L.; Lopes, J.E.; Chong, M.M.; Ivanov, I.I.; Min, R.; Victora, G.D.; Shen, Y.; Du, J.; Rubtsov, Y.P.; Rudensky, A.Y.; et al. TGF-beta-induced Foxp3 inhibits T(H)17 cell differentiation by antagonizing RORgammat function. *Nature* **2008**, *453*, 236–240. [CrossRef]
91. Shevach, E.M. Regulatory T cells in autoimmmunity*. *Annu Rev. Immunol.* **2000**, *18*, 423–449. [CrossRef] [PubMed]
92. Takahashi, T.; Tagami, T.; Yamazaki, S.; Uede, T.; Shimizu, J.; Sakaguchi, N.; Mak, T.W.; Sakaguchi, S. Immunologic self-tolerance maintained by CD25(+)CD4(+) regulatory T cells constitutively expressing cytotoxic T lymphocyte-associated antigen 4. *J. Exp. Med.* **2000**, *192*, 303–310. [CrossRef] [PubMed]
93. Fontenot, J.D.; Gavin, M.A.; Rudensky, A.Y. Foxp3 programs the development and function of CD4+CD25+ regulatory T cells. *Nat. Immunol.* **2003**, *4*, 330–336. [CrossRef]
94. Roychoudhuri, R.; Eil, R.L.; Restifo, N.P. The interplay of effector and regulatory T cells in cancer. *Curr. Opin. Immunol.* **2015**, *33*, 101–111. [CrossRef] [PubMed]
95. Castano, A.P.; Mroz, P.; Wu, M.X.; Hamblin, M.R. Photodynamic therapy plus low-dose cyclophosphamide generates antitumor immunity in a mouse model. *Proc. Natl. Acad. Sci. USA* **2008**, *105*, 5495–5500. [CrossRef] [PubMed]
96. Reginato, E.; Mroz, P.; Chung, H.; Kawakubo, M.; Wolf, P.; Hamblin, M.R. Photodynamic therapy plus regulatory T-cell depletion produces immunity against a mouse tumour that expresses a self-antigen. *Br J. Cancer* **2013**, *109*, 2167–2174. [CrossRef]
97. Oh, D.S.; Kim, H.; Oh, J.E.; Jung, H.E.; Lee, Y.S.; Park, J.H.; Lee, H.K. Intratumoral depletion of regulatory T cells using CD25-targeted photodynamic therapy in a mouse melanoma model induces antitumoral immune responses. *Oncotarget* **2017**, *8*, 47440–47453. [CrossRef] [PubMed]
98. Gao, L.; Zhang, C.; Gao, D.; Liu, H.; Yu, X.; Lai, J.; Wang, F.; Lin, J.; Liu, Z. Enhanced Anti-Tumor Efficacy through a Combination of Integrin alphavbeta6-Targeted Photodynamic Therapy and Immune Checkpoint Inhibition. *Theranostics* **2016**, *6*, 627–637. [CrossRef] [PubMed]
99. O'Shaughnessy, M.J.; Murray, K.S.; La Rosa, S.P.; Budhu, S.; Merghoub, T.: Somma, A.; Monette, S.; Kim, K.; Corradi, R.B.; Scherz, A.; et al. Systemic Antitumor Immunity by PD-1/PD-L1 Inhibition Is Potentiated by Vascular-Targeted Photodynamic Therapy of Primary Tumors. *Clin. Cancer Res.* **2018**, *24*, 592–599. [CrossRef]
100. Li, Y.; Du, Y.; Liang, X.; Sun, T.; Xue, H.; Tian, J.; Jin, Z. EGFR-targeted liposomal nanohybrid cerasomes: Theranostic function and immune checkpoint inhibition in a mouse model of colorectal cancer. *Nanoscale* **2018**, *10*, 16738–16749. [CrossRef]
101. Hwang, H.S.; Cherukula, K.; Bang, Y.J.; Vijayan, V.; Moon, M.J.; Thiruppathi, J.; Puth, S.; Jeong, Y.Y.; Park, I.K.; Lee, S.E.; et al. Combination of Photodynamic Therapy and a Flagellin-Adjuvanted Cancer Vaccine Potentiated the Anti-PD-1-Mediated Melanoma Suppression. *Cells* **2020**, *9*, 2432. [CrossRef] [PubMed]
102. Trempolec, N.; Doix, B.; Degavre, C.; Brusa, D.; Bouzin, C.; Riant, O.; Feron, O. Photodynamic Therapy-Based Dendritic Cell Vaccination Suited to Treat Peritoneal Mesothelioma. *Cancers* **2020**, *12*, 545. [CrossRef] [PubMed]
103. Zhang, R.; Zhu, Z.; Lv, H.; Li, F.; Sun, S.; Li, J.; Lee, C.S. Immune Checkpoint Blockade Mediated by a Small-Molecule Nanoinhibitor Targeting the PD-1/PD-L1 Pathway Synergizes with Photodynamic Therapy to Elicit Antitumor Immunity and Antimetastatic Effects on Breast Cancer. *Small* **2019**, *15*, e1903881. [CrossRef]
104. Bao, R.; Wang, Y.; Lai, J.; Zhu, H.; Zhao, Y.; Li, S.; Li, N.; Huang, J.; Yang, Z.; Wang, F.; et al. Enhancing Anti-PD-1/PD-L1 Immune Checkpoint Inhibitory Cancer Therapy by CD276-Targeted Photodynamic Ablation of Tumor Cells and Tumor Vasculature. *Mol. Pharm.* **2019**, *16*, 339–348. [CrossRef]
105. Muchowicz, A.; Wachowska, M.; Stachura, J.; Tonecka, K.; Gabrysiak, M.; Wolosz, D.; Pilch, Z.; Kilarski, W.W.; Boon, L.; Klaus, T.J.; et al. Inhibition of lymphangiogenesis impairs antitumour effects of photodynamic therapy and checkpoint inhibitors in mice. *Eur. J. Cancer* **2017**, *83*, 19–27. [CrossRef] [PubMed]
106. Kleinovink, J.W.; Fransen, M.F.; Lowik, C.W.; Ossendorp, F. Photodynamic-Immune Checkpoint Therapy Eradicates Local and Distant Tumors by CD8(+) T Cells. *Cancer Immunol. Res.* **2017**, *5*, 832–838. [CrossRef]
107. Xu, J.; Xu, L.; Wang, C.; Yang, R.; Zhuang, Q.; Han, X.; Dong, Z.; Zhu, W.; Peng, R.; Liu, Z. Near-Infrared-Triggered Photodynamic Therapy with Multitasking Upconversion Nanoparticles in Combination with Checkpoint Blockade for Immunotherapy of Colorectal Cancer. *ACS Nano* **2017**, *11*, 4463–4474. [CrossRef] [PubMed]
108. Lu, K.; He, C.; Guo, N.; Chan, C.; Ni, K.; Weichselbaum, R.R.; Lin, W. Chlorin-Based Nanoscale Metal-Organic Framework Systemically Rejects Colorectal Cancers via Synergistic Photodynamic Therapy and Checkpoint Blockade Immunotherapy. *J. Am. Chem. Soc.* **2016**, *138*, 12502–12510. [CrossRef]
109. Wachowska, M.; Stachura, J.; Tonecka, K.; Fidyt, K.; Braniewska, A.; Sas, Z.; Kotula, I.; Rygiel, T.P.; Boon, L.; Golab, J.; et al. Inhibition of IDO leads to IL-6-dependent systemic inflammation in mice when combined with photodynamic therapy. *Cancer Immunol. Immunother.* **2020**, *69*, 1101–1112. [CrossRef] [PubMed]
110. Grunwald, V.; Voss, M.H.; Rini, B.I.; Powles, T.; Albiges, L.; Giles, R.H.; Jonasch, E. Axitinib plus immune checkpoint inhibitor: Evidence- and expert-based consensus recommendation for treatment optimisation and management of related adverse events. *Br. J. Cancer* **2020**, *123*, 898–904. [CrossRef]
111. Powles, T.; Plimack, E.R.; Soulieres, D.; Waddell, T.; Stus, V.; Gafanov, R.; Nosov, D.; Pouliot, F.; Melichar, B.; Vynnychenko, I.; et al. Pembrolizumab plus axitinib versus sunitinib monotherapy as first-line treatment of advanced renal cell carcinoma (KEYNOTE-426): Extended follow-up from a randomised, open-label, phase 3 trial. *Lancet Oncol.* **2020**, *21*, 1563–1573. [CrossRef]

112. Wilky, B.A.; Trucco, M.M.; Subhawong, T.K.; Florou, V.; Park, W.; Kwon, D.; Wieder, E.D.; Kolonias, D.; Rosenberg, A.E.; Kerr, D.A.; et al. Axitinib plus pembrolizumab in patients with advanced sarcomas including alveolar soft-part sarcoma: A single-centre, single-arm, phase 2 trial. *Lancet Oncol.* **2019**, *20*, 837–848. [CrossRef]
113. Mo, D.C.; Luo, P.H.; Huang, S.X.; Wang, H.L.; Huang, J.F. Safety and efficacy of pembrolizumab plus lenvatinib versus pembrolizumab and lenvatinib monotherapies in cancers: A systematic review. *Int. Immunopharmacol.* **2021**, *91*, 107281. [CrossRef] [PubMed]
114. Solban, N.; Selbo, P.K.; Sinha, A.K.; Chang, S.K.; Hasan, T. Mechanistic investigation and implications of photodynamic therapy induction of vascular endothelial growth factor in prostate cancer. *Cancer Res.* **2006**, *66*, 5633–5640. [CrossRef] [PubMed]
115. Kosharskyy, B.; Solban, N.; Chang, S.K.; Rizvi, I.; Chang, Y.; Hasan, T. A mechanism-based combination therapy reduces local tumor growth and metastasis in an orthotopic model of prostate cancer. *Cancer Res.* **2006**, *66*, 10953–10958. [CrossRef]
116. Li, X.; Lovell, J.F.; Yoon, J.; Chen, X. Clinical development and potential of photothermal and photodynamic therapies for cancer. *Nat. Rev. Clin. Oncol.* **2020**, *17*, 657–674. [CrossRef] [PubMed]
117. Balakrishnan, P.B.; Sweeney, E.E.; Ramanujam, A.S.; Fernandes, R. Photothermal therapies to improve immune checkpoint blockade for cancer. *Int. J. Hyperthermia* **2020**, *37*, 34–49. [CrossRef]
118. Hu, Q.; Huang, Z.; Duan, Y.; Fu, Z.; Bin, L. Reprogramming Tumor Microenvironment with Photothermal Therapy. *Bioconjug Chem* **2020**, *31*, 1268–1278. [CrossRef] [PubMed]
119. Hou, Y.J.; Yang, X.X.; Liu, R.Q.; Zhao, D.; Guo, C.X.; Zhu, A.C.; Wen, M.N.; Liu, Z.; Qu, G.F.; Meng, H.X. Pathological Mechanism of Photodynamic Therapy and Photothermal Therapy Based on Nanoparticles. *Int. J. Nanomed.* **2020**, *15*, 6827–6838. [CrossRef]
120. Saneja, A.; Kumar, R.; Arora, D.; Kumar, S.; Panda, A.K.; Jaglan, S. Recent advances in near-infrared light-responsive nanocarriers for cancer therapy. *Drug Discov. Today* **2018**, *23*, 1115–1125. [CrossRef]
121. Zhu, H.; Liu, W.; Cheng, Z.; Yao, K.; Yang, Y.; Xu, B.; Su, G. Targeted Delivery of siRNA with pH-Responsive Hybrid Gold Nanostars for Cancer Treatment. *Int. J. Mol. Sci.* **2017**, *18*, 2029. [CrossRef]
122. Hu, Y.; Chi, C.; Wang, S.; Wang, L.; Liang, P.; Liu, F.; Shang, W.; Wang, W.; Zhang, F.; Li, S.; et al. A Comparative Study of Clinical Intervention and Interventional Photothermal Therapy for Pancreatic Cancer. *Adv. Mater.* **2017**, *29*. [CrossRef]
123. Rastinehad, A.R.; Anastos, H.; Wajswol, E.; Winoker, J.S.; Sfakianos, J.P.; Doppalapudi, S.K.; Carrick, M.R.; Knauer, C.J.; Taouli, B.; Lewis, S.C.; et al. Gold nanoshell-localized photothermal ablation of prostate tumors in a clinical pilot device study. *Proc. Natl. Acad. Sci. USA* **2019**, *116*, 18590–18596. [CrossRef] [PubMed]
124. Du, C.; Wu, X.; He, M.; Zhang, Y.; Zhang, R.; Dong, C.M. Polymeric photothermal agents for cancer therapy: Recent progress and clinical potential. *J. Mater. Chem. B* **2021**, *9*, 1478–1490. [CrossRef]
125. Ju, Y.; Zhang, H.; Yu, J.; Tong, S.; Tian, N.; Wang, Z.; Wang, X.; Su, X.; Chu, X.; Lin, J.; et al. Monodisperse Au-Fe2C Janus Nanoparticles: An Attractive Multifunctional Material for Triple-Modal Imaging-Guided Tumor Photothermal Therapy. *ACS Nano* **2017**, *11*, 9239–9248. [CrossRef] [PubMed]
126. Song, S.; Shen, H.; Yang, T.; Wang, L.; Fu, H.; Chen, H.; Zhang, Z. Indocyanine Green Loaded Magnetic Carbon Nanoparticles for Near Infrared Fluorescence/Magnetic Resonance Dual-Modal Imaging and Photothermal Therapy of Tumor. *ACS Appl. Mater. Interfaces* **2017**, *9*, 9484–9495. [CrossRef] [PubMed]
127. Mioc, A.; Mioc, M.; Ghiulai, R.; Voicu, M.; Racoviceanu, R.; Trandafirescu, C.; Dehelean, C.; Coricovac, D.; Soica, C. Gold Nanoparticles as Targeted Delivery Systems and Theranostic Agents in Cancer Therapy. *Curr. Med. Chem.* **2019**, *26*, 6493–6513. [CrossRef]
128. Li, Y.; Li, X.; Zhou, F.; Doughty, A.; Hoover, A.R.; Nordquist, R.E.; Chen, W.R. Nanotechnology-based photoimmunological therapies for cancer. *Cancer Lett.* **2019**, *442*, 429–438. [CrossRef] [PubMed]
129. Stern, J.M.; Kibanov Solomonov, V.V.; Sazykina, E.; Schwartz, J.A.; Gad, S.C.; Goodrich, G.P. Initial Evaluation of the Safety of Nanoshell-Directed Photothermal Therapy in the Treatment of Prostate Disease. *Int. J. Toxicol.* **2016**, *35*, 38–46. [CrossRef]
130. Ali, M.R.K.; Wu, Y.; El-Sayed, M.A. Gold-Nanoparticle-Assisted Plasmonic Photothermal Therapy Advances Toward Clinical Application. *J. Phys. Chem. C* **2019**, *123*, 15375–15393. [CrossRef]
131. Hildebrandt, B.; Wust, P.; Ahlers, O.; Dieing, A.; Sreenivasa, G.; Kerner, T.; Felix, R.; Riess, H. The cellular and molecular basis of hyperthermia. *Crit. Rev. Oncol. Hematol.* **2002**, *43*, 33–56. [CrossRef]
132. Movahedi, M.M.; Alamzadeh, Z.; Hosseini-Nami, S.; Shakeri-Zadeh, A.; Taheripak, G.; Ahmadi, A.; Zare-Sadeghi, A.; Ghaznavi, H.; Mehdizadeh, A. Investigating the mechanisms behind extensive death in human cancer cells following nanoparticle assisted photo-thermo-radiotherapy. *Photodiagnosis Photodyn. Ther.* **2020**, *29*, 101600. [CrossRef]
133. Richter, K.; Haslbeck, M.; Buchner, J. The heat shock response: Life on the verge of death. *Mol. Cell* **2010**, *40*, 253–266. [CrossRef]
134. Knavel, E.M.; Brace, C.L. Tumor ablation: Common modalities and general practices. *Tech. Vasc. Interv. Radiol.* **2013**, *16*, 192–200. [CrossRef] [PubMed]
135. Sweeney, E.E.; Cano-Mejia, J.; Fernandes, R. Photothermal Therapy Generates a Thermal Window of Immunogenic Cell Death in Neuroblastoma. *Small* **2018**, *14*, e1800678. [CrossRef] [PubMed]
136. Wang, Z.; Guo, B.; Middha, E.; Huang, Z.; Hu, Q.; Fu, Z.; Liu, B. Microfluidics-Prepared Uniform Conjugated Polymer Nanoparticles for Photo-Triggered Immune Microenvironment Modulation and Cancer Therapy. *ACS Appl. Mater. Interfaces* **2019**, *11*, 11167–11176. [CrossRef]
137. Yu, G.T.; Rao, L.; Wu, H.; Yang, L.L.; Bu, L.L.; Deng, W.W.; Wu, L.; Nan, X.L.; Zhang, W.F.; Zhao, X.Z.; et al. Myeloid-Derived Suppressor Cell Membrane-Coated Magnetic Nanoparticles for Cancer Theranostics by Inducing Macrophage Polarization and Synergizing Immunogenic Cell Death. *Adv. Funct. Mater.* **2018**, *28*. [CrossRef]

138. Han, Q.; Wang, X.; Jia, X.; Cai, S.; Liang, W.; Qin, Y.; Yang, R.; Wang, C. CpG loaded MoS2 nanosheets as multifunctional agents for photothermal enhanced cancer immunotherapy. *Nanoscale* **2017**, *9*, 5927–5934. [CrossRef] [PubMed]
139. Liu, Y.; Wei, C.; Lin, A.; Pan, J.; Chen, X.; Zhu, X.; Gong, Y.; Yuan, G.; Chen, L.; Liu, J.; et al. Responsive functionalized MoSe2 nanosystem for highly efficient synergistic therapy of breast cancer. *Colloids Surf. B Biointerfaces* **2020**, *189*, 110820. [CrossRef]
140. Guo, L.; Yan, D.D.; Yang, D.; Li, Y.; Wang, X.; Zalewski, O.; Yan, B.; Lu, W. Combinatorial photothermal and immuno cancer therapy using chitosan-coated hollow copper sulfide nanoparticles. *ACS Nano* **2014**, *8*, 5670–5681. [CrossRef] [PubMed]
141. Zhou, B.; Song, J.; Wang, M.; Wang, X.; Wang, J.; Howard, E.W.; Zhou, F.; Qu, J.; Chen, W.R. BSA-bioinspired gold nanorods loaded with immunoadjuvant for the treatment of melanoma by combined photothermal therapy and immunotherapy. *Nanoscale* **2018**, *10*, 21640–21647. [CrossRef] [PubMed]
142. Zhou, F.; Wu, S.; Song, S.; Chen, W.R.; Resasco, D.E.; Xing, D. Antitumor immunologically modified carbon nanotubes for photothermal therapy. *Biomaterials* **2012**, *33*, 3235–3242. [CrossRef]
143. Selby, M.J.; Engelhardt, J.J.; Quigley, M.; Henning, K.A.; Chen, T.; Srinivasan, M.; Korman, A.J. Anti-CTLA-4 antibodies of IgG2a isotype enhance antitumor activity through reduction of intratumoral regulatory T cells. *Cancer Immunol. Res.* **2013**, *1*, 32–42. [CrossRef] [PubMed]
144. Salmaninejad, A.; Valilou, S.F.; Shabgah, A.G.; Aslani, S.; Alimardani, M.; Pasdar, A.; Sahebkar, A. PD-1/PD-L1 pathway: Basic biology and role in cancer immunotherapy. *J. Cell. Physiol.* **2019**, *234*, 16824–16837. [CrossRef]
145. Liu, Y.; Maccarini, P.; Palmer, G.M.; Etienne, W.; Zhao, Y.; Lee, C.T.; Ma, X.; Inman, B.A.; Vo-Dinh, T. Synergistic Immuno Photothermal Nanotherapy (SYMPHONY) for the Treatment of Unresectable and Metastatic Cancers. *Sci. Rep.* **2017**, *7*, 8606. [CrossRef]
146. Wang, C.; Xu, L.; Liang, C.; Xiang, J.; Peng, R.; Liu, Z. Immunological responses triggered by photothermal therapy with carbon nanotubes in combination with anti-CTLA-4 therapy to inhibit cancer metastasis. *Adv. Mater.* **2014**, *26*, 8154–8162. [CrossRef]
147. Chen, Q.; Xu, L.; Liang, C.; Wang, C.; Peng, R.; Liu, Z. Photothermal therapy with immune-adjuvant nanoparticles together with checkpoint blockade for effective cancer immunotherapy. *Nat. Commun.* **2016**, *7*, 13193. [CrossRef]
148. Luo, L.; Zhu, C.; Yin, H.; Jiang, M.; Zhang, J.; Qin, B.; Luo, Z.; Yuan, X.; Yang, J.; Li, W.; et al. Laser Immunotherapy in Combination with Perdurable PD-1 Blocking for the Treatment of Metastatic Tumors. *ACS Nano* **2018**, *12*, 7647–7662. [CrossRef]
149. Ye, X.; Liang, X.; Chen, Q.; Miao, Q.; Chen, X.; Zhang, X.; Mei, L. Surgical Tumor-Derived Personalized Photothermal Vaccine Formulation for Cancer Immunotherapy. *ACS Nano* **2019**, *13*, 2956–2968. [CrossRef] [PubMed]
150. Liu, X.; Feng, Y.; Xu, J.; Shi, Y.; Yang, J.; Zhang, R.; Song, J.; Bai, X.; Wu, X.; Bao, Y.; et al. Combination of MAPK inhibition with photothermal therapy synergistically augments the anti-tumor efficacy of immune checkpoint blockade. *J. Control. Release* **2021**, *332*, 194–209. [CrossRef] [PubMed]
151. Zhao, W.; Hu, X.; Li, W.; Li, R.; Chen, J.; Zhou, L.; Qiang, S.; Wu, W.; Shi, S.; Dong, C. M2-Like TAMs Function Reversal Contributes to Breast Cancer Eradication by Combination Dual Immune Checkpoint Blockade and Photothermal Therapy. *Small* **2021**, *17*, e2007051. [CrossRef]
152. Yang, Q.; Peng, J.; Shi, K.; Xiao, Y.; Liu, Q.; Han, R.; Wei, X.; Qian, Z. Rationally designed peptide-conjugated gold/platinum nanosystem with active tumor-targeting for enhancing tumor photothermal-immunotherapy *J. Control. Release* **2019**, *308*, 29–43. [CrossRef]
153. Yan, M.; Liu, Y.; Zhu, X.; Wang, X.; Liu, L.; Sun, H.; Wang, C.; Kong, D.; Ma, G. Nanoscale Reduced Graphene Oxide-Mediated Photothermal Therapy Together with IDO Inhibition and PD-L1 Blockade Synergistically Promote Antitumor Immunity. *ACS Appl. Mater. Interfaces* **2019**, *11*, 1876–1885. [CrossRef]
154. Ge, R.; Liu, C.; Zhang, X.; Wang, W.; Li, B.; Liu, J.; Liu, Y.; Sun, H.; Zhang, D.; Hou, Y.; et al. Photothermal-Activatable Fe3O4 Superparticle Nanodrug Carriers with PD-L1 Immune Checkpoint Blockade for Anti-metastatic Cancer Immunotherapy. *ACS Appl. Mater. Interfaces* **2018**, *10*, 20342–20355. [CrossRef] [PubMed]
155. Cano-Mejia, J.; Burga, R.A.; Sweeney, E.E.; Fisher, J.P.; Bollard, C.M.; Sandler, A.D.; Cruz, C.R.Y.; Fernandes, R. Prussian blue nanoparticle-based photothermal therapy combined with checkpoint inhibition for photothermal immunotherapy of neuroblastoma. *Nanomedicine* **2017**, *13*, 771–781. [CrossRef] [PubMed]
156. Orth, M.; Lauber, K.; Niyazi, M.; Friedl, A.A.; Li, M.; Maihofer, C.; Schuttrumpf, L.; Ernst, A.; Niemoller, O.M.; Belka, C. Current concepts in clinical radiation oncology. *Radiat. Environ. Biophys.* **2014**, *53*, 1–29. [CrossRef]
157. Citrin, D.E. Recent Developments in Radiotherapy. *N. Engl. J. Med.* **2017**, *377*, 1065–1075. [CrossRef]
158. Bartzsch, S.; Corde, S.; Crosbie, J.C.; Day, L.; Donzelli, M.; Krisch, M.; Lerch, M.; Pellicioli, P.; Smyth, L.M.L.; Tehei, M. Technical advances in x-ray microbeam radiation therapy. *Phys. Med. Biol* **2020**, *65*, 02TR01. [CrossRef]
159. Amaravadi, R.K.; Thompson, C.B. The roles of therapy-induced autophagy and necrosis in cancer treatment. *Clin. Cancer Res.* **2007**, *13*, 7271–7279. [CrossRef]
160. Rouschop, K.M.; Ramaekers, C.H.; Schaaf, M.B.; Keulers, T.G.; Savelkouls, K.G.; Lambin, P.; Koritzinsky, M.; Wouters, B.G. Autophagy is required during cycling hypoxia to lower production of reactive oxygen species. *Radiother. Oncol.* **2009**, *92*, 411–416. [CrossRef]
161. McLaughlin, M.; Patin, E.C.; Pedersen, M.; Wilkins, A.; Dillon, M.T.; Melcher, A.A.; Harrington, K.J. Inflammatory microenvironment remodelling by tumour cells after radiotherapy. *Nat. Rev. Cancer* **2020**, *20*, 203–217. [CrossRef]
162. Wu, Q.; Allouch, A.; Martins, I.; Brenner, C.; Modjtahedi, N.; Deutsch, E.; Perfettini, J.L. Modulating Both Tumor Cell Death and Innate Immunity Is Essential for Improving Radiation Therapy Effectiveness. *Front. Immunol.* **2017**, *8*, 613. [CrossRef] [PubMed]

163. Demaria, S.; Bhardwaj, N.; McBride, W.H.; Formenti, S.C. Combining radiotherapy and immunotherapy: A revived partnership. *Int. J. Radiat. Oncol. Biol. Phys.* **2005**, *63*, 655–666. [CrossRef] [PubMed]
164. Frey, B.; Rubner, Y.; Wunderlich, R.; Weiss, E.M.; Pockley, A.G.; Fietkau, R.; Gaipl, U.S. Induction of abscopal anti-tumor immunity and immunogenic tumor cell death by ionizing irradiation—Implications for cancer therapies. *Curr. Med. Chem.* **2012**, *19*, 1751–1764. [CrossRef] [PubMed]
165. Burnette, B.; Weichselbaum, R.R. Radiation as an immune modulator. *Semin. Radiat. Oncol.* **2013**, *23*, 273–280. [CrossRef] [PubMed]
166. Stone, H.B.; Peters, L.J.; Milas, L. Effect of host immune capability on radiocurability and subsequent transplantability of a murine fibrosarcoma. *J. Natl. Cancer Inst.* **1979**, *63*, 1229–1235. [PubMed]
167. Ngwa, W.; Irabor, O.C.; Schoenfeld, J.D.; Hesser, J.; Demaria, S.; Formenti, S.C. Using immunotherapy to boost the abscopal effect. *Nat. Rev. Cancer* **2018**, *18*, 313–322. [CrossRef]
168. Burnette, B.C.; Liang, H.; Lee, Y.; Chlewicki, L.; Khodarev, N.N.; Weichselbaum, R.R.; Fu, Y.X.; Auh, S.L. The efficacy of radiotherapy relies upon induction of type i interferon-dependent innate and adaptive immunity. *Cancer Res.* **2011**, *71*, 2488–2496. [CrossRef]
169. Lim, J.Y.; Gerber, S.A.; Murphy, S.P.; Lord, E.M. Type I interferons induced by radiation therapy mediate recruitment and effector function of CD8(+) T cells. *Cancer Immunol. Immunother.* **2014**, *63*, 259–271. [CrossRef]
170. Bloy, N.; Garcia, P.; Laumont, C.M.; Pitt, J.M.; Sistigu, A.; Stoll, G.; Yamazaki, T.; Bonneil, E.; Buque, A.; Humeau, J.; et al. Immunogenic stress and death of cancer cells: Contribution of antigenicity vs adjuvanticity to immunosurveillance. *Immunol. Rev.* **2017**, *280*, 165–174. [CrossRef]
171. Galluzzi, L.; Buque, A.; Kepp, O.; Zitvogel, L.; Kroemer, G. Immunogenic cell death in cancer and infectious disease. *Nat. Rev. Immunol.* **2017**, *17*, 97–111. [CrossRef]
172. Lhuillier, C.; Rudqvist, N.P.; Yamazaki, T.; Zhang, T.; Charpentier, M.; Galluzzi, L.; Dephoure, N.; Clement, C.C.; Santambrogio, L.; Zhou, X.K.; et al. Radiotherapy-exposed CD8+ and CD4+ neoantigens enhance tumor control. *J. Clin. Investig.* **2021**, *131*. [CrossRef]
173. Goldszmid, R.S.; Idoyaga, J.; Bravo, A.I.; Steinman, R.; Mordoh, J.; Wainstok, R. Dendritic cells charged with apoptotic tumor cells induce long-lived protective CD4+ and CD8+ T cell immunity against B16 melanoma. *J. Immunol.* **2003**, *171*, 5940–5947. [CrossRef]
174. Casares, N.; Pequignot, M.O.; Tesniere, A.; Ghiringhelli, F.; Roux, S.; Chaput, N.; Schmitt, E.; Hamai, A.; Hervas-Stubbs, S.; Obeid, M.; et al. Caspase-dependent immunogenicity of doxorubicin-induced tumor cell death. *J. Exp. Med.* **2005**, *202*, 1691–1701. [CrossRef]
175. Vandenberk, L.; Belmans, J.; Van Woensel, M.; Riva, M.; Van Gool, S.W. Exploiting the Immunogenic Potential of Cancer Cells for Improved Dendritic Cell Vaccines. *Front. Immunol.* **2015**, *6*, 663. [CrossRef] [PubMed]
176. Schaue, D.; Ratikan, J.A.; Iwamoto, K.S.; McBride, W.H. Maximizing tumor immunity with fractionated radiation. *Int. J. Radiat. Oncol. Biol. Phys.* **2012**, *83*, 1306–1310. [CrossRef]
177. Chee, C.A.; Ilbery, P.L.; Rickinson, A.B. Depression of lymphocyte replicating ability in radiotherapy patients. *Br. J. Radiol.* **1974**, *47*, 37–43. [CrossRef] [PubMed]
178. Order, S.E. The effects of therapeutic irradiation on lymphocytes and immunity. *Cancer* **1977**, *39*, 737–743. [CrossRef]
179. Klug, F.; Prakash, H.; Huber, P.E.; Seibel, T.; Bender, N.; Halama, N.; Pfirschke, C.; Voss, R.H.; Timke, C.; Umansky, L.; et al. Low-dose irradiation programs macrophage differentiation to an iNOS(+)/M1 phenotype that orchestrates effective T cell immunotherapy. *Cancer Cell* **2013**, *24*, 589–602. [CrossRef] [PubMed]
180. Hasmim, M.; Noman, M.Z.; Messai, Y.; Bordereaux, D.; Gros, G.; Baud, V.; Chouaib, S. Cutting edge: Hypoxia-induced Nanog favors the intratumoral infiltration of regulatory T cells and macrophages via direct regulation of TGF-beta1. *J. Immunol.* **2013**, *191*, 5802–5806. [CrossRef]
181. Carvalho, H.A.; Villar, R.C. Radiotherapy and immune response: The systemic effects of a local treatment. *Clinics (Sao Paulo)* **2018**, *73*, e557s. [CrossRef]
182. Herter-Sprie, G.S.; Koyama, S.; Korideck, H.; Hai, J.; Deng, J.; Li, Y.Y.; Buczkowski, K.A.; Grant, A.K.; Ullas, S.; Rhee, K.; et al. Synergy of radiotherapy and PD-1 blockade in Kras-mutant lung cancer. *JCI Insight* **2016**, *1*, e87415. [CrossRef]
183. Gong, X.; Li, X.; Jiang, T.; Xie, H.; Zhu, Z.; Zhou, F.; Zhou, C. Combined Radiotherapy and Anti-PD-L1 Antibody Synergistically Enhances Antitumor Effect in Non-Small Cell Lung Cancer. *J. Thorac. Oncol.* **2017**, *12*, 1085–1097. [CrossRef] [PubMed]
184. Wu, L.; Wu, M.O.; De la Maza, L.; Yun, Z.; Yu, J.; Zhao, Y.; Cho, J.; de Perrot, M. Targeting the inhibitory receptor CTLA-4 on T cells increased abscopal effects in murine mesothelioma model. *Oncotarget* **2015**, *6*, 12468–12480. [CrossRef]
185. Young, K.H.; Baird, J.R.; Savage, T.; Cottam, B.; Friedman, D.; Bambina, S.; Messenheimer, D.J.; Fox, B.; Newell, P.; Bahjat, K.S.; et al. Optimizing Timing of Immunotherapy Improves Control of Tumors by Hypofractionated Radiation Therapy. *PLoS ONE* **2016**, *11*, e0157164. [CrossRef] [PubMed]
186. Demaria, S.; Kawashima, N.; Yang, A.M.; Devitt, M.L.; Babb, J.S.; Allison, J.P.; Formenti, S.C. Immune-mediated inhibition of metastases after treatment with local radiation and CTLA-4 blockade in a mouse model of breast cancer. *Clin. Cancer Res.* **2005**, *11*, 728–734. [PubMed]
187. Pfannenstiel, L.W.; McNeilly, C.; Xiang, C.; Kang, K.; Diaz-Montero, C.M.; Yu, J.S.; Gastman, B.R. Combination PD-1 blockade and irradiation of brain metastasis induces an effective abscopal effect in melanoma. *Oncoimmunology* **2019**, *8*, e1507669. [CrossRef]
188. Wang, F.; Luo, Y.; Tian, X.; Ma, S.; Sun, Y.; You, C.; Gong, Y.; Xie, C. Impact of Radiotherapy Concurrent with Anti-PD-1 Therapy on the Lung Tissue of Tumor-Bearing Mice. *Radiat. Res.* **2019**, *191*, 271–277. [CrossRef]

189. Ruckert, M.; Deloch, L.; Frey, B.; Schlucker, E.; Fietkau, R.; Gaipl, U.S. Combinations of Radiotherapy with Vaccination and Immune Checkpoint Inhibition Differently Affect Primary and Abscopal Tumor Growth and the Tumor Microenvironment. *Cancers* **2021**, *13*, 714. [CrossRef]
190. Chuong, M.; Chang, E.T.; Choi, E.Y.; Mahmood, J.; Lapidus, R.G.; Davila, E.; Carrier, F. Exploring the Concept of Radiation "Booster Shot" in Combination with an Anti-PD-L1 mAb to Enhance Anti-Tumor Immune Effects in Mouse Pancreas Tumors. *J. Clin. Oncol. Res.* **2017**, *5*, 1058.
191. Oweida, A.; Lennon, S.; Calame, D.; Korpela, S.; Bhatia, S.; Sharma, J.; Graham, C.; Binder, D.; Serkova, N.; Raben, D.; et al. Ionizing radiation sensitizes tumors to PD-L1 immune checkpoint blockade in orthotopic murine head and neck squamous cell carcinoma. *Oncoimmunology* **2017**, *6*, e1356153. [CrossRef]
192. Kim, K.J.; Kim, J.H.; Lee, S.J.; Lee, E.J.; Shin, E.C.; Seong, J. Radiation improves antitumor effect of immune checkpoint inhibitor in murine hepatocellular carcinoma model. *Oncotarget* **2017**, *8*, 41242–41255. [CrossRef]
193. Son, C.H.; Bae, J.H.; Shin, D.Y.; Lee, H.R.; Choi, Y.J.; Jo, W.S.; Ho Jung, M.; Kang, C.D.; Yang, K.; Park, Y.S. CTLA-4 blockade enhances antitumor immunity of intratumoral injection of immature dendritic cells into irradiated tumor in a mouse colon cancer model. *J. Immunother.* **2014**, *37*, 1–7. [CrossRef]
194. Belcaid, Z.; Phallen, J.A.; Zeng, J.; See, A.P.; Mathios, D.; Gottschalk, C.; Nicholas, S.; Kellett, M.; Ruzevick, J.; Jackson, C.; et al. Focal radiation therapy combined with 4-1BB activation and CTLA-4 blockade yields long-term survival and a protective antigen-specific memory response in a murine glioma model. *PLoS ONE* **2014**, *9*, e101764. [CrossRef] [PubMed]
195. Johnsrud, A.J.; Jenkins, S.V.; Jamshidi-Parsian, A.; Quick, C.M.; Galhardo, E.P.; Dings, R.P.M.; Vang, K.B.; Narayanasamy, G.; Makhoul, I.; Griffin, R.J. Evidence for Early Stage Anti-Tumor Immunity Elicited by Spatially Fractionated Radiotherapy-Immunotherapy Combinations. *Radiat. Res.* **2020**, *194*, 688–697. [CrossRef] [PubMed]
196. Helm, A.; Tinganelli, W.; Simoniello, P.; Kurosawa, F.; Fournier, C.; Shimokawa, T.; Durante, M. Reduction of Lung Metastases in a Mouse Osteosarcoma Model Treated with Carbon Ions and Immune Checkpoint Inhibitors. *Int. J Radiat Oncol. Biol Phys.* **2021**, *109*, 594–602. [CrossRef] [PubMed]
197. Gray, J.E.; Villegas, A.; Daniel, D.; Vicente, D.; Murakami, S.; Hui, R.; Kurata, T.; Chiappori, A.; Lee, K.H.; Cho, B.C.; et al. Three-Year Overall Survival with Durvalumab after Chemoradiotherapy in Stage III NSCLC-Update from PACIFIC. *J. Thorac. Oncol.* **2020**, *15*, 288–293. [CrossRef]
198. Santos, L.L.; Oliveira, J.; Monteiro, E.; Santos, J.; Sarmento, C. Treatment of Head and Neck Cancer with Photodynamic Therapy with Redaporfin: A Clinical Case Report. *Case Rep. Oncol.* **2018**, *11*, 769–776. [CrossRef] [PubMed]

MDPI
St. Alban-Anlage 66
4052 Basel
Switzerland
Tel. +41 61 683 77 34
Fax +41 61 302 89 18
www.mdpi.com

Pharmaceuticals Editorial Office
E-mail: pharmaceuticals@mdpi.com
www.mdpi.com/journal/pharmaceuticals

www.ingramcontent.com/pod-product-compliance
Lightning Source LLC
Chambersburg PA
CBHW040833250926
51850CB00007BA/1304
* 9 7 8 3 0 3 6 5 5 3 5 2 8 *